ADVANCES IN
NUMERICAL HEAT TRANSFER

Volume 3

T0225399

Computational and Physical Processes in Mechanics and Thermal Science

ADVANCES IN NUMERICAL HEAT TRANSFER

Volume 3

Edited by

W. J. Minkowycz
Mechanical and Industrial Engineering
University of Illinois at Chicago
Chicago, Illinois

E. M. Sparrow
Mechanical Engineering
University of Minnesota, Twin Cities
Minneapolis, Minnesota

Guest Editor

J. P. Abraham
School of Engineering
University of St. Thomas
St. Paul, Minnesota

CRC Press
Taylor & Francis Group
Boca Raton   London   New York

CRC Press is an imprint of the
Taylor & Francis Group, an **informa** business

CRC Press
Taylor & Francis Group
6000 Broken Sound Parkway NW, Suite 300
Boca Raton, FL 33487-2742

First issued in paperback 2017

© 2009 by Taylor & Francis Group, LLC
CRC Press is an imprint of Taylor & Francis Group, an Informa business

No claim to original U.S. Government works

ISBN 13: 978-1-138-11219-3 (pbk)
ISBN 13: 978-1-4200-9521-0 (hbk)

Visit the Taylor & Francis Web site at
http://www.taylorandfrancis.com

and the CRC Press Web site at
http://www.crcpress.com

Contents

Preface

The bioheat and fluid flow focus of this volume of *Advances in Numerical Heat Transfer* is motivated by the marked upwelling of current interest in these subjects that are critical to human health. Progress in these areas requires both ingenious modeling and innovative numerical simulation. These issues are at the heart of the compilation of knowledge that has been assembled in this volume.

The 10 chapters that comprise Volume 3 range widely over both fundamentals and applications. The modeling of thermal transport by perfusion within the framework of porous-media theory is the focus of Chapter 1. Other perfusion models are reviewed and synthesized in Chapter 2. Chapter 3 reviews different forms of the bioheat equation that are appropriate to several thermal therapies, including laser irradiation. In contrast to the continuum tissue models of the preceding chapters, Chapter 4 focuses on thermal transport in individual blood vessels.

Thermal methods of tumor detection and treatment are described in Chapter 5. Lengthy surgeries may require waste extraction from the blood with concomitant issues of blood heating and cooling, as exposited in Chapter 6. In Chapter 7, the enhancement of heat conduction in tumor tissue by intruded nanoparticles is demonstrated to improve the efficacy of thermal destruction of the tumor.

Although the current bioheat and fluid emphasis is on localized anatomies, whole-body thermal models remain of critical importance, as detailed in Chapter 8. Overarching issues in the thermal treatment of cancer are the focus of Chapter 9. Chapter 10 is a detailed case study describing the thermal ablation of an enlarged prostate.

The editors were vastly aided in the creation of Volume 3 by Professor John P. Abraham, who served as guest editor. All the editors owe a profound debt of gratitude to the editorial staff of Taylor & Francis Group for their splendid cooperation.

W. J. Minkowycz
E. M. Sparrow

Contributors

J. P. Abraham
School of Engineering
University of St. Thomas
St. Paul, Minnesota

P. S. Ayyaswamy
Department of Mechanical
  Engineering and Applied
  Mechanics
School of Engineering and Applied
  Science
University of Pennsylvania
Philadelphia, Pennsylvania

I. Babuška
Institute for Computational
  Engineering and Sciences
University of Texas at Austin
Austin, Texas

J. W. Baish
Department of Mechanical
  Engineering
Bucknell University
Lewisburg, Pennsylvania

C. Bajaj
Institute for Computational
  Engineering and Sciences
University of Texas at Austin
Austin, Texas

J. Bass
Institute for Computational
  Engineering and Sciences
University of Texas at Austin
Austin, Texas

L. Bidaut
Department of Imaging Physics
M. D. Anderson Cancer Center
University of Texas
Houston, Texas

J. C. Browne
Institute for Computational
  Engineering and Sciences
University of Texas at Austin
Austin, Texas

C. Cordero-Tumangday
Neurocritical Care and Acute Stroke
  Section
Departments of Neurology
  and Neurosurgery
Pritzker School of Medicine
University of Chicago
Chicago, Illinois

L. Demkowicz
Institute for Computational
  Engineering and Sciences
University of Texas at Austin
Austin, Texas

Z.-S. Deng
Technical Institute of Physics
    and Chemistry
Chinese Academy of Sciences
Beijing, People's Republic of China

K. R. Diller
Department of Biomedical Engineering
University of Texas at Austin
Austin, Texas

A. Elliott
Department of Imaging Physics
M. D. Anderson Cancer Center
University of Texas
Houston, Texas

Y. Feng
Computational Bioengineering
    and Nanotechnology Lab
Department of Mechanical
    Engineering
University of Texas at San Antonio
San Antonio, Texas

D. Fuentes
Institute for Computational
    Engineering and Sciences
University of Texas at Austin
Austin, Texas

S. Goswami
Institute for Computational
    Engineering and Sciences
University of Texas at Austin
Austin, Texas

A. Hawkins
Institute for Computational
    Engineering and Sciences
University of Texas at Austin
Austin, Texas

J. Hazle
Department of Imaging Physics
M. D. Anderson Cancer Center
University of Texas
Houston, Texas

K. Khanafer
Vascular Mechanics Lab
Biomedical Engineering
    Department
University of Michigan
Ann Arbor, Michigan

S. Khoshnevis
Department of Biomedical
    Engineering
University of Texas at Austin
Austin, Texas

A. N. T. J. Kotte
Department of Radiotherapy
University Medical Center
    Utrecht
Utrecht, The Netherlands

B. Kwon
Institute for Computational
    Engineering and Sciences
University of Texas at Austin
Austin, Texas

J. J. W. Lagendijk
Department of Radiotherapy
University Medical Center Utrecht
Utrecht, The Netherlands

J. Liu
Department of Biomedical
    Engineering
School of Medicine
Tsinghua University
Beijing, People's Republic of China

K. Mukundakrishnan
Department of Mechanical
 Engineering and Applied
 Mechanics
School of Engineering and Applied
 Science
University of Pennsylvania
Philadelphia, Pennsylvania

J. T. Oden
Institute for Computational
 Engineering and Sciences
University of Texas at Austin
Austin, Texas

S. Prudhomme
Institute for Computational
 Engineering and Sciences
University of Texas at Austin
Austin, Texas

B. W. Raaymakers
Department of Radiotherapy
University Medical Center Utrecht
Utrecht, The Netherlands

S. Ramadhyani
EnteroMedics Inc.
St. Paul, Minnesota

A. J. Rosengart
Neurocritical Care and Acute Stroke
 Section
Departments of Neurology
 and Neurosurgery
Pritzker School of Medicine
University of Chicago
Chicago, Illinois

T. Schappeler
Department of Mechanical
 Engineering
University of Maryland
Baltimore, Maryland

E. M. Sparrow
Department of Mechanical
 Engineering
University of Minnesota
Minneapolis, Minnesota

R. J. Stafford
Department of Imaging Physics
M. D. Anderson Cancer Center
University of Texas
Houston, Texas

Z.-Q. Sun
Technical Institute of Physics
 and Chemistry
Chinese Academy of Sciences
Beijing, People's Republic
 of China

K. Vafai
Mechanical Engineering
 Department
University of California
Riverside, California

E. H. Wissler
Department of Chemical
 Engineering
University of Texas at Austin
Austin, Texas

L. X. Xu
Department of Biomedical
 Engineering
School of Life Science
 and Biotechnology
Med-X Research Institute
Shanghai Jiao Tong University
Shanghai, People's Republic
 of China

J.-F. Yan
Technical Institute of Physics
   and Chemistry
Chinese Academy of Sciences
Beijing, People's Republic of China

A. Zhang
Department of Biomedical
   Engineering
School of Life Science
   and Biotechnology
Shanghai Jiao Tong University
Shanghai, People's Republic of China

Y.-X. Zhou
Technical Institute of Physics
   and Chemistry
Chinese Academy of Sciences
Beijing, People's Republic
   of China

L. Zhu
Department of Mechanical
   Engineering
University of Maryland
Baltimore, Maryland

1

Synthesis of Mathematical Models Representing Bioheat Transport

K. Khanafer and K. Vafai

CONTENTS

1.1 INTRODUCTION

The application of computational methods in modeling biological systems has been a topic of interest for various physicians and engineers. This interest stems from the rapid advancement of computational technology. Many medical operations have sought the help of engineering methods to ascertain the safety and to determine the risk levels involved in any surgery. Further, the accurate description of the thermal interaction between vasculature and tissues is essential for the advancement of medical technology in treating fatal diseases such as tumors and breast cancer. At present, mathematical models have been used significantly in the analysis of hyperthermia in treating tumors, cryosurgery, laser eye surgery, fetal-placental studies, and many other applications. For example, the success of hyperthermia treatment strongly depends on knowledge of the heat transfer processes in blood-perfused tissues. As such, accurate thermal modeling is essential for effective hyperthermia treatment.

1.1.1 Hyperthermia

Hyperthermia treatment has been demonstrated effective as a cancer therapy in recent years. Its objective is to raise the temperature of pathological tissues above cytotoxic temperatures (41°C to 45°C) without overexposing healthy tissues [1–4]. Conventional hyperthermia in conjunction with radiation has demonstrated increased effectiveness in the treatment of certain types of cancer, such as those of liver metastases (the spread of a disease from one organ or part to another noncontiguous organ or part) [5–7]. Uniform temperature distributions are significant to achieve and maintain during hyperthermia treatment [8] since the use of temperatures above 55°C may directly destroy tissues through thermal coagulation, as was illustrated by Beacco et al. [9]. For safety consideration in clinics, it is essential to ensure necrosis (the death of living cells or tissues) of the total tumor cells within the desired volume of treatment while minimizing the thermal damage to healthy tissues surrounding the tumor. Temperature variations, which may be associated with the mechanisms of heat removal by the body and inadequate heating technologies, are often heterogeneous, and can lead to an undesired heating of the tissues, hot spots, and potential burning.

An important source of temperature nonuniformity is the presence of large vessels entering the heated volume and carrying blood at a low systemic temperature (37°C). Blood flow is found to have a profound influence on the efficiency of thermal therapy treatment. The design of delivered power devices and numerous theoretical, experimental, and clinical studies have demonstrated that large blood vessels may produce localized cooling regions within heated tissues during hyperthermia treatment [10–15]. Thus, for process control, it is essential to obtain a temperature field of the entire treatment region in order to deliver an adequate amount of energy to the treatment target volume and raise its minimum temperature above 42°C, while controlling the temperatures in the normal tissue to prevent damage. Since it is important to determine accurately the temperature field over the entire affected region, many numerical and experimental methods have been developed to solve the bioheat equation. Tang et al. [16] and Dai et al. [17] developed a numerical method for obtaining an optimal temperature distribution in a triple-layered skin structure embedded with two countercurrent, multilevel blood vessels: an artery and a vein. The authors concluded that their results could be useful for certain types of hyperthermia cancer treatments, such as for skin cancer.

He et al. [18] developed a two-dimensional (2D) finite element thermal model of a human breast with a tumor to study the variation of the blood perfusion rate and distribution of oxygen partial pressure (PO_2) in human tumors. Laser irradiation was used as an adjunct method in the treatment of cancer. The blood circulation inside the breast was modeled using one-dimensional nonlinear equations of pulsatile fluid flow. The distribution of PO_2 inside the capillaries, tumor vessels, and surrounding tissue was obtained by the Krogh analysis model (the Krogh model predicts a biphasic relationship between O_2 delivery and the rate of O_2 uptake per unit tissue volume). Shih et al. [19] used the explicit finite difference method to solve the transient equation for the temperature field within a perfused tumor tissue encompassing a blood vessel in an axisymmetric configuration during thermal therapy. Their results illustrated that short-duration high-intensity heating was more effective for treating a tumor with a blood vessel of 200 μm or less diameter, while neither longer heating duration nor higher heating power density was sufficient for complete necrosis of a tumor with a blood vessel with a diameter larger than 2 mm. Zhou and Liu [20] developed a three dimensional (3D) time-dependent heat transfer model coupled with the Navier-Stokes equation-based blood flow model to solve for temperature distributions in laser-irradiated tissues embedded with large blood vessels and the flow field within the vessels. A better understanding of the role of a large vessel in laser-induced thermotherapy (LITT) was obtained.

Khanafer et al. [21] conducted a numerical study to determine the influence of pulsatile laminar flow and heating protocol on the temperature distribution in a single blood vessel and tumor tissue receiving hyperthermia treatment using physiological velocity waveforms. Their results showed that the presence of large vessels had a significant effect on temperature distributions. Further, a uniform heating scheme was found to generate larger temperatures compared to the pulsed heating scheme, which may induce areas of overheating (beyond the therapeutic regions) that could damage normal tissues (Figure 1.1).

Craciunescu and Clegg [22] analyzed the effects of pulsating blood flow on the temperature distribution of a heated tissue. They found that the pulsation of blood flow rate yields an obvious change of the energy transport between the vessel wall and the blood flow within large blood vessels. Their results were based on the assumption that the vessel wall was a perfect thermal sink.

1.1.2 Bioheat Transfer in the Human Eye

With the growing interest in the bioengineering field, the area of ophthalmology, in particular laser eye surgery, has become better known in the last 20 years [23]. This field has gained increasing popularity with the advancement of computational technology. The popular employment of laser technology for surgical applications has given rise to a new area of burn studies, that is, the change in

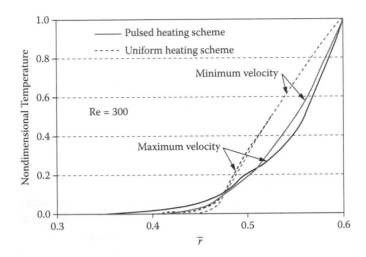

Figure 1.1   Influence of the heating protocol on the temperature distribution at various flow conditions. (Reprinted from Khanafer et al. [21]. With permission from Elsevier.)

temperature of the bio-organ associated with the absorption of high-intensity irradiation of light. As early as the 1960s, when lasers were first introduced in the medical field, there was immediate concern over the potential of incurring injury to the eye owing to the absorption of energy causing elevated temperatures. Injury to the eye can be severe when the blood flow cannot regulate the heat loading within the ocular tissues. The tissues most vulnerable in the eye are the cornea and the aqueous humor, as the infrared radiation raises the overall temperature of the aqueous eye [23,24]. Invasive or direct-contact techniques initially used in measuring the eye temperature are now confined to animal experiments due to the damaging nature of the test procedures [25]. The application of military technology in medical sciences has a way for measuring human body temperature utilizing infrared imaging. Infrared (IR)-imaging techniques have been widely used in measuring the temperature of the eye ever since. The human eye is very sensitive, and any direct contact with foreign objects is intolerable. Computational modeling of the human eye is thus very important for estimating the eye temperature during an eye procedure. A mathematical model can be useful for the doctors in enabling them to optimize their surgical protocol. This will lead to a reduction in intraocular tissue thermal damage.

Ooi et al. [26] used the boundary element method to analyze the 2D steadystate bioheat model of the human eye. The human eye was modeled as comprising four distinct homogeneous regions. The boundary condition on the outer surface of the cornea was nonlinear due to heat radiation. An iterative approach was used to treat the nonlinear heat radiation term. The authors showed that the calculated heat flux results were more accurately obtained using the boundary element method than the finite element method on the corneal surface. Ng and

Ooi [27] developed a 2D finite element model to simulate the thermal steady-state conditions of the human eye based on the properties and parameters reported in the open literature. Their results were verified against previous studies on human as well as animal eyes. Their results compared favorably with images from IR screening and another finite element model. Extending earlier work, Ng and Ooi [28] also developed a 3D model of the human eye to simulate the steady-state temperature distribution during standard conditions and during electromagnetic (EM) wave radiation. Their results were in good agreement with the experimental results in the open literature.

A mathematical model of the human eye based on the bioheat transfer equation was developed by Scott [29]. The intraocular temperature distribution was calculated using the Galerkin finite element method. A mathematical model to predict the temperature distribution within the human eye when subjected to a laser source was presented by Chuak et al. [30]. The model was developed by employing the Pennes bioheat transfer equation. The intraocular temperature distributions were calculated using the finite volume method. To compute the intraocular temperature distribution, Amara [31] presented a thermal model of the human eye exposed to laser irradiation. The physical system was described by a set of partial differential equations consisting of the heat equation that included the laser heat source, and the boundary and initial conditions. The analytical system was transformed to an integral formulation where a Galerkin function was applied. The results illustrated that decreasing the laser wavelength increases the adverse effects on the eye. This was due to the production of higher temperatures that can lead to the denaturation of the ocular tissues.

Long-term industrial exposure to low levels of infrared radiation has for many years been associated with the development of cataracts, which is considered to be a thermally related injury. A finite element model of the human eye was employed to calculate the temperature rise experienced by the intraocular media when exposed to infrared radiation [32]. The model was used to calculate transient and steady-state temperature distributions for various exposure times and a range of incident irradiances. The effect of the eye's natural cooling mechanisms on the heating was investigated. Hirata et al. [33–35] applied the finite difference time domain method to study the temperature rise in the human eye exposed to electromagnetic waves. Hirata [33] investigated the effect of frequency, polarization, and angle of incidence of an electromagnetic (EM) wave on the specific absorption rate (SAR) and maximum temperature increase in the human eye at 900 MHz, 1.5 GHz, and 1.9 GHz. In particular, the temperature increase in the eye was compared for near-field and far-field exposures. The results illustrated that the SAR and temperature increase in the eye were found to be largely dependent on the separation between the eye and a source, and the frequency, polarization, and angle of incidence of the EM wave. Lagendijk [36] conducted measurements on rabbit eyes and used the results to predict the thermal properties of the rabbit eye using a finite difference method. The measured temperature at the cornea surface was in good agreement with the calculated temperature using the finite difference method (FDM) mathematical model at the same location.

1.2   THERMAL MODELS FOR BLOOD-PERFUSED TISSUES

Heat transport in biological tissues, which is usually expressed by the bioheat equation, is a complicated process since it involves thermal conduction in tissues, convection and perfusion of blood (delivery of the arterial blood to a capillary bed in tissues), and metabolic heat generation. Therefore, several authors have developed mathematical models of bioheat transfer as extended and modified versions of the original work of Pennes [37], as reported by Charny [38] and Arkin et al. [39]. An example of the applications of the bioheat equation exists in the fetal-placental studies. The existence of a thermal gradient between fetal and maternal tissue has been considered a medical subject of research interest. This gradient is found to play a significant role in dissipating heat produced by the fetus during its metabolic processes. The magnitude of this temperature difference is determined by the fetal metabolic rate and the rate of heat exchange from fetal to maternal tissues [40–46].

Another example of the bioheat equation is related to the presence of the global system mobile (GSM) electromagnetic fields in the environment due to cellular phones and base stations, which have been causing increasing public concern regarding the possible adverse health effects of these fields. Electromagnetic-thermal analysis of human exposure to base station antenna radiation was presented by Poljak et al. [47]. The formulation was based on a simplified cylindrical representation of the human body. The electromagnetic analysis involved incident and internal field dosimetry, while the thermal model was based on the Pennes bioheat transfer equation for solving thermal processes inside the human body. In what follows, a concise summary of the pertinent thermal models and their limitations for blood-perfused tissues that best categorize different approaches in modeling the bioheat transfer is presented.

1.2.1   The Pennes Bioheat Equation

The Pennes model [37] was initially developed for predicting heat transfer in the human forearm. Due to the simplicity of the Pennes bioheat model (it assumes uniform thermal conductivity, perfusion rate, and metabolic heating), it was implemented in various biological research works such as for therapeutic hyperthermia for the treatment of cancer [48–50]. The equation that Pennes developed is expressed in its simplest form as

$$(\rho c_p)_t \frac{\partial T_t}{\partial t} = \nabla \cdot (k_t \nabla T_t) + q_p + q_m \tag{1.1}$$

where ρ, c_p, T_t, k_t, and q_m are tissue density, tissue-specific heat, tissue temperature, tissue thermal conductivity, and uniform rate of metabolic heat generation
in the tissue layer per unit volume, respectively. The heat transfer from the blood
to the tissue, q_p, is assumed to be proportional to the temperature difference
between the arterial blood entering the tissue and the venous blood leaving the
tissue. This quantity is presented as

$$q_p = \omega \rho_b c_b (T_{a,in} - T_{v,out}) \tag{1.2}$$

where $T_{a,in}$ and $T_{v,out}$ are the temperature of the blood upon entering and leaving
the tissue via the arteriole–venule network, respectively; ρ_b is the blood density;
c_b is the blood-specific heat; and ω is the volumetric rate of blood perfusion in
the tissue per unit volume. The Pennes model assumed thermal equilibrium
between the venous blood and the tissue temperatures (i.e., $T_{v,out} = T_t$), yielding
the familiar Pennes perfusion heat source term:

$$q_p = \omega \rho_b c_b (T_{a,in} - T_t) \tag{1.3}$$

1.2.2 Wulff Continuum Model

Due to the simplicity of the Pennes model, many authors have looked into the
validity of the assumptions used to develop the Pennes bioheat equation. Wulff's
study [51] was one of the first that questioned the assumptions of the Pennes
model. Wulff [51] assumed that the heat transfer between flowing blood and tissue should be modeled to be proportional to the temperature difference between
these two media rather than between the two bloodstream temperatures (i.e., the
temperature of the blood entering and leaving the tissue). Thus, the energy flux at
any point in the tissue should be expressed by

$$q = -k_t \nabla T_t + \rho_b h_b v_h \tag{1.4}$$

where v_h is the local mean blood velocity and T_t is the tissue temperature. The
specific enthalpy of the blood h_b, which accounts for both the sensible enthalpy
plus the enthalpy of reaction, is given by

$$h_b = \int_{T_o}^{T_b} c_b(T_b^*) dT_b^* + \frac{P}{\rho_b} + \Delta H_f (1 - \phi) \tag{1.5}$$

where P is the system pressure, ΔH_f is the enthalpy of formation of the metabolic reaction, and ϕ is the extent of reaction, respectively. T_o and T_b are the

reference and blood temperatures, respectively. Thus, the energy balance equation can be written as

$$\rho c_p \frac{\partial T_t}{\partial t} = -\nabla \bullet q = -\nabla \bullet (-k_t \nabla T_t + \rho_b h_b v_h) = \nabla \bullet \left[k_t \nabla T_t - \rho_b v_h \left(\int_{T_o}^{T_b} c_b(T_b^*) dT_b^* \right. \right.$$

$$\left. \left. + \frac{P}{\rho_b} + \Delta H_f (1-\phi) \right) \right] \tag{1.6}$$

Neglecting the mechanical work term (P/ρ_b), setting the divergence of the product $(\rho_b v_h)$ to zero, and assuming constant physical properties, Equation (1.6) can be simplified as follows:

$$\rho c_p \frac{\partial T_t}{\partial t} = k_t \nabla^2 T_t - \rho_b v_h (c_b \nabla T_b - \Delta H_f \nabla \phi) \tag{1.7}$$

Since blood is effectively microcirculating within the tissue, it will likely be in thermal equilibrium with the surrounding tissue. As such, Wulff [51] assumed that T_b is equivalent to the tissue temperature T_t. The metabolic reaction term $(\rho_b v_h \Delta H_f \nabla \phi)$ is equivalent to q_m; therefore, the final form of the bioheat equation that was derived by Wulff [51] is

$$(\rho c_p)_t \frac{\partial T_t}{\partial t} = k_t \nabla^2 T_t - (\rho c)_b v_h \bullet \nabla T_t + q_m \tag{1.8}$$

The main challenge in solving this bioheat equation is in the evaluation of the local blood mass flux $\rho_b v_h$.

1.2.3 Klinger Continuum Model

Since the Pennes model [37] neglected the effect of blood flow inside the tissues, Klinger [52] considered the convective heat transfer caused by the blood flow inside the tissue. Taking into account the spatial and temporal variations of the velocity (v) and heat source, and assuming constant physical properties of tissue and incompressible blood flow, the modified Pennes model was expressed as

$$(\rho c_p)_t \frac{\partial T_t}{\partial t} + (\rho c)_b v \bullet \nabla T_t = k_t \nabla^2 T_t + q_m \tag{1.9}$$

1.2.4 Continuum Model of Chen and Holmes (CH)

Similar to the analysis of Wulff [51] and Klinger [52], the bioheat transfer analysis of Chen and Holmes [53] is a microvascular model. The Chen and Holmes

(CH) model [53] assumes that the total tissue control volume is composed of the solid-tissue subvolume (V_s) and blood subvolume (V_b). Using a simplified volume-averaging technique, the energy balance equations for both the solid-tissue space and vascular spaces can be written as follows:

Solid Phase

$$\delta V_s(\rho c)_s \frac{\partial T_s}{\partial t} = dQ_{ks} + dQ_{bs} + dQ_m \tag{1.10}$$

where ρ_s and c_s are the solid-tissue density and specific heat, respectively; δV_s is the differential volume of the solid phase; dQ_{ks} is the energy transferred by conduction; dQ_{bs} is the heat gain from the blood subvolume; and dQ_m is the metabolic heating energy.

The energy balance equation for the vascular space is similar to Equation (1.10) except with an additional term associated with the bulk fluid flow in this space:

Fluid Phase

$$\delta V_b(\rho c)_b \frac{\partial T_b}{\partial t} = dQ_{kb} - dQ_{bs} + \int_S (\rho c)_b \, T \mathbf{v} \bullet d\mathbf{s} \tag{1.11}$$

where ρ_b and c_b are the blood density and specific heat, respectively; δV_b is the differential volume of the blood in the vascular space; dQ_{kb} is the conductive contribution; and the integral term in Equation (1.11) denotes the energy transfer by convection as the blood flows across the surface area S at velocity $\mathbf{v}$. Therefore, the energy balance for the tissue space is derived by the addition of Equations (1.10) and (1.11) and division of the result by the total control volume δV, which yields the following:

$$(\rho c) \frac{\partial T_t}{\partial t} = q_k + q_m + q_p \tag{1.12}$$

where ρ and c are

$$\rho = (1 - \varepsilon_b)\rho_s + \varepsilon_b \rho_b \quad \text{and} \quad c = \frac{1}{\rho}[(1 - \varepsilon_b)\rho_s c_s + \varepsilon_b \rho_b c_b] \tag{1.13}$$

where ε_b is the porosity of the tissue where blood flows and T_t is the local mean tissue temperature expressed as

$$T_t = \frac{1}{\rho c}[(1 - \varepsilon_b)(\rho c)_s T_s + \varepsilon_b (\rho c)_b T_b] \tag{1.14}$$

The quantity q_k denotes the heat transfer by conduction per unit volume, q_m is the metabolic heat generation per unit volume, and q_p is the perfusion energy generated per unit volume. The total heat transfer by conduction per unit volume (q_k) in the tissue control volume is expressed by

$$q_k = \frac{Q_{ks} + Q_{kb}}{\delta V} = \nabla \bullet (k_{eff} \nabla T_t) \tag{1.15}$$

where k_{eff} is the effective thermal conductivity of the combined tissue and vascular spaces. The effective thermal conductivity is written as

$$k_{eff} = \varepsilon_b k_b + (1 - \varepsilon_b) k_s \tag{1.16}$$

Since $\varepsilon_b = \frac{\delta V_b}{\delta V} \sim \frac{\delta V_b}{\delta V_s} \ll 1$, it follows that k_{eff} is independent of blood flow and equal to the conductivity of the solid tissue ($k_{eff} \cong k_s$).

In Wulff's formulation [51], thermal equilibrium was assumed between blood and the solid-tissue medium at all locations within the control volume. However, Chen and Holmes [53] allowed for the blood within the tissue matrix to flow at a temperature different than the tissue temperature. Therefore, the convective heat transfer across the surface (q_p) due to blood flow was written as the sum of the perfusion heating term (originated by Pennes), a contribution proportional to local blood perfusion velocity as represented by the Wulff model [51], and a contribution due to the perfusion thermal conductivity. Thus, the perfusion term including all these terms is

$$q_p = \frac{1}{\delta V} \int_S \rho_b c_b T \mathbf{v} \bullet ds \cong (\rho c)_b \omega^* (T_a^* - T_s) - (\rho c)_b \mathbf{v}_p \bullet \nabla T_s + \nabla \bullet k_p \nabla T_s \tag{1.17}$$

where $\mathbf{v}_p$ is the mean perfusion velocity; ω^* is the total perfusion bleed-off to the tissue only from the microvessels, while the Pennes term ω includes bleed-off from all generations of the vasculature; T_a^* is the blood temperature; and T_s is the solid-tissue temperature. Note that T_a^* is different from T_a (temperature of blood entering the tissue), which was used in the Pennes model; see Equation (1.2). The second term in Equation (1.17) is indicative of blood convective perfusion, and the third term on the right-hand side of Equation (1.17) indicates the enhancement of thermal conductivity in a tissue due to the blood flow within microvessels (i.e., the thermal dispersion effect).

Therefore, the bioheat equation based on the Chen and Holmes [53] model can be written as

$$(\rho c_p)_t \frac{\partial T_t}{\partial t} = \nabla \bullet (k_{eff} \nabla T_t) + (\rho c)_b \omega^* (T_a^* - T_t) - (\rho c)_b \mathbf{v}_p \bullet \nabla T_t + \nabla \bullet k_p \nabla T_t + q_m \tag{1.18}$$

where T_s is replaced by the volume-weighted continuum temperature (T_t). This is reasonable as long as $\varepsilon_b \ll 1$.

1.2.5 The Weinbaum, Jiji, and Lemons (WJL) Bioheat Equation Model

Weinbaum and colleagues [54–56] modified the thermal conductivity in the Pennes equation by means of an "effective conductivity," which is a function of the blood flow rate and vascular geometry. The modified bioheat equation was obtained based on a hypothesis that small arteries and veins are parallel and the flow direction is countercurrent, resulting in counterbalanced heating and cooling effects. It should be noted that this assumption is mainly applicable within the intermediate tissue of the skin. Neglecting axial conduction, the artery and vein energy balances are written as

$$(\rho c)_b \frac{d}{ds}\left(n\pi a^2 \bar{u} T_a\right) = -n q_a - (\rho c)_b (2\pi a n g) T_a \qquad (1.19)$$

$$(\rho c)_b \frac{d}{ds}\left(n\pi a^2 \bar{u} T_v\right) = -n q_v - (\rho c)_b (2\pi a n g) T_v \qquad (1.20)$$

where q_a is the heat loss from the artery by conduction through its wall, q_v is the heat gain by conduction per unit length through the vein wall into the vein, T_a and T_v are the bulk mean temperatures inside the blood vessel, n is the number of arteries or veins, $\bar{u}$ is the mean velocity in either the artery or vein, a is the radius, and g is the perfusion bleed-off velocity per unit vessel surface area. The second term on the right-hand side of Equation (1.19) indicates heat loss from the arterial blood due to perfusion bleed-off, while in Equation (1.20) it represents the heat gain by the venous blood from perfusion drainage. For an equal-size artery–vein pair, subtracting Equation (1.20) from Equation (1.19) yields

$$(\rho c)_b \left[\frac{d}{ds}\left(n\pi a^2 \bar{u} T_a\right) - \frac{d}{ds}\left(n\pi a^2 \bar{u} T_v\right) \right] = -n(q_a - q_v) - (\rho c)_b (2\pi a n g)(T_a - T_v) \qquad (1.21)$$

where the first term on the right-hand side is the net heat transfer by conduction from the tissue into the paired vessels, and the second term is the net heat deposited in the tissue due to the perfusion bleed-off. The first term on the left-hand side represents the total blood heat exchange in the countercurrent vessels and the surrounding tissue. This term can be balanced by conduction and metabolic heating as follows:

$$(\rho c)_b \left[\frac{d}{ds}\left(n\pi a^2 \bar{u} T_a\right) - \frac{d}{ds}\left(n\pi a^2 \bar{u} T_v\right) \right] = \nabla \bullet (k_t \nabla T_t) + q_m \qquad (1.22)$$

If the continuity equation $[\frac{d}{ds}(na^2 \bar{u}) = -2nag]$ is used, the second terms in the right-hand side of Equations (1.19) and (1.20) can be eliminated, yielding the following equations:

$$q_a = -(\rho c)_b (\pi a^2 \bar{u}) \frac{dT_a}{ds} \ \& \ q_v = -(\rho c)_b (\pi a^2 \bar{u}) \frac{dT_v}{ds} \qquad (1.23)$$

Therefore, the rate of the energy entering and leaving the tissue control volume can be expressed as

$$q_a - q_v = (\rho c)_b (\pi a^2 \bar{u}) \frac{d}{ds} [T_v - T_a] \qquad (1.24)$$

Thus, the final form of the bioheat equation can be obtained by substituting Equation (1.22) for the left-hand side of Equation (1.21) and substituting Equation (1.24) for the first term on the right-hand side of Equation (1.21) as

$$(\rho c)_b (n\pi a^2 \bar{u}) \frac{d}{ds} [T_a - T_v] - (\rho c)_b (n 2\pi a\ g)(T_a - T_v) = \nabla \bullet (k_t \nabla T_t) + q_m \qquad (1.25a)$$

or

$$(\rho c)_b (n\pi a^2 \bar{u}) \frac{d}{ds} [T_a - T_v] = \nabla \bullet (k_t \nabla T_t) + (\rho c)_b \omega'(T_a - T_v) + q_m \qquad (1.25b)$$

where $\omega' = (n 2\pi a\ g)$. Equation (1.25) includes a perfusion bleed-off term that apparently resembles the Pennes perfusion term. This term is proportional to $(T_a - T_v)$ rather than $(T_a - T_t)$.

1.2.6 The Weinbaum-Jiji Bioheat Equation Model

Since both T_a and T_v are unknowns in Equation (1.23), the tissue temperature T_t cannot be determined. Therefore, Weinbaum and Jiji [57] derived a simplified single equation to study the influence of blood flow on the tissue temperature distribution. The mean tissue temperature can be approximated as

$$T_t \cong \frac{T_a + T_v}{2} \qquad (1.26)$$

Thus, the magnitude of the difference $(q_a - q_v)$ is much smaller than the magnitude of either q_a or q_v. Moreover, Weinbaum and Jiji [57] assumed that the tissue surrounding the vessel pair is a pure conduction region such that

$$q_a \cong q_v = \sigma k_t (T_a - T_v) \qquad (1.27)$$

where σ is a geometrical shape factor given by

$$\sigma = \frac{\pi}{\cosh^{-1}(L_s/a)} \qquad (1.28)$$

The ratio (L_s/a) indicates the ratio of the vessel spacing to vessel diameter. Equations (1.23), (1.26), and (1.27) are solved to obtain an equation for the artery–vein temperature difference and the tissue temperature gradient by adding the q_a and q_v terms from Equation (1.23):

$$T_a - T_v = -\frac{\pi a^2 \bar{u} (\rho c)_b}{\sigma k_t} \frac{dT_t}{ds}$$

(1.29)

Substituting Equation (1.29) in the original WJL model, Equation (1.25a), yields a new bioheat equation proposed by Weinbaum and Jiji [57] as follows:

$$\frac{n\pi^2 ak_b}{4k_t} Pe \left(\frac{2gPe}{\sigma \bar{u}} \frac{dT_t}{ds} - \frac{d}{ds} \left[\frac{aPe}{\sigma} \frac{dT_t}{ds} \right] \right) = \nabla \bullet (k_t \nabla T_t) + q_m$$

(1.30)

where Pe is the Peclet number; which is defined as $Pe = \frac{2a(\rho c)_b \bar{u}}{k_b}$.

Table 1.1 illustrates a summary of variants within the previously discussed models.

1.3 MATHEMATICAL MODELING OF BIOHEAT EQUATION USING POROUS-MEDIA THEORY

1.3.1 Energy Equation

Although Pennes's bioheat equation is considered to be a useful model to predict temperature distribution in the human body due to its simplicity, it is still questionable. Based on the previous section, one can note that the summarized bioheat transfer models in Table 1.1 are extended or modified versions of the original work of Pennes's model. They were based on improving the main flaws of Pennes's equation. An accurate description of the thermal interaction between vasculature and tissues is essential for the advancement of medical technology through effective modeling of arteries, tissues, and organs. Therefore, it is crucial to develop a more robust bioheat model that incorporates the effects of blood thermal dispersion, porosity variation, effective tissue conductivity, and effective tissue capacitance, and a more precise representation of the heat exchange between the blood and the tissue. Since the compound matrix of tissues, arteries, veins, and capillary tubes can be considered as a porous medium, porous media theory is very well suitable for developing a rigorous model of a bioheat equation.

Transport phenomena in porous media have received continuing interest in the past five decades. This interest stems from their importance in many industrial and clinical applications [58–60]. Examples include computational biology, tissue replacement production, drug delivery, advanced medical imaging, porous scaffolds for tissue engineering, and transport in biological tissues [61–65]. Porous media theory may also be utilized in biosensing systems [66–69]. Some aspects of transport in porous media were also discussed in the

Table 1.1 Analysis of Variants within Bioheat Transfer Models

Model	Transient Term	Conduction Term	Perfusion Term	Metabolic Heat Source
General[a]	$(\rho c_p)_t \dfrac{\partial T_t}{\partial t}$	$\nabla \bullet (k_{eff} \nabla T_t)$	q_p	q_m
Pennes	$(\rho c_p)_t \dfrac{\partial T_t}{\partial t}$	$k_t \nabla^2 T_t$	$(\rho c)_b \omega (T_{a,in} - T_{v,out})$	q_m
Wulff	$(\rho c_p)_t \dfrac{\partial T_t}{\partial t}$	$k_t \nabla^2 T_t$	$-(\rho c)_b v_h \nabla T_t$	$\rho_b v_h \Delta H_f \nabla \phi$
Klinger	$(\rho c_p)_t \dfrac{\partial T_t}{\partial t}$	$k_t \nabla^2 T_t$	$-(\rho c)_b \mathbf{v} \nabla T_t$	q_m
Chen and Holmes	$(\rho c_p)_t \dfrac{\partial T_t}{\partial t}$	$\nabla \bullet (k_{eff} \nabla T_t)$	$-(\rho c)_b \mathbf{v}_p \bullet \nabla T_t + (\rho c)_b \omega^*(T_a^* - T_t)$ $+ \nabla \bullet (k_p \nabla T_t)$	q_m
Weinbaum, Jiji, and Lemons	NA	$\nabla \bullet (k_t \nabla T_t)$	$(\rho c)_b (n2\pi a\, g)(T_a - T_v)$ $-(\rho c)_b (n\pi a^2 \bar{u}) \dfrac{d}{ds}[T_a - T_v]$	q_m
Weinbaum and Jiji	NA	$\nabla \bullet (k_t \nabla T_t)$	$\dfrac{n\pi^2 a k_b}{4 k_t} Pe \left(-\dfrac{2gPe}{\sigma \bar{u}} \dfrac{dT_t}{ds} + \dfrac{d}{ds}\left[\dfrac{aPe}{\sigma} \dfrac{dT_t}{ds} \right] \right)$	q_m

$^a\ (\rho c_p)_t \dfrac{\partial T_t}{\partial t} = \nabla \bullet (k_{eff} \nabla T_t) + q_p + q_m.$

two editions of the *Handbook of Porous Media* [69,70] and in Hadim and Vafai [71] and Vafai and Hadim [72]. Complicated and interesting biomedical aspects can be modeled using the porous media concept. Xuan and Roetzel [73,74] utilized the porous media approach to model a tissue–blood system composed mainly of tissue cells and interconnected voids that contain either arterial or venous blood. The thermal energy exchange between the tissue and blood was modeled using the principle of local thermal nonequilibrium as described in the works of Amiri and Vafai [75,76]; Alazmi and Vafai [77]; Lee and Vafai [78]; Vafai and Sozen [79–81]; and Sozen and Vafai [82,83]. Thus, two energy equations were derived that represent the blood phase and the solid matrix phase, as given below:

Blood Phase

$$\varepsilon(\rho c)_b \left(\frac{\partial <T>^b}{\partial t} + <\vec{V}>^b \cdot \nabla <T>^b \right) = \nabla \cdot \left(\mathbf{k}_b^a \cdot \nabla <T>^b \right) + h_{bs} [<T>^s - <T>^b]$$

(1.31)

Solid Matrix Phase

$$(1-\varepsilon)(\rho c)_s \frac{\partial <T>^s}{\partial t} = \nabla \cdot \left(\mathbf{k}_s^a \cdot \nabla <T>^s \right) - h_{bs} [<T>^s - <T>^b] + (1-\varepsilon) q_m$$

(1.32)

where $<T>^b, <T>^s, \mathbf{k}_b^a, \mathbf{k}_s^a, <\vec{V}>^b$ and h_{bs}, and ε are the local volume-averaged arterial blood temperature, local volume-averaged solid-tissue temperature, blood effective thermal conductivity tensor, solid-tissue effective thermal conductivity tensor, blood velocity vector, and interstitial convective heat transfer coefficient, respectively. The interstitial convective heat transfer coefficient is a function of blood velocity and properties and geometric structure of the solid phase. The heat exchange between the blood and the tissue is expressed as $h_{bs} [<T>^s - <T>^b]$. For isotropic conduction, the effective thermal conductivity $\mathbf{k}_b^a$ of blood and solid tissue $\mathbf{k}_s^a$ can be expressed as

$$\mathbf{k}_b^a = \varepsilon \mathbf{k}_b + \mathbf{k}_b^t \quad \text{and} \quad \mathbf{k}_s^a = (1-\varepsilon) \mathbf{k}_s$$

(1.33)

where $\mathbf{k}_b^t$ is the thermal dispersion conductivity. The concept of thermal dispersion is well established in the theory of porous media as presented in the works of Amiri and Vafai [75,76]. Due to insufficient knowledge about the thermal and anatomic properties of tissue, the velocity field of the blood, and interstitial convective heat transfer coefficients, the local thermal equilibrium model represents a good approximation for determining the temperature field in applications involving small-sized blood vessels $(\varepsilon \ll 1)$. This implies that blood flowing in these small vessels will be completely equilibrated with the surrounding tissue.

Therefore, Equations (1.31) and (1.32) reduce to the following equation [21,85]:

$$\left[(\rho c)_b \varepsilon + (1-\varepsilon)(\rho c)_s\right]\frac{\partial <T>}{\partial t} + \varepsilon(\rho c)_b <\vec{V}>^b \cdot \nabla <T> = \nabla\left[\left(\mathbf{k}_s^a + \mathbf{k}_b^a\right)\cdot\nabla <T>\right]$$
$$+ q_m(1-\varepsilon)$$

(1.34)

The second term on the left-hand side of Equation (1.34) represents the heat transfer due to the blood perfusion. Note that the perfusion source term in the Pennes model was derived based on a uniform blood perfusion assumption and was equal to $\omega\rho_b c_b(T_{a,in} - T_{v,out})$. The representation of the blood perfusion in Equation (1.34) is more consistent with representation in the Klinger and Wulff models. In hyperthermia applications, tissue may absorb energy from an external source such as electromagnetic or ultrasonic radiation, and, therefore, another heat source term should be added to the right side of Equation (1.34) as follows:

$$[(\rho c)_b \varepsilon + (1-\varepsilon)(\rho c)_s]\frac{\partial <T>}{\partial t} + \varepsilon(\rho c)_b <\vec{V}>^b \cdot \nabla <T> = \nabla\left[\left(\mathbf{k}_s^a + \mathbf{k}_b^a\right)\cdot\nabla <T>\right]$$
$$+ q_m(1-\varepsilon) + q_h(1-\varepsilon)$$

(1.35)

Table 1.2 and Table 1.3 summarize the previously discussed bioheat transfer models in this work.

Table 1.2   Main Characteristics of Bioheat Models Using the Porous Media Approach

Bioheat Model	Main Characteristics
Porous media model (local thermal equilibrium principle)[a]	This model modifies the Pennes equation by accounting for the following effects: Variable tissue porosity Effective tissue conductivity Effective tissue capacitance Blood dispersion
Porous media model (local thermal nonequilibrium principle)[b]	This model requires more knowledge about the thermal and anatomic properties of the tissue, the velocity field of the blood, and interstitial convective heat transfer coefficients. This model considers the following effects: Variable tissue porosity Blood dispersion Effective tissue conductivity Effective tissue capacitance

[a] Amiri and Vafai [75,76]; Marafie and Vafai [84].
[b] Amiri and Vafai [75,76]; Alazmi and Vafai [77]; Lee and Vafai [78]; Vafai and Sozen [79–81]; Sozen and Vafai [82,83].

Table 1.3   Main Characteristics of Bioheat Models Using a Simplified Approach

Bioheat Model	Assumptions	Main Characteristics
Pennes [37]	Uniform physical properties and metabolic heating Heat transfer from the blood to the tissue is proportional to the temperature difference between the arterial blood entering the tissue and the venous blood leaving the tissue	Simple model Not valid for all tissues
Wulff [51]	Thermal equilibrium between flowing blood and the surrounding tissue Uniform mean blood velocity inside the tissue	Modified version of the Pennes model
Klinger [52]	Constant physical properties	Considers the convective heat transfer caused by the blood flow inside the tissue Considers the spatial and temporal variations of the velocity field and heat source
Chen and Holmes [53]	Utilized two separate volumes: one for solid tissue and one for blood in the vascular space	The total heat transfer by conduction relates to heat transfer by conduction in the solid tissue and in the vascular space The total perfusion term corresponds to the effect of blood flow on tissue temperature around large vessels, heat transfer that takes place as a result of the blood flow, and heat transfer due to the small temperature changes (microvessels) Introduce perfusion conductivity tensor in the bioheat equation. Allows for the blood within the tissue matrix to flow at a temperature different than that of the tissue temperature

(Continued)

Table 1.3 Main Characteristics of Bioheat Models Using a Simplified Approach (Continued)

Bioheat Model	Assumptions	Main Characteristics
Weinbaum, Jiji, and Lemons [54–56]	Based on a hypothesis that small arteries and veins are parallel and the flow direction is countercurrent, resulting in counterbalanced heating and cooling effects	Coupled energy equations for artery–vein pair and tissue
	Isotropic blood perfusion between the countercurrent vessels	Utilizes the effective conductivity
Weinbaum and Jiji [57]	The mean tissue temperature is approximated by an average temperature of the bulk mean temperatures inside the blood vessel	Valid when arteries and veins are close, leading to negligible blood perfusion effects
	Assumes that the tissue around the vessel pair is a pure conduction region	Utilizes the effective conductivity

1.3.2 Mathematical Model of Velocity Field and Macromolecule Transport within the Arterial Wall and Arterial Lumen

In order to solve Equations (1.31), (1.34), and (1.35), the velocity distribution within a tissue should be determined first. Khanafer et al. [21] modeled the arterial wall as a homogeneous porous medium. The velocity distribution within the tissue was determined using the volume-averaged governing equations of porous media coupled with the velocity field in the blood vessel. Vafai and coworkers [61–65] had developed a new fundamental and comprehensive four-layer model for the description of the velocity field and mass transport in the arterial wall coupled with the velocity field and mass transport in the arterial lumen. The endothelium, intima, internal elastic lamina (IEL), and media layers were all treated as macroscopically homogeneous porous media and mathematically modeled using proper types of the volume-averaged porous media equations, with the Staverman filtration and osmotic reflection coefficients employed to account for selective permeability of each porous layer to certain solutes. The typical anatomical structure of an arterial wall is shown schematically in Figure 1.2.

1.3.2.1 Lumen

Assuming incompressible and Newtonian fluid flow, blood flow in the arterial lumen was modeled using Navier-Stokes and continuity equations as follows:

$$\nabla \cdot \vec{V} = 0 \tag{1.36}$$

$$\frac{\partial \vec{V}}{\partial t} + \vec{V} \cdot \nabla \vec{V} = -\frac{1}{\rho} \nabla P + \nu \nabla^2 \vec{V} \tag{1.37}$$

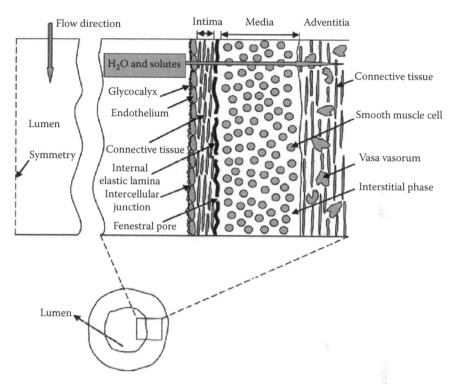

Figure 1.2   Schematic illustration of the geometric artery wall. (Reprinted from Yang and Vafai [62]. With permission from Elsevier.)

The concentration field in the arterial lumen is computed using the mass transport equation:

$$\frac{\partial c}{\partial t} + \vec{V} \cdot \nabla c = D \nabla^2 c \tag{1.38}$$

1.3.2.2 Endothelium and Internal Elastic Lamina

The endothelium and IEL were modeled as biological porous membranes [61–65]. The Staverman filtration and osmotic reflection coefficients were employed to account for selective rejection of species by the membranes and for the effects of osmotic pressure. The volume-averaged governing equations were given by

$$\nabla \cdot <\vec{V}> = 0 \tag{1.39}$$

$$\frac{\rho}{\varepsilon} \frac{\partial <\vec{V}>}{\partial t} = -\nabla <p>^f + \frac{\mu}{\varepsilon} \nabla^2 <\vec{V}> - \frac{\mu <\vec{V}>}{K} + R_u T \sigma_d \nabla <c> \tag{1.40}$$

$$\frac{\partial <c>}{\partial t} + (1 - \sigma_f) <\vec{V}> \cdot \nabla <c> = D_e \nabla^2 <c> \tag{1.41}$$

where K is the permeability, and D_e is the effective macromolecule diffusivity in the medium. The symbol $<>$ denotes the local volume average of a quantity [86,87], and the superscript f refers to the local volume average inside the fluid. The parameters σ_f and σ_d are the Staverman filtration and osmotic reflection coefficients (to account for the selective permeability of biological membranes to certain solutes), respectively; T is the absolute temperature of the medium; and R_u is the universal gas constant.

1.3.2.3 Intima and Media

The intima and media were also modeled as macroscopically homogeneous porous media. Since the layers comprising the arterial wall are selectively permeable to certain species such as low-density lipoprotein (LDL), the Staverman filtration reflection coefficient has to be introduced here as well to account for this effect. The osmotic effect in the transport modeling is not included in this part since the maximum osmotic pressure gradient in the media layer is far below the hydraulic pressure gradient [88]. Therefore, the volume-averaged governing equations of the intima and media layers are as follows [61–65]:

$$\nabla \cdot <\vec{V}> = 0 \tag{1.42}$$

$$\frac{\rho}{\varepsilon}\frac{\partial <\vec{V}>}{\partial t} = -\nabla <p>^f + \frac{\mu}{\varepsilon}\nabla^2 <\vec{V}> - \frac{\mu <\vec{V}>}{K} \tag{1.43}$$

$$\frac{\partial <c>}{\partial t} + (1-\sigma_f)<\vec{V}>\cdot\nabla <c> = D_e\nabla^2 <c> + k(c) \tag{1.44}$$

where k is the effective volumetric first-order reaction rate coefficient. Table 1.4 shows various momentum equations used in modeling flow in the arterial wall. Yang and Vafai [62] presented an analytical solution for a robust four-layer porous model for description of the LDL transport in the arterial wall coupled with the transport in the lumen. The analytical results were found consistent with the numerical data for different physiological conditions, as depicted in Figures 1.3 and 1.4.

1.4 CONCLUSIONS

Most of the previous studies in the literature have assumed rigid walls for the arteries, Newtonian blood flow, and steady flow when analyzing flow and heat transfer characteristics in human tissues. Proper analysis of the arterial wall is critical in accurate modeling of arterial transport. This must be done through the use of a multilayer model that accounts for various physical attributes of the vessel and interfacial aspects between the layers. The multilayer model describes the arterial anatomy most accurately. In this model, the arterial wall is composed

Table 1.4   Momentum Equations Using the Modeling Flow in the Arterial Wall

Model	Remarks
$$\vec{V} - \nabla \cdot \left(\frac{K}{\mu} P \right) = 0$$	Arterial wall modeled as single-layer porous medium Darcy model Constant permeability
$$\frac{\partial \vec{V}}{\partial t} + \vec{V} \cdot \nabla \vec{V} = -\frac{1}{\rho} \nabla P + \nu \nabla^2 \vec{V} - \frac{\nu \vec{V}}{K}$$	Arterial wall modeled as single-layer porous medium Brinkman's model Constant permeability Neglect the Staverman filtration and osmotic reflection coefficients

$$\frac{\rho}{\varepsilon} \frac{\partial <\vec{V}>}{\partial t} = -\nabla <p>^f + \frac{\mu}{\varepsilon} \nabla^2 <\vec{V}> - \frac{\mu <\vec{V}>}{K}$$ $$+ R_u T \sigma_d \nabla <c>$$	Endothelium and internal elastic lamina
	More realistic
	The osmotic reflection coefficients were employed to
	  account for selective rejection of species by the
	  membranes and for the effects of osmotic pressure

$$\frac{\rho}{\varepsilon} \frac{\partial <\vec{V}>}{\partial t} = -\nabla <p>^f + \frac{\mu}{\varepsilon} \nabla^2 <\vec{V}> - \frac{\mu <\vec{V}>}{K}$$	Intima and media More realistic Neglects the osmotic effect in the transport modeling

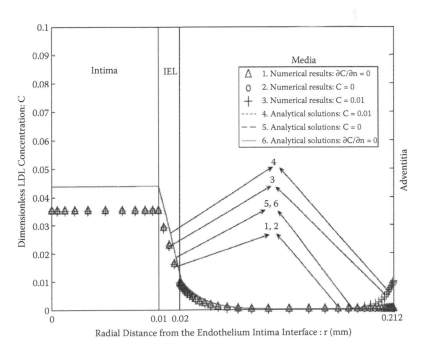

Figure 1.3   Calculated species profiles across the intima, IEL, and media layers. (Reprinted from Yang and Vafai [62]. With permission from Elsevier.)

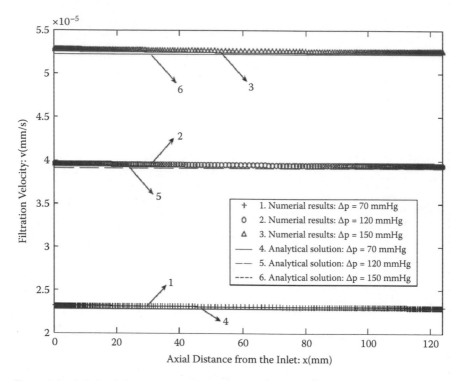

Figure 1.4 Calculated filtration velocity profiles at the lumen endothelium interface along the axial direction. (Reprinted from Yang and Vafai [62]. With permission from Elsevier.)

of four porous layers with different physiological characteristics. The multilayer model requires a number of transport parameters (properties) for each layer and in return provides an accurate profile for macromolecule distribution across the arterial wall, illuminating the role and behavior of each porous layer in the transport of macromolecules across the arterial wall. Another important factor in the accurate modeling of arterial transport is the use of a proper set of governing equations that take into account the dominant processes involved in the transport phenomenon. The Staverman filtration and osmotic reflection coefficients must be included to account for the selective rejection of species by the endothelium and IEL porous membranes as well as the effects of osmotic pressure.

The interaction between blood flow and wall can involve a wide range of fluid-mechanical phenomena. When blood flows through the lumen, the flow may deform the arterial walls and consequently alter the properties of the wall, which in turn affect flow and heat transfer characteristics in the lumen and arterial wall. Also, it will be beneficent to consider variations in the porosity and permeability of an arterial wall in future studies. In addition, the variations in the physical properties of the arterial walls such as Young's modulus and Poisson's ratio can be considered in future studies since the materials of the walls are nonlinear and nonhomogeneous. However, these variations at this time cannot be properly described due to lack of pertinent data. For an accurate analysis of heat transfer during a hyperthermia procedure in the treatment of fetal diseases such as tumors, the porous media approach is highly recommended in order to deliver the required amount of heat source to the target volume of cancer without destroying the surrounding healthy tissues. This is because the porous media approach takes into account many pertinent effects that were neglected or simplified in the existing bioheat equations. Therefore, the mathematical models based on the porous media approach presented in the previous section are applicable in various biomedical applications such as laser eye surgery.

NOMENCLATURE

a:	radius
c:	concentration
c_b:	blood-specific heat
c_p:	tissue-specific heat
c_s:	solid-specific heat
D_e:	effective macromolecule diffusivity in the medium
g:	perfusion bleed-off per unit vessel surface area
ΔH_f:	specific enthalpy of the metabolic reaction
h_b:	specific enthalpy of the blood
h_{bs}:	interstitial convective heat transfer coefficient
K:	permeability
$\mathbf{k}_b^a$:	blood effective thermal conductivity tensor
k_{eff}:	effective thermal conductivity
$\mathbf{k}_s^a$:	solid-tissue effective thermal conductivity tensor
k_t:	tissue thermal conductivity
n:	number of arteries or veins
P:	system pressure
Pe:	Peclet number
q_a:	heat loss from the artery by conduction through its wall
q_k:	heat transfer by conduction per unit volume
q_m:	rate of metabolic heat generation in the tissue layer per unit volume
q_p:	heat transfer from the blood to the tissue
q_v:	heat gain by conduction per unit length through the vein wall into the vein
R_u:	universal gas constant
$T_{a,in}$:	temperature of the blood entering the tissue
T_a^*:	blood temperature in the vascular space
T_b:	temperature of blood
$<T>^b$:	local volume-averaged arterial blood temperature

T_o: reference temperature
T_s: solid temperature
$<T>^s$:   local volume-averaged solid tissue temperature
T_t: tissue temperature
$T_{v,out}$:   temperature of the blood leaving the tissue
$<\vec{V}>^b$:   blood velocity vector
v_h: local mean blood velocity
v_p: mean perfusion velocity

GREEK SYMBOLS

ρ: tissue density
ρ_b: blood density
ρ_s: solid-tissue density
ω: volumetric rate of blood perfusion in the tissue per unit volume
ω^*:   total perfusion bleed-off to the tissue only from the microvessels
ϕ: extent of reaction
ε: porosity
σ: geometrical factor
σ_f: osmotic reflection
σ_d: Staverman filtration

REFERENCES

1. J. Overgaard, D. G. Gonzales, M. C. Hulshof, G. Arcangeli, O. Dahl, O. Mella, and S. M. Bentzen, Hyperthermia as an Adjuvant to Radiation Therapy of Recurrent or Metastatic Melanoma: A Multicenter Randomized Trial by the European Society for Hyperthermic Oncology, *Int. J. Hypertherm.* vol. 12, pp. 3–20, 1996.
2. J. R. Oleson, D. A. Sim, and M. R. Manning, Analysis of Prognostic Variables in Hyperthermia Treatment of 161 Patients, *Int. J. Radiat. Oncol. Biol. Phys.* vol. 10, pp. 2231–2239, 1994.
3. M. W. Dewhirst and T. V. Samulski, *Hyperthermia in the Treatment for Cancer*, Kalamazoo, MI: Upjohn, 1988.
4. S. B. Field and J. W. Hand, *An Introduction to the Practical Aspects of Hyperthermia*, New York: Taylor & Francis, 1990.
5. V. Muralidharan, C. Malconti-Wilson, and C. Cristophi, Interstitial Laser Hyperthermia for Colorectal Liver Metastases: The Effect of Thermal Sensitization and the Use of a Cylindrical Diffuser Tip on Tumor Necrosis, *J. Clin. Laser Med. Surg.* vol. 20, pp. 189–196, 2002.
6. E. J. Hall and L. Roizin-Towle, Biological Effects of Heat, *Cancer Res.* vol. 44, pp. 4708s–4713s, 1984.
7. C. Streffer, Biological Basis for the Use of Hyperthermia in Tumor Therapy, *Strahlentherapie und Onkologie* vol. 163, pp. 416–419, 1987.
8. J. Crezee and J. W. Lagendijk, Temperature Uniformity during Hyperthermia: The Impact of Large Vessels, *Phys. Med. Biol.* vol. 37, pp. 1321–1337, 1992.

9. C. M. Beacco, S. R. Mordon, and J. M. Bruneaud, Development and Experimental in Vivo Validation of Mathematical Modeling of Laser Coagulation, *Las. Surg. Med.* vol. 14, pp. 362–373, 1994.

10. J. W. Baish, Formulation of a Statistical Model of Heat Transfer in Perfused Tissue, *J. Biomech. Engin.* vol. 116, pp. 521–527, 1994.

11. D. E. Lemons, S. Chien, L. I. Crawshaw, S. Weinbaum, and L. M. Jiji, Significance of Vessel Size and Type in Vascular Heat Transfer, *Am. J. Physiol.* vol. 253, pp. R128–R135, 1987.

12. W. Levin, M. D. Shem, B. Cooper, R. E. Hill, J. W. Hunt, and E. F. Liu, The Effect of Vascular Occlusion on Tumor Temperatures during Superficial Hyperthermia, *Int. J. Hyperthermia* vol. 10, pp. 495–505, 1994.

13. R. B. Roemer, The Local Tissue Cooling Coefficient: A Unified Approach to Thermal Washout and Steady-State Perfusion Calculations, *Int. J. Hyperthermia* vol. 6, pp. 421–430, 1990.

14. M. C. Kolios, M. D. Sherar, and J. W. Hunt, Large Blood Vessel Cooling in Heated Tissues: A Numerical Study, *Phys. Med. Biol.* vol. 40, pp. 477–494, 1995.

15. M. C. Kolios, A. E. Worthington, D. W. Holdsworth, M. D. Sherar, and J. W. Hunt, An Investigation of the Flow Dependence of Temperature Gradients near Large Vessels during Steady State and Transient Tissue Heating, *Phys. Med. Biol.* vol. 44, pp. 1479–1497, 1999.

16. X. Tang, W. Dai, R. Nassar, and A. Bejan, Optimal Temperature Distribution in a Three-Dimensional Triple-Layered Skin Structure Embedded with Artery and Vein Vasculature, *Num. Heat Trans.: Pt. A: Appl.* vol. 50, pp. 809–834, 2006.

17. W. Dai, A. Bejan, X. Tang, L. Zhang, and R. Nassar, Optimal Temperature Distribution in a Three Dimensional Triple-Layered Skin Structure with Embedded Vasculature, *J. Appl. Phys.* vol. 99(104702), pp. 809–834, 2006.

18. Y. He, M. Shirazaki, H. Liu, R. Himeno, and Z. Sun, A Numerical Coupling Model to Analyze the Blood Flow, Temperature, and Oxygen Transport in Human Breast Tumor under Laser Irradiation, *Comp. Biol. Med.* vol. 36, pp. 1336–1350, 2006.

19. T. C. Shih, H. L. Liu, and A. Horng, Cooling Effect of Thermally Significant Blood Vessels in Perfused Tumor Tissue during Thermal Therapy, *Int. Commun. Heat Mass Trans.* vol. 33, pp. 135–141, 2006.

20. J. Zhou and J. Liu, Numerical Study on 3-D Light and Heat Transport in Biological Tissues Embedded with Large Blood Vessels during Laser-Induced Thermotherapy, *Numer. Heat Trans. A* vol. 45, pp. 415–449, 2004.

21. K. M. Khanafer, J. L. Bull, I. Pop, and R. Berguer, Influence of Pulsatile Blood Flow and Heating Scheme on the Temperature Distribution during Hyperthermia Treatment, *Int. J. Heat Mass Trans.* vol. 50, pp. 4883–4890, 2007.

22. O. I. Craciunescu and C. T. Clegg, Pulsatile Blood Flow Effects on Temperature Distribution and Heat Transfer in Rigid Vessels, *ASME J. Biomech. Engin.* vol. 123, pp. 500–505, 2001.

23. K. R. Diller, in Y. I. Cho, Editor, *Advances in Heat Transfer: Bioengineering Heat Transfer*, New York: Academic Press, 1992.

24. J. Voke, Lasers and Their Use in Ophthalmology, *Optom. Today: Pt. 3*, pp. 31–36, June 2001.

25. C. Purslow and J. Wolffsohn, Ocular Surface Temperature: A Review, *Eye Cont. Lens* vol. 31, pp. 117–123, 2005.

26. E. H. Ooi, W. T. Ang, and E. Y. K. Ng, Bioheat Transfer in the Human Eye: A Boundary Element Approach, *Engin. Anal. Bound. Elem.* vol. 31, pp. 494–500, 2007.

27. E. Y. K. Ng and E. H. Ooi, FEM Simulation of the Eye Structure with Bioheat Analysis, *Comp. Met. Prog. Biomed.* vol. 82, pp. 268–276, 2006.

28. E. Y. K. Ng and E. H. Ooi, Ocular Surface Temperature: A 3D FEM Prediction Using Bioheat Equation, *Comp. Biol. Med.* vol. 37, pp. 829–835, 2007.
29. J. A. Scott, A Finite Element Model of Heat Transport in the Human Eye, *Phys. Med. Biol.* vol. 33, pp. 227–242, 1988.
30. K. J. Chuak, J. C. Ho, S. K. Chou, and M. R. Islam, The Study of the Temperature Distribution within a Human Eye Subjected to a Laser Source, *Int. Commun. Heat Mass Trans.* vol. 32, pp. 1057–1065, 2005.
31. E. H. Amara, Numerical Investigations on Thermal Effects of Laser-Ocular Media Interaction, *Int. J. Heat Mass Trans.* vol. 38, pp. 2479–2488, 1995.
32. J. A. Scott, The Computation of Temperature Rises in the Human Eye Induced by Infrared Radiation, *Phys. Med. Biol.* vol. 33, pp. 243–257, 1988.
33. A. Hirata, Temperature Increase in Human Eyes Due to Near-Field and Far-Field Exposures at 900 MHz, 1.5 GHz and 1.9 GHz, *IEEE Trans. Electromagn. Compat.* vol. 47, pp. 68–76, 2005.
34. A. Hirata, S. Matsuyama, and T. Shiozawa, Temperature Rises in the Human Eye Exposure to EM Waves in the Frequency Range 0.6–6 GHz, *IEEE Trans. Electromagn. Compat.* vol. 42, pp. 386–393, 2000.
35. A. Hirata, S. Watanabe, M. Kojima, I. Hata, K. Wake, M. Taki, K. Sasaki, O. Fujiwara, and T. Shiozawa, Computational Verification of Anesthesia Effect on Temperature Variations in Rabbit Eyes Exposed to 2.45 GHz Microwave Energy, *Bioelectromagnetics* vol. 27, pp. 602–612, 2006.
36. J. J. W. Lagendijk, A Mathematical Model to Calculate Temperature Distributions in Human and Rabbit Eyes during Hyperthermic Treatment, *Phys. Med. Biol.* vol. 27, pp. 1301–1311, 1982.
37. H. H. Pennes, Analysis of Tissue and Arterial Blood Temperature in the Resting Human Forearm, *J. Appl. Phys.* vol. 1, pp. 93–122, 1948.
38. C. K. Charny, Mathematical Models of Bioheat Transfer, *Adv. Heat Trans.* vol. 22, pp. 19–152, 1992.
39. H. Arkin, L. X. Xu, and K. R. Holmes, Recent Developments in Modeling Heat Transfer in Blood Perfused Tissues, *IEEE Trans. Biomed. Engin.* vol. 41, pp. 97–107, 1994.
40. T. W. McGrail and R. C. Seagrave, Application of the Bioheat Transfer Equation in Fetal-Placental Studies, *Annals N.Y. Acad. Sci.* vol. 335, pp. 161–172, 1980.
41. C. Wood and R. W. Beard, Temperature of the Human Fetus, *J. Obstet. Gynec. Brit. Commun.* vol. 71, pp. 768–769, 1964.
42. K. Adamsonsk and M. E. Towell, Thermal Homeostasis in the Fetus and Newborn, *Anesthesiology* vol. 26, pp. 531–540, 1965.
43. F. M. Hart and J. J. Faber, Fetal and Maternal Temperatures in Rabbits, *J. Appl. Phys.* vol. 20, pp. 737–741, 1965.
44. D. Walker, A. Walker, and C. Wood, Temperature of the Human Fetus, *J. Obstet. Gynec. Brit. Commun.* vol. 76, pp. 503–511, 1969.
45. D. Abramsr, D. Caton, L. B. Curetc, C. Crenshaw, L. Mann, and D. Barron, Fetal Brain-Maternal Aorta-Temperature Differences in Sheep, *Am. J. Physiol.* vol. 217, pp. 1619–1622, 1969.
46. D. Abramsr, D. Caton, J. Clapp, and D. Barron, Thermal and Metabolic Features of Life in Utero, *Clin. Obstet. Gynec.* vol. 13, pp. 549–564, 1970.
47. D. Poljak, A. Peratta, and C. Brebbia, The Boundary Element Electromagnetic-Thermal Analysis of Human Exposure to Base Station Antennas Radiation, *Engin. Anal. Bound. Elem.* vol. 28, pp. 763–770, 2004.

48. C. K. Charny and R. L. Levin, Heat Transfer Normal to Paired Arteries and Veins Embedded in Perfused Tissue during Hyperthermia, *Trans. ASME J. Biomech. Engin.* vol. 110, pp. 277–282, 1988.

49. C. K. Charny and R. L. Levin, Bioheat Transfer in a Branching Countercurrent Net-work during Hyperthermia, *ASME J. Biomech. Engin.* vol. 111, pp. 263–270, 1989.

50. R. B. Roemer and T. C. Cetas, Applications of Bioheat Transfer Simulations in Hyperthermia, *Cancer Res.* vol. 44, pp. 4788–4798, 1984.

51. W. Wulff, The Energy Conservation Equation for Living Tissues, *IEEE Trans. Biomed. Engin.* vol. 21, pp. 494–495, 1974.

52. H. G. Klinger, Heat Transfer in Perfused Tissue—I: General Theory, *Bull. Math. Biol.* vol. 36, pp. 403–415, 1974.

53. M. M. Chen and K. R. Holmes, Microvascular Contributions in Tissue Heat Transfer, *Annals N.Y. Acad. Sci.* vol. 335, pp. 137–150, 1980.

54. S. Weinbaum and L. M. Jiji, A Two Phase Theory for the Influence of Circulation on the Heat Transfer in Surface Tissue, in M. K. Wells, Editor, *Advances in Bioengineering*, pp. 179–182, New York: ASME, 1979.

55. S. Weinbaum, L. M. Jiji, and D. E. Lemons, Theory and Experiment for the Effect of Vascular Microstructure on Surface Tissue Heat Transfer: Part I: Anatomical Foundation and Model Conceptualization, *ASME J. Biomech. Engin.* vol. 106, pp. 321–330, 1984.

56. S. Weinbaum, L. M. Jiji, and D. E. Lemons, Theory and Experiment for the Effect of Vascular Microstructure on Surface Tissue Heat Transfer: Part II: Model Formulation and Solution, *ASME J. Biomech. Engin.* vol. 106, pp. 331–341, 1984.

57. S. Weinbaum and L. M. Jiji, A New Simplified Equation for the Effect of Blood Flow on Local Average Tissue Temperature, *ASME J. Biomech. Engin.* vol. 107, pp. 131–139, 1985.

58. A. Bejan, I. Dincer, S. Lorente, A. Miguel, and A. Reis, *Porous and Complex Flow Structures in Modern Technologies*, New York: Springer-Verlag, 2004.

59. K. Khanafer and K. Vafai, The Role of Porous Media in Biomedical Engineering as Related to Magnetic Resonance Imaging and Drug Delivery, *Heat Mass Trans.* vol. 42, pp. 939–953, 2006.

60. A.-R. A. Khaled and K. Vafai, The Role of Porous Media in Modeling Flow and Heat Transfer in Biological Tissues, *Int. J. Heat Mass Trans.* vol. 46, pp. 4989–5003, 2003.

61. L. Ai and K. Vafai, A Coupling Model for Macromolecule Transport in a Stenosed Arterial Wall, *Int. J. Heat Mass Trans.* vol. 49, pp. 1568–1591, 2006.

62. N. Yang and K. Vafai, Low Density Lipoprotein (LDL) Transport in an Artery: A Simplified Analytical Solution, *Int. J. Heat Mass Trans.* vol. 51, pp. 497–505, 2008.

63. M. Khakpour and K. Vafai, A Critical Assessment of Arterial Transport Models, *Int. J. Heat Mass Trans.* vol. 51, pp. 807–822, 2008.

64. M. Khakpour and K. Vafai, A Complete Analytical Solution for Mass Transport within a Multilayer Arterial Wall, *Int. J. Heat Mass Trans.* vol. 51, pp. 2905–2913, 2008.

65. K. Khanafer, A. R. A. Khaled, and K. Vafai, Spatial Optimization of an Array of Aligned Microcantilever Biosensors, *J. Micromech. Microengin.* vol. 14, pp. 1328–1336, 2004.

66. A. R. A. Khaled and K. Vafai, Optimization Modeling of Analyte Adhesion over an Inclined Microcantilever-Based Biosensor, *J. Micromech. Microengin.* vol. 14, pp. 1220–1229, 2004.

67. A. R. A. Khaled and K. Vafai, Analysis of Oscillatory Flow Disturbances and Thermal Characteristics inside Fluidic Cells Due to Fluid Leakage and Wall Slip Conditions, *J. Biomech.* vol. 3, pp. 721–729, 2004.

68. K. Khanafer and K. Vafai, Geometrical and Flow Configurations for Enhanced Microcantilever Detection within a Fluidic Cell, *Int. J. Heat Mass Trans.* vol. 48, pp. 2886–2895, 2005.

69. K. Vafai, *Handbook of Porous Media*, New York: Marcel Dekker, 2000.

70. K. Vafai, *Handbook of Porous Media*, 2nd ed., New York: Taylor & Francis Group, 2005.

71. H. Hadim and K. Vafai, Overview of Current Computational Studies of Heat Transfer in Porous Media and Their Applications: Forced Convection and Multiphase Transport, *Adv. Numer. Heat Trans.* vol. 2, pp. 291–330, 2000.

72. K. Vafai and H. Hadim, Overview of Current Computational Studies of Heat Transfer in Porous Media and Their Applications: Natural Convection and Mixed Convection, *Adv. Numer. Heat Trans.* vol. 2, pp. 331–371, 2000.

73. Y. M. Xuan and W. Roetzel, Bioheat Equation of the Human Thermal System, *Chem. Eng. Technol.* vol. 20, pp. 268–276, 1997.

74. Y. M. Xuan and W. Roetzel, Transfer Response of the Human Limb to an External Stimulus, *Int. J. Heat Mass Trans.* vol. 41, pp. 229–239, 1998.

75. A. Amiri and K. Vafai, Analysis of Dispersion Effects and Nonthermal Equilibrium, Non-Darcian, Variable Porosity Incompressible-Flow through Porous Media, *Int. J. Heat Mass Trans.* vol. 37, pp. 939–954, 1994.

76. A. Amiri and K. Vafai, Transient Analysis of Incompressible Flow through a Packed Bed, *Int. J. Heat Mass Trans.* vol. 41, pp. 4259–4279, 1998.

77. B. Alazmi and K. Vafai, Constant Wall Heat Flux Boundary Conditions in Porous Media under Local Thermal Non-Equilibrium Conditions, *Int. J. Heat Mass Trans.* vol. 45, pp. 3071–3087, 2002.

78. D. Y. Lee and K. Vafai, Analytical Characterization and Conceptual Assessment of Solid and Fluid Temperature Differentials in Porous Media, *Int. J. Mass Trans.* vol. 42, pp. 423–435, 1999.

79. K. Vafai and M. Sozen, Analysis of Energy and Momentum Transport for Fluid Flow through a Porous Bed, *ASME J. Heat Trans.* vol. 112, pp. 690–699, 1990.

80. K. Vafai and M. Sozen, An Investigation of a Latent Heat Storage Packed Bed and Condensing Flow through It, *ASME J. Heat Trans.* vol. 112, pp. 1014–1022, 1990.

81. K. Vafai and M. Sozen, A Comparative Analysis of Multiphase Transport Models in Porous Media, *Ann. Rev. Heat Trans.* vol. 3, pp. 145–162, 1990.

82. M. Sozen and K. Vafai, Analysis of Oscillating Compressible Flow through a Packed Bed, *Int. J. Heat Fluid Flow* vol. 12, pp. 130–136, 1991.

83. M. Sozen and K. Vafai, Analysis of the Non-Thermal Equilibrium Condensing Flow of a Gas through a Packed Bed, *Int. J. Heat Mass Trans.* vol. 33, pp. 1247–1261, 1990.

84. A. Marafie and K. Vafai, Analysis of Non-Darcian Effects on Temperature Differentials in Porous Media, *Int. J. Heat Mass Trans.* vol. 44, pp. 4401–4411, 2001.

85. K. Vafai and C. L. Tien, Boundary and Inertia Effects on Convective Mass Transfer in Porous Media, *Int. J. Heat Mass Trans.* vol. 25, pp. 1183–1190, 1981.

86. K. Vafai and C. L. Tien, Boundary and Inertia Effects on Flow and Heat Transfer in Porous Media, *Int. J. Heat Mass Trans.* vol. 24, pp. 195–203, 1980.

87. Z. J. Huang and J. M. Tarbell, Numerical Simulation of Mass Transfer in Porous Media of Blood Vessel Walls, *Am. J. Physiol.* vol. 273, pp. H464–H477, 1997.

2

Numerical Models of Blood Flow Effects in Biological Tissues

J. W. Baish, K. Mukundakrishnan,
and P. S. Ayyaswamy

CONTENTS

2.1 INTRODUCTION

This chapter reviews the numerical methods used to predict temperatures in perfused tissues for temperatures just above or below normal body temperature where the effects of phase change are unimportant. Our emphasis is on the effects of blood flow that make tissue different from a simple conducting solid. The range of temperatures considered covers those for thermal therapies such as hyperthermia treatment (seeking temperatures from 43°C to 50°C) and moderate forms of thermal ablation that exceed this range but neither char nor boil away tissue. Likewise, cooling will be discussed, but little emphasis will be given to tissue freezing. The reader is directed elsewhere [1–3] for information on these topics.

Heat transfer in perfused tissues is shaped by processes on many scales from the nanoscale of cell membranes and organelles through the macroscale of the organism as a whole. For most clinical applications, the scales of most interest range from that of the smallest blood vessels, the capillaries, with dimensions of several microns through those of specific anatomical structures such as organs measuring several centimeters or more. Between these two limits, the intended purpose of the model will be a major factor in deciding whether a blood vessel will need to be considered individually or whether its effects might be modeled collectively with those of numerous other vessels. For example, the question of whether underheating during hyperthermia treatment occurs near a 2 mm diameter blood vessel will likely include a geometric description of the vessel, its blood flow, and the local heating pattern, whereas a question of whether the same hyperthermia treatment will increase the average temperature in the same or an adjacent organ might be addressed with a continuum model of the vicinity of the heat source. Such a model would likely include the net effects of blood flow in each organ, but would not represent individual blood vessels. Between these two extremes, a question of whether there might be a small fraction of tissue within an organ that is underheated due to numerous blood vessels measuring a few hundred microns in diameter or smaller will require a model that links the vascular scale to the organ scale.

The formulation of suitable continuum models that account for the effects of many blood vessels in a collective sense has been a lively subject of debate for many decades. While some such models have been proposed on purely empirical grounds, the goal of putting these models on a firm theoretical foundation has spurred efforts to model heat transfer due to numerous blood vessels in realistic geometrical arrangements. In the following sections, we will review the basic physics of heat transfer in perfused tissues and the most widely used approaches to account for the thermal effects of blood flow in individual vessels or collections of vessels.

2.2 BASIC CONCEPTS

The most common formulation of heat transfer in perfused tissue begins with a division into two subvolumes that share an internal boundary, the vascular wall. Inside the vessel walls, the blood subvolume contains liquid blood moving under the action of the upstream pressures produced in the heart (and to a lesser extent by smooth muscle in the vascular wall). The surrounding tissue subvolume is considered to be a solid through which no flow occurs. For heat transfer purposes, the small quantity of interstitial flow that arises from permeability of the vessel walls is generally neglected.

Heat transfer in the tissue subvolume is represented by a standard heat conduction equation as follows [4]:

$$\nabla \cdot k_t \nabla T_t(\mathbf{r},t) + q_t'''(\mathbf{r},t) = \rho_t c_t \frac{\partial T_t(\mathbf{r},t)}{\partial t} \tag{2.1}$$

where T_t is the local tissue temperature, k_t is the thermal conductivity of the tissue, q_t''' is the rate of volumetric heat generation from metabolism or external sources, ρ_t is the tissue density, and c_t is the tissue specific heat. The material properties are effective values averaged over any heterogeneities that exist on the cellular scale. In the blood subvolume, heat transfer may occur by advection as well, yielding the following [4]:

$$\nabla \cdot k_b \nabla T_b(\mathbf{r},t) - \rho_b c_b \mathbf{u}_b(\mathbf{r},t) \cdot \nabla T_b(\mathbf{r},t) + q_b'''(\mathbf{r},t) = \rho_b c_b \frac{\partial T_b(\mathbf{r},t)}{\partial t} \tag{2.2}$$

where $\mathbf{u}_b$ is the local blood velocity and all other parameters pertain to the local properties of the blood. Potential energy, kinetic energy, and viscous dissipation effects are typically negligible in living tissue.

The blood vessel walls form the internal boundary between the tissue and blood subvolumes. On this surface, we expect a continuity of heat flux:

$$k_b \nabla T_b(\mathbf{r}_b,t) = k_t \nabla T_t(\mathbf{r}_b,t) \tag{2.3}$$

and temperature:

$$T_b(\mathbf{r}_b,t) = T_t(\mathbf{r}_b,t) \tag{2.4}$$

where $\mathbf{r}_b$ represents points on the vessel wall.

Determining the geometry and velocity field inside a single, particular blood vessel is a challenging but tractable problem, but the major obstacle in modeling perfused tissues is that the total number of blood vessels in even a cubic centimeter of tissues can number 10,000 or more with the vascular architecture being extraordinarily complex and specific to each organ. The issue then is not with merely formulating the coupled convection–advection problem above, but with having adequate information on the blood velocity and the geometry of the boundary linking the tissue and blood subvolumes. In Section 2.2.1, we review some of the basic approaches used to model heat transfer to individual blood vessels when they are well characterized, followed by discussion of the continuum models used to represent large numbers of vessels collectively.

2.2.1 Heat Transfer to Blood Vessels

Inside each blood vessel, the heat transfer problem is generally a question of how the temperature of the blood changes along the length of the vessel as heat is exchanged with the surrounding tissue. Figure 2.1 shows the immediate surroundings of a typical blood vessel. Each vessel is assumed to be surrounded by similar vessels so that the outer surface of the tissue cylinder forms an approximate boundary of symmetry with adjacent tissue cylinders. The tissue surrounding each vessel can be expected to include capillaries or other small vessels, but most analyses of individual blood vessels neglect advective effects from blood vessels on a smaller scale. A full transient analysis of a typical blood vessel is readily accomplished, but the focus in here is on steady-state processes. This is justifiable

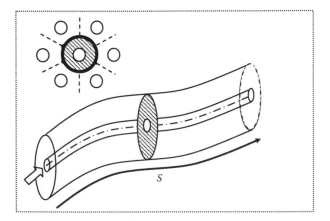

Figure 2.1   Geometry near a typical blood vessel. The shaded area represents the approximate circular area of solid tissue over which averaging is done to determine the local average tissue temperature. The diameter of the tissue cylinder ranges from 10 to 40 times the diameter of the blood vessel for thermally significant blood vessels. Large vessels tend to be more widely spaced than small vessels.

because the distance over which heat conducts in most clinically relevant problems is much greater than the vascular spacing. Recognizing the diffusion time scales as $t_{Diffusion} \sim \delta^2/\alpha$, where δ is the distance over which diffusion takes place and α is the thermal diffusivity, we see that gradients near an individual vessel decay much faster than those over several vessel spacings. Heat generation, axial conduction, and pulsatility in the blood are also neglected.

The rate of heat transfer into the vessel under the prescribed conditions can be represented by the following [5]:

$$\dot{m}_b c_b \frac{d\bar{T}_b(s)}{ds} = q'(s) \tag{2.5}$$

where $\bar{T}_b(s)$ is the mixed mean temperature of the blood for a given vessel cross section, $q'(s)$ is the rate at which heat conducts into the vessel per unit length, s is the spatial coordinate along the vessel axis, and $\dot{m}_b$ is the mass flowrate of the blood in the vessel. Assuming a circular cross section for the blood vessel, we can relate the mass flowrate of the blood to the velocity with

$$\dot{m}_b = \pi r_b^2 \rho_b \bar{u}_b \tag{2.6}$$

where r_b is the vessel radius and $\bar{u}_b$ is the mean speed of the blood across the vessel cross section. The mixed mean temperature of the blood is defined as follows [4]:

$$\bar{T}_b(s) = \frac{1}{\pi r_b^2 \bar{u}_b} \int_0^{r_b} u(r)T(r)2\pi r\, dr \tag{2.7}$$

The rate of heat transfer into the blood vessels may be related to the local average tissue temperature $\bar{T}_t(s)$ (see Figure 2.1) and the blood temperatures as follows:

$$q'(s) = U' 2\pi r_b \left(\bar{T}_t(s) - \bar{T}_b(s)\right) \tag{2.8}$$

where U' is the overall heat transfer coefficient between the tissue and the blood. The thermal resistance due to conduction through the surrounding solid tissue and that between the vessel wall and the mixed mean temperature of the blood are included in the calculation of U'. Local heat transfer rates are typically calculated from a steady-state analysis of the vicinity of the vessels. Convection between the vessel wall at $T_t(r_b)$ and the mixed mean temperature of the blood can be represented by

$$q' = h 2\pi r_b \left(T_t(r_b) - \bar{T}_b\right) \tag{2.9}$$

where the convective heat transfer coefficient h is related to the Nusselt number Nu by

$$h = 2r_b Nu/k_b \tag{2.10}$$

The convective heat transfer coefficient h may be found from Victor and Shah's [6] recommendation that the Nusselt number may be obtained from

$$Nu = \frac{2r_b h}{k_b} = 4 + 0.155 \exp(1.58 \log_{10} Gz) \quad Gz < 103 \tag{2.11}$$

where Gz is the Graetz number, defined as

$$Gz = \frac{4\rho_b c_b \bar{u}_b r_b^2}{k_b L} \tag{2.12}$$

where L is the vessel length. See also Barozzi and Dumas [7]. Many studies use approximations based on established flow conditions that yield $Nu \approx 4$ [8,9].

Neglecting advection in smaller vessels in the tissue nearby, the rate of heat conduction into the vessel of interest can often be expressed as

$$q' = k_t \sigma (\bar{T}_t - T_t(r_b)) \tag{2.13}$$

where the conduction shape factor σ depends only on the geometry of the vessel and surrounding tissue in cross section. For a vessel centered in a cylinder of tissue, the shape factor can be approximated by

$$\sigma \approx \frac{2\pi}{\ln\left(\frac{r_t}{r_b}\right)} \tag{2.14}$$

where r_t is the radius of the tissue cylinder. Recognizing that the thermal resistances inside and outside the vessel are in series, we have

$$\frac{1}{U'} = \frac{2\pi r_b}{k_t \sigma} + \frac{2r_b}{Nu k_b} \tag{2.15}$$

Numerous other configurations of the blood vessels such as vessels near surfaces, near vascular junctions, or in countercurrent pairs have been considered in the literature [10,11]. For example, closely spaced arteries and veins in counterflow can be modeled by the following (Figure 2.2):

$$q'(s) = k_t \sigma_\Delta (\bar{T}_v(s) - \bar{T}_a(s)) \tag{2.16}$$

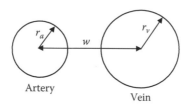

Figure 2.2   Cross section of an artery–vein pair with countercurrent flow. The diameter of the vein is often two to three times that of the artery.

where $\overline{T}_a(s)$ is the mixed mean temperature in the artery, $\overline{T}_v(s)$ is the mixed mean temperature in the adjacent vein, and the conduction shape factor (neglecting thermal resistance inside the vessels) is given approximately by the following [4]:

$$\sigma_\Delta \approx \frac{2\pi}{\cosh^{-1}\left(\frac{w^2 - r_a^2 - r_v^2}{2r_a r_v}\right)} \qquad (2.17)$$

where w is the distance between the vessel axes, r_a is the radius of the artery, and r_v is the radius of the vein.

2.2.2 Equilibration Lengths

A valuable application of the simplified vascular models above is in determination of the anatomical scale at which blood exchanges the greatest amount of heat with the tissue and as a result where the blood reaches equilibrium with the surrounding tissue temperature [12]. Consider an idealized situation in which a single blood vessel passes through a representative tissue cylinder such as that shown in Figure 2.1, which is held at a constant temperature $\overline{T}_t$. The difference between the blood and the tissue temperature will decay along the length of the vessel according to

$$\frac{\overline{T}_b(s) - \overline{T}_t}{\overline{T}_b(0) - \overline{T}_t} = \exp\left(-\frac{s}{L_e}\right) \qquad (2.18)$$

where the equilibration length is given by

$$L_e = \frac{\dot{m}_b c_b}{U' 2\pi r_b} \qquad (2.19)$$

The anatomical site of equilibration can be obtained by comparing the equilibration length as calculated in Equation (2.19) to the actual length of the vessels. Numerous studies show that blood passes through the largest vessels in the body with little change in temperature $(L_e \gg L)$ [11,12]. In contrast, the equilibration length in the capillaries is so short that blood entering these vessels must be assumed to be already in equilibrium with the surrounding tissue ($L_e \ll L$). Blood in vessels of an intermediate size, ranging from 50 to several hundred microns in diameter, can be expected to undergo the most active thermal equilibration ($L_e \approx L$). Such vessels have been identified as "thermally significant." Vessels in the thermally significant range are frequently found in a countercurrent arrangement that has been shown to influence their net effect on tissue heat transfer [11,13,14].

The configurations considered in equilibration length studies are highly idealized, suitable for making order-of-magnitude estimates of blood temperatures. More detailed and realistic models of the vascular architecture and the interaction of large numbers of blood vessels with the tissue will be described in later Section 2.5. Nonetheless, equilibration length estimates are valuable for assessing when vessels must be analyzed explicitly rather than lumped together in a continuum formulation such as those in Section 2.3.

2.3 MODELS OF PERFUSED TISSUES

2.3.1 Continuum Models

Continuum models of perfused tissues are used to predict a local average tissue [12]:

$$\bar{T}_t(\mathbf{r},t) = \frac{1}{\delta V} \int_{\delta V} T_t(\mathbf{r}',t)dV' \tag{2.20}$$

where δV is a volume that is assumed to be large enough to encompass a reasonable number of thermally significant blood vessels, but is much smaller than the scale of the tissue as a whole. These models are based on modifications to the standard heat diffusion equation such that the collective effects of blood flow can be approximately incorporated without having to solve the full, coupled conduction–advection problem defined in Equations (2.1) through (2.4). The various forms of the continuum models arise from the differing assumptions used with regard to the effects of the blood subvolume. These models all have the advantages that they require relatively little or no information on the vascular architecture and the flow in any particular blood vessels. Blood flow is modeled with only a few adjustable parameters that may be obtained empirically or from analysis of simplified vascular geometries. Moreover, the governing equations are all readily solved by a variety of well-established numerical methods, which will be summarized later in this chapter.

2.3.2 Pennes Heat Sink Model

The most widely used continuum model of perfused tissue was introduced in 1948 by Harry Pennes [15]. The Pennes formulation is so widely used that it is sometimes known as the "bioheat equation" despite the existence of several alternatives. Pennes modified the heat diffusion equation by introducing a source or sink term as follows [15]:

$$\nabla \cdot k \nabla \overline{T}_t(\mathbf{r},t) + \varepsilon(\mathbf{r},t)\omega_b(\mathbf{r},t)\rho_b c_b (T_a - \overline{T}_t(\mathbf{r},t)) + q'''(\mathbf{r},t) = \rho c \frac{\partial \overline{T}_t(\mathbf{r},t)}{\partial t} \qquad (2.21)$$

where ε is a heat transfer effectiveness $(0 \le \varepsilon \le 1)$, ω_b is taken to be the blood perfusion rate in units of volume of blood per unit volume of tissue per unit time, and T_a is an arterial supply temperature. The properties k, ρ, and c are composite properties taken to incorporate the tissue and blood subvolumes. The key assumption in this model is that blood is delivered to each capillary (and, as a result, the vicinity of every cell in the body) at a constant arterial temperature, which is generally taken to equal the core temperature of the body, nominally 37°C. After delivery to the capillary bed, the blood is then assumed to be removed through veins that no longer influence the tissue temperature. Pennes originally introduced ε to allow for the possibility that blood never reached full equilibration with the tissue. Later equilibration length studies showed that full equilibration takes place long before blood reaches the capillary bed, implying that $\varepsilon = 1$. A more modern interpretation of the heat transfer effectiveness accounts for heat exchange between closely spaced arteries and veins in the thermally significant size range. Studies of countercurrent exchange yield estimates in the range $0.5 \le \varepsilon \le 0.8$ [14,16,17]. The issues related to numerical solutions of the Pennes equation are unchanged by whether the combination $\varepsilon\omega_b$ is known or ε and ω_b are known independently.

The actual size of the volume δV over which $\overline{T}_t(\mathbf{r},t)$ is averaged when interpreting the Pennes equation is somewhat ambiguous. In order for blood to be available to each δV of tissue at near-core temperature, we would expect that δV should be large enough to consistently include nonequilibrated vessels, which will typically have diameters greater than several hundred microns. Equilibration length studies point to a scale of about $\delta V > 1$ cm³. The tissue temperature $\overline{T}_t(\mathbf{r},t)$ would then be a sort of moving average over this scale. Unfortunately, such a moving average does not yield information on the range of temperatures present in the averaging volume. Inherent to the Pennes approach is the assumption that each averaging volume contains at least one vessel carrying blood near the core temperature. When used to model a thermal therapy such as hyperthermia in which the heating of the entire tumor to > 43°C is desirable, the Pennes formulation implicitly carries an assumption that a small volume of tissue remains at 37°C. While such an undertreated bit of tissue may or may not actually exist, its possible presence cannot be assessed with the Pennes formulation alone.

2.3.3 Directed Perfusion Model

In 1974, Wulff criticized the Pennes formulation on the grounds that it contains an awkward assumption of two temperatures at each point in space ($\overline{T}_t(r,t)$ and T_a)

and that it neglects the directional nature of blood flow. As an alternative to the Pennes equation, he proposed the following [18]:

$$\nabla \cdot k \nabla \overline{T_t}(\mathbf{r},t) - \rho c \mathbf{u}(\mathbf{r},t) \cdot \nabla \overline{T_t}(\mathbf{r},t) + q''(\mathbf{r},t) = \rho c \frac{\partial \overline{T_t}(\mathbf{r},t)}{\partial t} \qquad (2.22)$$

where $\mathbf{u}(\mathbf{r},t)$ is a velocity averaged over both the tissue and blood subvolumes. Wulff could have resolved the apparent contradiction in the Pennes formulation of two temperatures at each spatial point by interpreting the tissue temperature as a local average over δV rather than as a temperature at a point, but he raises a more complex issue with regard to the directional nature of blood flow. Baish et al. [9] showed that the Wulff formulation is valid when the tissue and blood subvolumes are in near equilibrium $(\overline{T_t}(\mathbf{r},t) \approx \overline{T_b}(\mathbf{r},t))$, as is commonly true in nonbiological porous media. In small volumes of tissue, including only thermally insignificant blood vessels, this assumption might be valid, but even on such a small scale many vessels are found in countercurrent pairs. The net contribution of a countercurrent pair with equal supply and return flows to $\mathbf{u}(\mathbf{r},t)$ will cancel out, leading to the misleading conclusion that blood flow does not contribute to heat transfer in perfused tissue. Despite its limited practical applicability, the Wulff formulation prompted a more careful examination of the vascular contribution to tissue heat transfer.

2.3.4 Effective Conductivity Model

Historically, the oldest continuum model simply added an enhancement to the thermal conductivity to account for the blood flow such that [19]:

$$\nabla \cdot k_{eff} \nabla \overline{T_t}(\mathbf{r},t) + q'''(\mathbf{r},t) = \rho c \frac{\partial \overline{T_t}(\mathbf{r},t)}{\partial t} \qquad (2.23)$$

where the effective conductivity k_{eff} is composed of the intrinsic thermal conductivity of the tissue and a perfusion-dependent increment. An empirical value of k_{eff} can be defined in any situation in which a heat flux can be related to temperature difference between parallel surfaces, that is,

$$k_{eff} \equiv \frac{q''}{\Delta T} \qquad (2.24)$$

An important application of this approach is in the use of small inserted thermistor probes to measure the blood perfusion. The heat loss from such a heated probe can be modeled with an effective or apparent conductivity that can, in turn, be correlated analytically with the perfusion rate in the Pennes equation [20–28].

While use of such an effective conductivity sometimes provides a convenient way of correlating experimental data, caution is advised unless the value of k_{eff} is independent of the geometry of the problem, as would be the case for an intrinsic thermodynamic property such as k_t.

An alternative perspective on the effective conductivity model was developed in the 1980s by several investigators [12,13]. This approach views the effective conductivity as being analogous to an eddy diffusivity. Estimates of how the effective conductivity depends on the vascular architecture and flow were then developed by analytical means. For example, the Weinbaum-Jiji formulation [13] considered a tissue that had closely spaced, countercurrent artery–vein pairs as its dominant vascular feature. Under the condition that the tissue temperature along the vessel axes satisfied

$$\bar{T}_t \approx \frac{1}{2}(\bar{T}_a + \bar{T}_v) \tag{2.25}$$

they were able to show that the effective conductivity could be calculated with

$$k_{eff} = k_t \left(1 + \frac{\pi^2 \rho_b^2 c_b^2 n r_a^4 \bar{u}_b^2 \cos^2 \phi}{k_t^2 \sigma_\Delta} \right) \tag{2.26}$$

where the enhancement is in the direction of the vessel axes, n is the number of artery–vein pairs per unit area, and ϕ is the angle of the vessel axes relative to the temperature gradient. Further studies extended the model to vessels of unequal size and better identified the limitations of the model. The Weinbaum-Jiji formula for the effective conductivity is most appropriate for tissue without intense volumetric heating and containing vessels no larger than 300 μm in diameter. When the blood vessels are not in near equilibrium with the tissue, as required above, Baish showed that a countercurrent pair of vessels could still be modeled as a highly conductive pathway through the tissue, but one that was not necessarily in local equilibrium, much like metal fibers in a low-conductivity composite [29]. The radius of such a fiber was found to satisfy

$$r_{fiber} = (w r_b)^{1/2} \tag{2.27}$$

while its conductivity is given by

$$k_{fiber} = \frac{\rho_b^2 c_b^2 \bar{u}_b^2 r_b^3 \cosh^{-1}(w/r_b)}{w k_t} \tag{2.28}$$

2.3.5 Combination Models

A landmark study by Chen and Holmes in 1980, showed that the effective conductivity, heat sink, and directed perfusion mechanism are likely to be simultaneously present in perfused tissue [12]. They proposed a combination model expressed as

$$\nabla \cdot k_{e\!f\!f} \nabla \overline{T}_t(\mathbf{r},t) + \omega_b(\mathbf{r},t)\rho_b c_b(T_a^* - \overline{T}_t(\mathbf{r},t))$$

$$-\rho c \mathbf{u}(\mathbf{r},t) \cdot \nabla \overline{T}_t(\mathbf{r},t) + q'''(\mathbf{r},t) = \rho c \frac{\partial \overline{T}_t(\mathbf{r},t)}{\partial t} \qquad (2.29)$$

where T_a^* is the temperature exiting the last artery that is individually modeled. On a practical level, the combination formulation presents little new difficulty for numerical solution, but requires knowledge of many adjustable parameters that are seldom known simultaneously with high precision. In particular, the directed perfusion term similar to that found in Equation (2.22) is rarely known well enough to be used.

Roemer and Dutton [30] extended the work of Chen and Holmes by introducing additional terms that more completely represent the many ways in which the tissue and blood subvolumes can interact in tissue. Roemer and Dutton's formulation yields valuable insight into the meaning of the various terms, but does not reduce to a single differential equation as in the other continuum formulations. Instead, their model is now of the integro-differential form, requiring simultaneous evaluation of integrals over the blood subvolume and differential equations for the averaged tissue temperature.

2.3.6 Parameter Values

An extensive review of the parameter values used in the various preceding formulations may be found elsewhere [31]. The intrinsic thermal properties of tissues such as the density, specific heat, and thermal conductivity depend primarily on the water, fat, and protein content of the tissue. More complex, dynamic relationships may exist between the tissue temperature and the perfusion rate and metabolism due to thermoregulation. A few representative examples follow.

Changes in blood flow due to thermoregulation may be modeled by the following dependence of the effective conductivity on the tissue temperature [32]:

$$k_{e\!f\!f} = 4.82 - 4.44833[1.00075^{-1.575(\overline{T}_t - 25)}]\,W/m - K \qquad (2.30)$$

Alternatively, the blood perfusion term in the Pennes equation ($\omega_b \rho_b$ in kg/m³–s) can be assigned a temperature dependence on the tissue temperature (°C) as follows for muscle [33,34]:

$$\omega_b \rho_b = \begin{cases} 0.45 + 3.55\exp\left[-\dfrac{(\overline{T}_t - 45.0)^2}{12.0}\right], & \overline{T}_t \le 45.0 \\[2mm] 4.00, & \overline{T}_t > 45.0 \end{cases} \qquad (2.31)$$

for fat:

$$
\omega_b \rho_b = \begin{cases} 0.36 + 0.36\exp\left[-\dfrac{(\bar{T_t} - 45.0)^2}{12.0}\right], & \bar{T_t} \leq 45.0 \\ \\ 0.72, & \bar{T_t} > 45.0 \end{cases}
\tag{2.32}
$$

and for tumors:

$$
\omega_b \rho_b = \begin{cases} 0.833, & \bar{T_t} < 37.0 \\ 0.833 - (\bar{T_t} - 37.0)^{4.8}/5.438x10^3, & 37.0 \leq \bar{T_t} < 42.0 \\ 0.416, & \bar{T_t} > 42.0 \end{cases}
\tag{2.33}
$$

Metabolic rates also show a dependence on temperature that may be modeled by the following [35]:

$$
q'''_{met} = 170(2)^{[(T_o - \bar{T_t})/10]} \; W/m^3
\tag{2.34}
$$

where T_o is a reference temperature. Numerical solution of the continuum models will sometimes require iteration due to the nonlinearities introduced by the temperature dependence of the parameter values. Because these dependencies are relatively weak, the numerical difficulties are seldom problematic.

2.4 SOLUTIONS OF CONTINUUM MODELS

The majority of numerical methods for solving the equations of the continuum models can be broadly classified in three types, namely, (1) finite difference method, (2) finite element, and (3) finite volume methods.

The combination model of Chen and Holmes will be taken as the representative equation for our numerical analysis since all other continuum models are but limit cases of the combination model. For example, the Pennes equation can be recovered from the combination model by setting $\mathbf{u}_b = 0$. Hence, in this study, for generality purposes, the numerical solution of the Pennes equation for tissue heat transfer modeling will be discussed mainly in the context of the combination model of Chen and Holmes given by Equation (2.29).

Consider Equation (2.29):

$$
\rho c \frac{\partial \bar{T_t}(\mathbf{r},t)}{\partial t} = \nabla \cdot k_{eff} \nabla \bar{T_t}(\mathbf{r},t) + \omega_b(\mathbf{r},t)\rho_b c_b(T_a^* - \bar{T_t}(\mathbf{r},t))
$$

$$
- \rho c \mathbf{u}(\mathbf{r},t) \cdot \nabla \bar{T_t}(\mathbf{r},t) + q'''(\mathbf{r},t), \; \mathbf{r} \in \Omega
\tag{2.35}
$$

where Ω is the physical domain over which the problem is posed with the following boundary conditions,

$$\bar{T}_t = f(\mathbf{r},t) \text{ for } \mathbf{r} \text{ on } \Gamma_1, \ t>0 \tag{2.36a}$$

and

$$-k_{eff} \nabla \bar{T}_t(\mathbf{r},t) \cdot \mathbf{n} = g(\mathbf{r},t) \text{ for } \mathbf{r} \text{ on } \Gamma_2, \ t>0 \tag{2.36b}$$

where $\mathbf{n}$ is the unit normal vector on the boundary of Ω that consists of the union of Γ_1 and Γ_2. The initial condition for the problem is

$$\bar{T}_t = h(\mathbf{r}) \text{ for } \mathbf{r} \text{ on } \Omega, \ t=0 \tag{2.37}$$

2.4.1 Finite Difference Method

The bases of the finite difference method are the construction of a discrete grid, the replacement of continuous derivatives in the governing equation to equivalent finite difference expressions using Taylor series expansion techniques, and solving the resultant algebraic equations.

Consider a discrete point in the grid uniquely specified by the spatial indices (i, j, k) such that the coordinates are given by $(x,y,z)_{i,j,k} = (i\Delta x, j\Delta y, k\Delta z)$, where Δx, Δy, and Δz are the mesh discretization sizes. Let the time at which the temperature is evaluated be denoted by n such that $t^n = n\Delta t$, where Δt is the discrete time step size. Then, the Taylor series expansion for the temperature at the point $(i + 1, j, k)$ about the point (i, j, k) at a time level n, and the Taylor series expansion for the temperature at the point (i, j, k) at the time level $n + 1$ about the time level n, are given as follows:

$$\bar{T}_{i+1}^n = \sum_{m=0}^{\infty} \frac{\Delta x^m}{m!} \left[\frac{\partial^m \bar{T}}{\partial x^m} \right]_i^n, \qquad \bar{T}_i^{n+1} = \sum_{m=0}^{\infty} \frac{\Delta t^m}{m!} \left[\frac{\partial^m \bar{T}}{\partial t^m} \right]_i^n \tag{2.38}$$

Only the ith direction expansions are considered, and expansions along other directions follow from a change of the appropriate subscript indices.

A more methodical way of constructing general finite difference expressions for the pth derivative of $\bar{T}$ at point x_i is given as

$$\frac{\partial^p \bar{T}_i}{\partial x^p} = \sum_{j=-N1}^{N2} \beta_j \bar{T}_{i+j}, \tag{2.39}$$

where the coefficients β_j are determined by a Taylor series expansion of $\bar{T}_{i+j}$ using Equation (2.38) about the point i and equating coefficients of similar terms.

In Equation (2.39), $N1$ and $N2$ are integers depending on the accuracy of the approximation of the derivative. If $N1 = N2$, the finite difference approximation is centered in space. Some of the standard finite difference expressions for first-($p = 1$) and second-order ($p = 2$) derivatives obtained using Equations (2.38) and (2.39) are given below:

Centered:
$$\frac{\partial \overline{T}_i}{\partial x} = \frac{\overline{T}_{i+1} - \overline{T}_{i-1}}{2\Delta x}, \quad \text{error} = O(\Delta x^2)$$

$$\frac{\partial^2 \overline{T}_i}{\partial x^2} = \frac{\overline{T}_{i+1} - 2\overline{T}_i + \overline{T}_{i-1}}{\Delta x^2}, \quad \text{error} = O(\Delta x^2)$$

(2.40)

Backward:
$$\frac{\partial \overline{T}_i}{\partial x} = \frac{\overline{T}_i - \overline{T}_{i-1}}{\Delta x}, \quad \text{error} = O(\Delta x)$$

(2.41)

Forward:
$$\frac{\partial \overline{T}_i}{\partial x} = \frac{\overline{T}_{i+1} - \overline{T}_i}{\Delta x}, \quad \text{error} = O(\Delta x)$$

(2.42)

Using a second-order accurate centered difference scheme (Equation 2.40) for the discretization of spatial terms in Equation (2.35) and collecting coefficients for the temperature at a given node, we obtain

$$\rho c \frac{\partial \overline{T}_{i,j,k}}{\partial t} = \alpha_1 \overline{T}_{i,j,k} + \alpha_2 \overline{T}_{i-1,j,k} + \alpha_3 \overline{T}_{i+1,j,k} + \alpha_4 \overline{T}_{i,j-1,k}$$

$$+ \alpha_5 \overline{T}_{i,j+1,k} + \alpha_6 \overline{T}_{i,j,k-1} + \alpha_7 \overline{T}_{i,j,k+1} + \beta$$

(2.43)

where

$$\alpha_1 = -2\left(\frac{k_x}{\Delta x^2} + \frac{k_y}{\Delta y^2} + \frac{k_z}{\Delta z^2}\right) - \omega_b \rho_b c_b, \quad \alpha_2 = -\frac{k_x}{\Delta x^2} + \frac{\rho c u_x}{2\Delta x}, \quad \alpha_3 = -\alpha_2$$

$$\alpha_4 = -\frac{k_y}{\Delta y^2} + \frac{\rho c u_y}{2\Delta y}, \quad \alpha_5 = -\alpha_4, \quad \alpha_6 = -\frac{k_z}{\Delta z^2} + \frac{\rho c u_z}{2\Delta z}, \quad \alpha_7 = -\alpha_6, \quad \beta = \omega_b \rho_b c_b T_a^* + q'''$$

(2.44)

In the above, u_x, u_y, and u_z are the components of the velocity **u**. Equation (2.43) can be written in compact linear algebraic form as follows:

$$\mathbf{M}\dot{\overline{T}} + \mathbf{K}\overline{T} = \mathbf{Q}$$

(2.45)

where **M** is a diagonal matrix with entries ρc, **K** is a matrix with entries $-\alpha_1$ through $-\alpha_7$, and **Q** is the source term matrix with entries β. The overdot denotes a time derivative. Since Equation (2.45) is a transient equation, one needs to integrate it with respect to time. Conventionally, time advancement from level n to $n+1$ is done using any of the following finite difference procedures:

Forward difference (explicit scheme): $\mathbf{M}\bar{\mathbf{T}}^{n+1} = (\mathbf{M} - \Delta t\mathbf{K})\bar{\mathbf{T}}^n + \Delta t\,\mathbf{Q}^n$

Backward difference (implicit scheme): $(\mathbf{M} + \Delta t\mathbf{K})\bar{\mathbf{T}}^{n+1} = \mathbf{M}\bar{\mathbf{T}}^n + \Delta t\,\mathbf{Q}^{n+1}$ (2.46)

Central difference: $\mathbf{M}\bar{\mathbf{T}}^{n+1} = \mathbf{M}\bar{\mathbf{T}}^{n-1} - 2\Delta t\mathbf{K}\bar{\mathbf{T}}^n + 2\Delta t\mathbf{Q}^n$

θ method: $(\mathbf{M} + \theta\Delta t\mathbf{K})\bar{\mathbf{T}}^{n+1} = (\mathbf{M} - (1-\theta)\Delta t\mathbf{K})\bar{\mathbf{T}}^n + \Delta t\,\mathbf{Q}^{n+\theta},\ \ 0 \le \theta \le 1$

In general, the central difference time discretization scheme is numerically unstable. A forward differencing in time and a central differencing in space are also referred to as the forward-time centered-space (FTCS) scheme. When $\theta = 0.5$, the time-stepping method is also called the Crank-Nicolson method and is second-order accurate in both time and space. A central differencing of the convective terms together with an explicit treatment of the spatial terms lead to numerical instabilities for convection-dominated problems (i.e., a large Peclet number $Pe = \rho c |\mathbf{u}| L/k_{eff} \gg 1$, where L is a characteristic length of the problem) if either the mesh size or the time step is large such that it violates the Courant-Friedrichs-Lewy (CFL) condition [36]. Stability of numerical solution is guaranteed if $\alpha_1 \le 0$ and the coefficients $\alpha_{2-7} \ge 0$. For large Pe, an upwind method of discretization can be employed to satisfy the above condition for the coefficients and hence avoid instabilities. If the velocity is positive (negative), the temperature gradient in the convective term is discretized using backward (forward) differencing (see Equations 2.41 and 2.42). It should be noted that such an upwind scheme is only first-order accurate and also introduces artificial diffusion due to the leading-order truncated terms in the backward (forward) differencing. Assuming the velocity $\mathbf{u}_b$ is positive and using a backward differencing for the convective terms and a central differencing for other terms, the following coefficients are obtained for $\alpha_1 - \alpha_7$ in Equation (2.43):

$$\alpha_1 = -2\left(\frac{k_x}{\Delta x^2} + \frac{k_y}{\Delta y^2} + \frac{k_z}{\Delta z^2}\right) - \omega_b\rho_b c_b - \frac{\rho c u_x}{\Delta x} - \frac{\rho c u_y}{\Delta y} - \frac{\rho c u_z}{\Delta z}$$

(2.47)

$$\alpha_2 = -\frac{k_x}{\Delta x^2} + \frac{\rho c u_x}{2\Delta x}, \quad \alpha_3 = \frac{k_x}{\Delta x^2}, \quad \alpha_4 = -\frac{k_y}{\Delta y^2} + \frac{\rho c u_y}{\Delta y}, \quad \alpha_5 = \frac{k_y}{\Delta y^2}$$

$$\beta = \omega_b\rho_b c_b T_a^* + q'''$$

Higher-order upwind discretizations can also be obtained from the general procedure outlined in Equation (2.39), and the coefficients $\alpha_1 - \alpha_7$ can be obtained in a straightforward manner.

The matrices $\mathbf{M}$ and $\mathbf{K}$ in Equation (2.45) are generally sparse due to relatively few nonzero elements. The overall computational cost for solving a given problem will be primarily determined by the method of solution employed for such sparse linear algebraic equations. Direct solution methods based on Gaussian elimination are usually not practical for large-sized problems due to the excessive memory and computational (central processing unit, or CPU) requirements. Iterative methods for sparse linear systems are the most preferred solution techniques because of lesser memory requirements even for fully three-dimensional (3D) problems and also because of their ease of implementation for parallel computations. An excellent review of sparse linear iterative methods including multigrid techniques [37] can be found in the book by Saad [38].

2.4.1.1 Treatment of Boundary Conditions

The oft-encountered boundary conditions in bioheat transfer can be classified into three types: (1) Dirichlet, where the temperature is specified at the boundary (Equation 2.36a); (2) Von Neumann, where the heat flux is prescribed at the boundary (Equation 2.36b); and (3) mixed or convective boundary conditions, where the heat flux at the boundary is specified as a function of the boundary temperature with a specified convective heat transfer coefficient h and an ambient temperature T_∞ (see Equation 2.51, below). Now, for the purposes of numerical discretization, consider Figure 2.3 for a representative boundary point (i,j,k) and assume $\mathbf{n} = (1,0,0)$. The following discussions can be suitably modified for points lying on different coordinate directions (with a different value of $\mathbf{n}$).

1. Dirichlet boundary condition. Here, the temperature is directly specified and Equation (2.43) is not solved for this node point.
2. Von Neumann boundary condition:

$$-k_{eff}\nabla \overline{T}_t(\mathbf{r},t)\cdot \mathbf{n} = g(\mathbf{r},t) \tag{2.48}$$

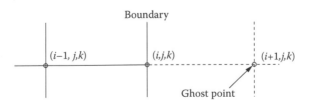

Figure 2.3 Boundary point (i, j, k) together with the interior point $(i-1, j, k)$ and a ghost point $(i+1, j, k)$.

For the boundary points, the gradient term in Equation (2.48) can be discretized using only the boundary and interior points (one-sided difference), say, a first-order backward-difference scheme (Equation 2.41). This yields the following relation between the boundary and interior points:

$$\bar{T}_{i,j,k} - \bar{T}_{i-1,j,k} + \frac{g\Delta x}{k_{eff}} = 0 \tag{2.49}$$

Now, Equation (2.43) is replaced by Equation (2.49) for all the boundary points, and hence the corresponding entries for $\bar{T}_{i,j,k}$ of matrix **K** in Equation (2.45) are altered accordingly. Higher-order discretization using more interior points can also be constructed using Equation (2.39) by setting $N2 = 0$ (which guarantees that only interior points are to be considered).

Another method of discretizing Equation (2.48) is with the introduction of ghost points or fictitious points outside the problem domain (see Figure 2.3). The advantage with such a procedure is that a centered difference procedure can still be used with the boundary points maintaining a second-order accuracy. The gradient term in Equation (2.48) is differenced using a centered scheme at the point (i, j, k) using the ghost point and is written as follows:

$$\left(\bar{T}_{i+1,j,k}\right)_g = \bar{T}_{i-1,j,k} - \frac{2g\Delta x}{k_{eff}} \tag{2.50}$$

where the subscript g on $\bar{T}_{i+1,j,k}$ denotes the ghost point. Equation (2.50) is then used in Equation (2.43) to replace $\bar{T}_{i+1,j,k}$ (term 3 of the right-hand side, or RHS) and then rearranged to yield the respective terms in matrix **K**.

3. Mixed or convective boundary condition:

$$-k_{eff}\nabla\bar{T}_t(\mathbf{r},t)\cdot\mathbf{n} = h(\bar{T}_t - T_\infty) \tag{2.51}$$

where h is the convective heat transfer coefficient and T_∞ is a known temperature of the surrounding fluid. For simplicity, the gradient term in Equation (2.51) is discretized using a first-order backward-difference scheme to yield the following relation between the interior and boundary points:

$$\bar{T}_{i,j,k}\left(1 + \frac{h\Delta x}{k_{eff}}\right) - \bar{T}_{i-1,j,k} - \frac{hT_\infty\Delta x}{k_{eff}} = 0 \tag{2.52}$$

Again, for such boundary points, Equation (2.43) is replaced with Equation (2.52). Higher-order one-sided differences or ghost point methods can also be used to discretize the gradient term in Equation (2.51).

2.4.1.2 Treatment of Multimaterial Contact Problems

Another physical situation of great interest in bioheat transfer is the presence of multiple materials within the same physical domain on which the problem is posed. The condition that is generally imposed at the interface of two materials is that both the temperatures and the heat flux are the same across the interface.

For example, consider Figure 2.4, where two materials of different conductivities, k_{eff_1} and k_{eff_2}, are in contact with each other. Let (i, j, k) denote the point that lies on the interface between the two materials. By the construction of the discrete grid, the temperature is continuous at (i, j, k), and hence temperature continuity is satisfied. The flux continuity condition to be imposed at (i, j, k) can be expressed as follows:

$$k_{eff_1}(\nabla \overline{T}_t(\mathbf{r},t)\cdot \mathbf{n})_1 = k_{eff_2}(\nabla \overline{T}_t(\mathbf{r},t)\cdot \mathbf{n})_2, \text{ on the interface} \qquad (2.53)$$

where subscripts 1 and 2 denote respective materials in contact with each other. Discretizing using a simple first-order finite difference using the points as shown in Figure 2.4, we get the following relation for the temperature at (i, j, k):

$$\overline{T}_{i,j,k} = \left(\frac{k_{eff_2}}{k_{eff_1} + k_{eff_2}}\right)\overline{T}_{i-1,j,k} + \left(\frac{k_{eff_1}}{k_{eff_1} + k_{eff_2}}\right)\overline{T}_{i+1,j,k} \qquad (2.54)$$

As with the treatment of other boundary conditions, Equation (2.54) replaces Equation (2.43). Higher-order differences considering more interior points in each material can also be considered.

Several numerical studies have used finite difference techniques to numerically analyze the problems relating to tissue heat transfer. Most of these studies typically consider the Pennes equation (see Equation 2.21) for modeling the heat transfer. As mentioned earlier, in our numerical analysis presented above, the Pennes equation can be recovered by setting $\mathbf{u} = 0$. Also, by setting the mass matrix $\mathbf{M} = 0$, a steady-state version of the governing equation can be obtained.

Kotte et al. [39] numerically studied the tissue heat transfer problem associated with the interstitial hyperthermia using ferromagnetic seeds. A one-dimensional

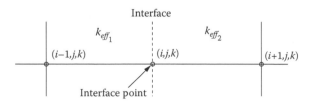

Figure 2.4   Interface point (i, j, k) together with point $(i - 1, j, k)$ in material 1 and point $(i + 1, j, k)$ in material 2.

(1D) Pennes equation was used to model the transient heat transfer behavior of both the ferromagnetic seed and the surrounding tissue. An explicit FTCS scheme finite difference in conjunction with a novel seed-modeling method was used in their numerical simulation. A simple von Neumann stability analysis of the governing equation assuming a constant k_{eff} was performed to evaluate the time step restriction in their problem. They showed that the time step Δt used in their numerical simulations had to satisfy the following inequality for achieving numerical stability:

$$\Delta t < 2\rho c \left(\omega_b \rho_b c_b + \frac{4k_{eff}}{\Delta x^2} \right)^{-1} \qquad (2.55)$$

Van Leeuwen et al. [40] carried out a 3D numerical model of the heat transfer in a neonatal that was also performed using the FTCS scheme. A 3D Pennes equation was used as their governing equation.

Dai et al. [41] used a fourth-order compact finite difference scheme as suggested in Reference [42] to solve a 1D Pennes equation in a triple-layered skin structure and compared their solution with a Crank-Nicolson finite difference scheme. They obtained highly accurate solutions with compact differencing with almost the same CPU time compared to the standard Crank-Nicolson finite difference scheme. For a 3D triple-layered skin structure, Dai et al. [43] developed a numerical method to obtain optimal temperature distributions embedded with multilevel blood vessels. The heat transfer in the tissue was solved using a 3D Pennes equation. A Crank-Nicolson-centered second-order finite difference scheme was used in the discretization of individual terms. A large number of grid points was used along the z-direction in their simulation due to the thinness of the epidermic layer. In view of the latter observation, the usage of a compact finite difference scheme as given in Reference [41] was suggested to increase the accuracy with fewer grid points, although results were not shown in that regard.

Karaa et al. [44] presented an implicit numerical solution of the 3D Pennes bioheat transfer equation and applied it to study some typical bioheat transfer processes often encountered in cancer hyperthermia, laser surgery, and thermal parameter estimation. A second-order Crank-Nicolson scheme in conjunction with an incomplete factorization (ILU) preconditioned generalized minimal residual (GMRES) scheme [38] was used to iteratively solve the resulting linear algebraic equation (Equation 2.45). Their study showed the acceleration of convergence due to the use of such preconditioners. It should, however, be noted that an extensive comparison of their preconditioned iterative solver with various other available iterative techniques was not undertaken in their study.

In Deng and Liu [45], the effects of large blood vessels on the transient tissue temperature distributions during cryosurgery treatment were studied. The model considered both the perfused tissue and large blood vessels. The Pennes bioheat transfer equation was used to describe the heat transfer in perfused tissues, while for a single or countercurrent blood vessel, the heat transfer was described by means of an energy equation (a convective heat transfer equation) with a constant Nusselt number. A finite difference method based on the effective heat capacity method was applied to solve the heat transfer. All terms in the Pennes equation were treated using an explicit scheme except the arterial source term, which was treated in a weighted semi-implicit manner (similar to the θ method in Equation 2.46, i.e., $\bar{T}_{i,j,k} \equiv \left[(1-\beta)\bar{T}_{i,j,k}^{n} + \beta\bar{T}_{i,j,k}^{n+1}\right]$, where β is the relaxation parameter and $0 \leq \beta \leq 1$). The instability due to the explicit treatment of diffusion terms in the Pennes equation was avoided by choosing a time step Δt such that it satisfies the following relation: $1 - W(1-\beta)\Delta t - 6Fo \geq 0$, where $W = \frac{\rho_b c_b \omega_b}{\rho c}$, $Fo = \frac{k\Delta t}{\rho c \Delta x^2}$. This relation was obtained by using the fact that, for stability, the coefficient $\alpha_1 \leq 0$ in Equation (2.43). The properties of the tissue were assumed to be isotropic and constant (with $k_{eff} = k$, a constant). A convective heat transfer equation was used to model the temperature variation along the flow direction for large blood vessels. The numerical analysis in Section 2.4.1, being a more general one, encompasses the energy equation, too. The convective terms were discretized using an implicit upwind scheme to alleviate the convective numerical instabilities.

2.4.2 Finite Element Method

The finite element method belongs to the class of integral techniques for solving the governing equation. The most commonly used integral approach method is the weighted residual method. This technique forms an integral equation (also known as weak formulation) by multiplying the governing equation by a weighting function W and then integrating it over a prescribed interval. A functional form of the independent variables with unknown coefficients is then assumed and substituted into the integral equations. The space over which the heat transfer equation is solved is divided into N intervals of integrations (finite elements) that give rise to a system of equations for N unknown temperatures at these nodal locations.

When $W = 0$ or 1, the method is known as the subdomain or integral-relation method; when $W = x^i$ (i = 0 ... N), the method is known as the method of moments; and when $W = F$, with F being a known function set that is assumed to represent the solution, the method is known as the Galerkin method.

The first step in finite elements is the construction of the weak form or the integral formulation of Equation (2.35). This is obtained by first multiplying the equation with a set of weighting functions W (which satisfies the condition $W = 0$ on Γ_1) followed by integration over the domain Ω.

$$\int_{\Omega} W\left(\rho c \frac{\partial \bar{T}_t}{\partial t}\right) d\Omega = \int_{\Omega} W[\nabla \cdot k_{eff} \nabla \bar{T}_t] d\Omega + \int_{\Omega} W[\omega_b \rho_b c_b (T_a^* - \bar{T}_t)] d\Omega$$

$$- \int_{\Omega} W[\rho c \mathbf{u} \cdot \nabla \bar{T}_t] d\Omega + \int_{\Omega} W q''' \, d\Omega \tag{2.56}$$

After integrating the second-order derivatives by parts using the Gauss divergence theorem and using the fact that $W = 0$ on Γ_1, we obtain the following weak formulation of Equation (2.56) on application of the boundary conditions given by Equation (2.36a,b):

$$\int_\Omega \rho c W \frac{\partial \overline{T}_t}{\partial t}\, d\Omega + \int_\Omega k_{eff} \nabla W \cdot \nabla \overline{T}_t\, d\Omega + \int_{\Gamma_2} g W\, d\Gamma_2 + \int_\Omega \omega_b \rho_b c_b W \overline{T}_t\, d\Omega$$

$$- \int_\Omega \rho c W [\mathbf{u} \cdot \nabla \overline{T}_t]\, d\Omega = \int_\Omega \omega_b \rho_b c_b T_a^* W\, d\Omega + \int_\Omega W q'''\, d\Omega \qquad (2.57)$$

Integrating by parts reduces the continuity requirements for $\overline{T}_t$ (hence called a weak formulation) while they are increased for the weighting functions. Equation (2.36b) is also called a "natural boundary condition."

The second step is the formulation of the Galerkin solution method. In this method, both $\overline{T}_t$ and W are expanded in terms of complete trial and weighting function sets Π and Θ, which are defined as follows:

$$\Pi = \left\{ \overline{T}(\mathbf{r},t) \mid \overline{T}_t = \sum_{j=1}^{\infty} T_j(t)\varphi_j(\mathbf{r}), \quad \overline{T}_t = f(\mathbf{r},t) \text{ on } \Gamma_1 \right.$$

$$\Theta = \left\{ W(\mathbf{r},t) \mid W = \sum_{j=1}^{\infty} b_j(t)\varphi_j(\mathbf{r}), \quad W = 0 \text{ on } \Gamma_1 \right. \qquad (2.58)$$

where φ_j's are the basis functions, and T_j and b_j are unknown coefficients.

The third step is to construct an approximate solution $\hat{T}_t$ for $\overline{T}_t$, obtained by using finite dimensional subsets $\Pi^N \subset \Pi$ and $\Theta^N \subset \Theta$, which are defined below:

$$\Pi^N = \left\{ \hat{T}_t(\mathbf{r},t) \mid \hat{T}_t = \sum_{j=1}^{N} T_j(t)\varphi_j(\mathbf{r}), \quad \hat{T} = f(\mathbf{r},t) \text{ on } \Gamma_1 \right.$$

$$\Theta^N = \left\{ \hat{W}_t(\mathbf{r},t) \mid \hat{W} = \sum_{j=1}^{N} b_j(t)\varphi_j(\mathbf{r}), \quad \hat{W} = 0 \text{ on } \Gamma_1 \right. \qquad (2.59)$$

The final step is the matrix formulation. The problem now is to find $\hat{T}_t$ in Π^N, such that the approximate form of Equation (2.57) is given by

$$\int_\Omega \rho c \hat{W} \frac{\partial \hat{T}_t}{\partial t}\, d\Omega + \int_\Omega k_{eff} \nabla \hat{W} \cdot \nabla \hat{T}_t\, d\Omega + \int_{\Gamma_2} g \hat{W}\, d\Gamma_2 + \int_\Omega \omega_b \rho_b c_b \hat{W} \hat{T}_t\, d\Omega$$

$$- \int_\Omega \rho c \hat{W} [\mathbf{u} \cdot \nabla \hat{T}_t]\, d\Omega = \int_\Omega \omega_b \rho_b c_b T_a^* \hat{W}\, d\Omega + \int_\Omega \hat{W} q'''\, d\Omega \qquad (2.60)$$

Using the definitions of $\hat{T}_t$ and $\hat{W}$ from Equation (2.59) in Equation (2.60) leads to the following set of N equations after some simplifications:

$$
\sum_{j=1}^{N} \left\{ \int_{\Omega} \rho c \varphi_i \varphi_j \, d\Omega \right\} \frac{\partial T_j}{\partial t} + \sum_{j=1}^{N} \left\{ \int_{\Omega} k_{eff} \nabla \varphi_i \cdot \nabla \varphi_j \, d\Omega + \int_{\Gamma_2} g \varphi_j \, d\Gamma_2 \right\} T_j
$$

$$
+ \sum_{j=1}^{N} \left\{ \int_{\Omega} \omega_b \rho_b c_b \varphi_i \varphi_j \, d\Omega - \int_{\Omega} \rho c \varphi_i [\mathbf{u} \cdot \nabla \varphi_j] \, d\Omega \right\} T_j = \int_{\Omega} (q''' + \omega_b \rho_b c_b T_a^*) \varphi_i \, d\Omega
$$

(2.61)

Using matrix notation, Equation (2.61) can also be written as

$$
\mathbf{M}\dot{\mathbf{T}} + \mathbf{K}\mathbf{T} = \mathbf{Q}
$$

(2.62)

where the overdot denotes a time derivative. The NxN matrices M, K and the Nx1 matrix Q are given as follows:

$$
\mathbf{M}_{ij} = \int_{\Omega} \rho c \varphi_i \varphi_j \, d\Omega
$$

$$
\mathbf{K}_{ij} = \int_{\Omega} k_{eff} \nabla \varphi_i \cdot \nabla \varphi_j \, d\Omega + \int_{\Gamma_2} g \varphi_j \, d\Gamma_2 + \int_{\Omega} \omega_b \rho_b c_b \varphi_i \varphi_j \, d\Omega - \int_{\Omega} \rho c \varphi_i [\mathbf{u} \cdot \nabla \varphi_j] \, d\Omega
$$

$$
\mathbf{Q}_i = \int_{\Omega} (q''' + \omega_b \rho_b c_b T_a^*) \varphi_i \, d\Omega
$$

(2.63)

If a mixed or convective boundary condition of the form $g(\mathbf{r},t) = h(\bar{T}_t - T_\infty)$ is prescribed on Γ_2, then in Equation (2.60), g is replaced with the above convective term and then integrated. Here, h and T_∞ are the known convective heat transfer coefficient and the temperature of the surrounding fluid. The entries in the matrices **K** and **Q** are modified as follows:

$$
\mathbf{K}_{ij} = \int_{\Omega} k_{eff} \nabla \varphi_i \cdot \nabla \varphi_j \, d\Omega + \int_{\Gamma_2} h \varphi_i \varphi_j \, d\Gamma_2 + \int_{\Omega} \omega_b \rho_b c_b \varphi_i \varphi_j \, d\Omega - \int_{\Omega} \rho c \varphi_i [\mathbf{u} \cdot \nabla \varphi_j] \, d\Omega
$$

$$
\mathbf{Q}_i = \int_{\Omega} \omega_b \rho_b c_b T_a^* \varphi_i \, d\Omega + \int_{\Omega} q''' \varphi_i \, d\Omega + \int_{\Gamma_2} h T_\infty \varphi_i \, d\Gamma_2
$$

(2.64)

For multimaterial contact problems, the interface boundary condition becomes a natural boundary condition in the finite element method. In such cases, it can be shown that [46]:

$$
\mathbf{K}_{ij} = \int_{\Omega} k_{\text{eff}} \nabla \varphi_i \cdot \nabla \varphi_j \, d\Omega + \int_{\Gamma_2} \left[\underbrace{k_{\text{eff}_2} (\nabla \overline{T} \cdot \mathbf{n})_1 - k_{\text{eff}_1} (\nabla \overline{T} \cdot \mathbf{n})_2}_{\text{jump in heat flux across the interface}} \right] \varphi_j \, d\Gamma_2
$$

$$
+ \int_{\Omega} \omega_b \rho_b c_b \varphi_i \varphi_j \, d\Omega - \int_{\Omega} \rho c \varphi_i [\mathbf{u} \cdot \nabla \varphi_j] \, d\Omega
$$

(2.65)

With heat flux continuity at the interface (Equation 2.53), the boundary flux term (second term) of $\mathbf{K}_{ij}$ in Equation (2.65) vanishes identically.

The solution of Equation (2.62) for the coefficients T_j constitutes the approximate solution $\hat{T}_t$ for $\overline{T}_t$. The coefficients T_j can also be interpreted as the nodal unknowns of the temperature $\hat{T}_t$, for which φ_j's serve as the interpolating functions. These interpolating functions are almost exclusively chosen from low-order piecewise polynomials, for example, piecewise linear or quadratic elements, whose support extends only over a few contiguous elements. This produces relatively few nonzero terms in the matrix $\mathbf{K}$ and thus helps in achieving the solution in an economical way. Usually, for higher-dimensional problems (two-dimensional [2D] or 3D), the numerical integrations in Equation (2.63) (and also Equations 2.64 and 2.65) are performed on an element-by-element basis. This is achieved by mapping every element in the physical space to a master element defined by an element-based coordinate system. The transformation from the physical element to the master element is done using elementary transformation techniques. The individual terms in the equation are then integrated to form elemental matrices with element nodes as unknown quantities. The individual elemental matrices are then assembled into a global matrix of the form given by Equation (2.62). More details on the steps involved in the above method can be obtained in References [46–48].

Equation (2.62) is a transient equation similar to Equation (2.45). The transient equation is advanced in time using a finite difference approach similar to the one explained in Section 2.4.1 (see Equation 2.46).

In order to avoid numerical stabilities for convection-dominated problems ($Pe = \rho c |\mathbf{u}| L/k_{\text{eff}} \gg 1$), Taylor-Galerkin or streamline upwinding methods can be used. The details of such methods are given in References [46,47,49].

In regard to finite element studies of tissue heat transfer, Torvi and Dale [50] studied the thermal effects of flash fires on skin to predict skin temperatures and the times taken to attain second- and third-degree burns. The effects of variations in thermal physical properties on skin temperature and burn predictions were carried out numerically using a multiple-layer, variable-property, finite element model. The multiple-layered skin consisted of the epidermis, the dermis, and the subcutaneous layer. Heat transfer in the skin was assumed to be transient and 1D and obeyed the Pennes equation. A Galerkin weighted-residual method with cubic Hermitian temperature interpolation functions were used (see Reference [48] for the definitions of cubic Hermitian interpolation functions). A Crank-Nicolson scheme was used to discretize the equations in time. It was noted from their studies that five Hermitian elements for their problem domain (one for the epidermis, and two each for the dermis and subcutaneous region) provided the same or better accuracy than nine elements using quadratic interpolation or 18 elements using linear elements. Chatterjee and Adams [51] studied the thermal response during hyperthermia treatment of the prostrate region of a human body using a 2D Pennes bio-heat transfer equation. A Galerkin finite element model in conjunction with automatic mesh generation capabilities of the commercial software ANSYS® (Canonsburg, Pennsylvania) was used to solve the equation. The limitations of the commercial software in modeling the thermal behavior of the human body were also discussed in their work. Dennis et al. [52] numerically studied the cooling profiles of the brain in response to various external cooling methods and protocols.

A transient simulation of the 3D Pennes equation was carried out assuming local tissue properties using a Galerkin finite element technique. The simulations performed considered ice packs applied to the head and neck as well as the use of a head-cooling helmet. Tetrahedral-shaped elements were used exclusively to discretize the complex 3D geometry of a human head. A Crank-Nicolson scheme was employed to discretize the equation in time, and the resulting linear algebraic equation was efficiently solved using a preconditioned conjugate gradient method. A typical time-accurate numerical simulation was performed on a notebook PC (PIII 700 MHz) in less than 20 minutes using less than 64 MB of in-core memory. From the simulations, it was found that both the cooling approaches resulted in insubstantial cooling within 30 minutes. It was suggested that additional cooling methods need to be explored, such as cooling of other pertinent parts of the human anatomy.

In a similar fashion, Wainwright [53] used finite element techniques to model the heat transfer in a human leg in the high-frequency domain of radiofrequency and microwave radiation. Galerkin-weighted residual methods were used to solve the partial differential equations for both the electromagnetic and thermal fields. Construction of the finite element mesh was done by processing the voxel data set using the automatic mesh generator as described in Reference [54]. This method was based on a modified Delaunay triangulation algorithm.

The mesh was refined in the region of the ankle to show the details of the specific absorption rate (SAR) distribution in the intricate anatomy of muscle, tendon, and bone.

2.4.3 Finite Volume Method

The basis of the finite volume method is that the integral form of the conservation equation (Equation 2.35) is used to derive the discretized approximations. Let us assume for the purposes of simplicity that the properties of tissue are constant and isotropic, and that the flow field ($\mathbf{u}$) is known everywhere.

The first step in the finite volume method is to divide the physical domain into a number of nonoverlapping control volumes such that there is one control volume surrounding each grid point. In the second step, the governing equation for the unknown variable $\bar{T}_t$ is integrated over each control volume. Finally, the integrals are evaluated by assuming a piecewise profile for the variation of $\bar{T}_t$, which results in a set of algebraic equations for the grid point and its neighbors under consideration. The important aspect of the finite volume formulation is that the resulting solution satisfies an integral conservation property on the whole domain.

There are two common ways of constructing finite volumes. The first one is to place the grid points and draw control volume faces midway between neighboring points (see Figure 2.5). This method is more accurate in evaluating the fluxes at the control volume faces. However, special treatment is required for evaluating

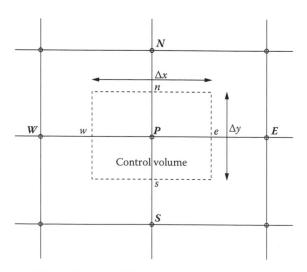

Figure 2.5 A typical 2D control volume with faces e, w, n, and s. Node points are denoted by E, W, N, S, and P.

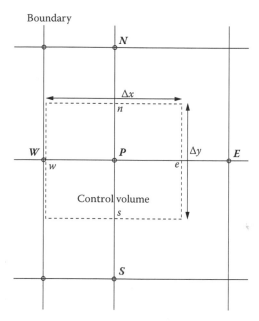

Figure 2.6   A typical 2D boundary control volume with boundary face w passing through the grid point W.

the boundary conditions at the boundary nodes. The second method of constructing finite volumes is to draw the control volumes that coincide with the domain and multimaterial boundaries and then place the grid point at the center of each control volume. Control volume sizes are generally nonuniform. In this method, the boundary grid points are located on the control volume faces (see Figure 2.6). The second scheme of constructing finite volumes is more common in practice, and hence all the discussions below pertain to this scheme.

Consider a representative control volume as shown in Figure 2.5. Only a 2D situation is considered for elucidation purposes, and an extension to 3D situations is straightforward. By convention, the control volume faces are denoted by e, w, n, and s for the east, west, north, and south faces, respectively. E, W, N, S, and P denote the grid points along the east, west, north, south, and center of the control volume, respectively. The unknown temperature is solved for these grid points as follows.

Integrating Equation (2.35) over a representative control volume, as given above, yields

$$\int_{dV} \rho c \frac{\partial \overline{T}_t}{\partial t}\, dV = \int_{dV} [\nabla \cdot k_{eff} \nabla \overline{T}_t + \omega_b \rho_b c_b (T_a^* - \overline{T}_t) - \rho c \mathbf{u} \cdot \nabla \overline{T}_t + q''']\, dV \qquad (2.66)$$

Application of Gauss divergence in Equation (2.66) yields Equation (2.67) for $\overline{T}$ (after dropping the subscript t on the variable $\overline{T}$):

$$\rho c \int_{dV} \frac{\partial \overline{T}}{\partial t} dV = \int_{d\Gamma} k_{eff} (\nabla \overline{T} \cdot \hat{\mathbf{n}}) d\Gamma - \rho c \int_{d\Gamma} \overline{T}(\mathbf{u} \cdot \hat{\mathbf{n}}) d\Gamma - \omega_b \rho_b c_b \int_{dV} \overline{T} dV + \int_{dV} Q''' dV$$

$$(2.67)$$

where $\hat{\mathbf{n}}$ is the unit normal vector on a given control volume face, and $Q''' = q''' + \omega_b \rho_b c_b T_a^*$ is the total volumetric source of heat. Here, $d\Gamma$ denotes surface integration of quantities, and dV denotes volume integration. The individual terms on the right-hand side of Equation (2.67) are integrated as follows:

$$\text{Diffusion term:} \quad \int_{d\Gamma} k_{eff} (\nabla \overline{T} \cdot \hat{\mathbf{n}}) d\Gamma$$

The above integral is evaluated on each face of the control volume. The integral is usually approximated as the product of the integrand at the cell face center and the cell face area. This gives

$$\int_{d\Gamma} k_{eff} (\nabla \overline{T} \cdot \hat{\mathbf{n}}) d\Gamma = k_{eff} \left(\frac{\partial \overline{T}}{\partial x}\right)_e \Delta y - k_{eff} \left(\frac{\partial \overline{T}}{\partial x}\right)_w \Delta y + k_{eff} \left(\frac{\partial \overline{T}}{\partial y}\right)_n \Delta x - k_{eff} \left(\frac{\partial \overline{T}}{\partial y}\right)_s \Delta x$$

$$(2.68)$$

where the surface areas are calculated assuming unit depth along the z-direction (for 2D problems). Higher-order approximations can be used, and the details can be obtained in Reference [36]. Assuming a linear profile for the variation of $\overline{T}$ between neighboring points, the following difference expression for the right-hand side of Equation (2.68) can be obtained:

$$\int_{d\Gamma} k_{eff} (\nabla \overline{T} \cdot \hat{\mathbf{n}}) d\Gamma = k_{eff} \frac{\overline{T}_E - \overline{T}_P}{\delta x_e} \Delta y - k_{eff} \frac{\overline{T}_P - \overline{T}_W}{\delta x_w} \Delta y + k_{eff} \frac{\overline{T}_N - \overline{T}_P}{\delta y_n} \Delta x - k_{eff} \frac{\overline{T}_P - \overline{T}_s}{\delta y_s} \Delta x$$

$$(2.69)$$

where $\delta x_e = x_E - x_P$, and so on. Equation (2.69) can be rearranged conveniently as follows:

$$\int_{d\Gamma} k_{eff} (\nabla \overline{T} \cdot \hat{\mathbf{n}}) d\Gamma = a_E^D \overline{T}_E + a_W^D \overline{T}_W + a_N^D \overline{T}_N + a_S^D \overline{T}_s - a_P^D \overline{T}_P \qquad (2.70)$$

where

$$a_E^D = \frac{k_{eff}\Delta y}{\delta x_e}, \ a_W^D = \frac{k_{eff}\Delta y}{\delta x_w}, \ a_N^D = \frac{k_{eff}\Delta x}{\delta y_n}, \ a_S^D = \frac{k_{eff}\Delta x}{\delta y_s}, \text{ and } a_P^D = \left(a_E^D + a_W^D + a_N^D + a_S^D\right)$$

Equation (2.70) is second-order accurate and is equivalent to the central differencing scheme (which is space centered) in finite difference methods.

$$\text{Convection term:} \quad \rho c \int\limits_{d\Gamma} \overline{T}(\mathbf{u}\cdot\hat{\mathbf{n}})d\Gamma$$

Depending on the relative magnitude of the convective terms, one can choose to evaluate the convective terms using upwind schemes. In this study, two schemes will be discussed, namely, a linear interpolation scheme and a first-order upwind scheme.

2.4.3.1 Linear Interpolation of Temperature

As with the diffusion terms, the convective terms are evaluated using the product of the integrand at the cell face center and the cell face area. Thus,

$$\rho c \int\limits_{d\Gamma} \overline{T}(\mathbf{u}\cdot\hat{\mathbf{n}})d\Gamma = (\rho c u_x)_e \Delta y\, \overline{T}_e - (\rho c u_x)_w \Delta y\, \overline{T}_w + (\rho c u_y)_n \Delta x\, \overline{T}_n - (\rho c u_y)_s \Delta x\, \overline{T}_s$$

$$(2.71)$$

Assuming a linear variation of $\overline{T}$ with $\overline{T}_{e,w,n,s} = \overline{T}_{E,W,N,S}\lambda_{e,w,n,s} + \overline{T}_P(1-\lambda_{e,w,n,s})$, where $\lambda_{e,w} = \frac{x_{e,w}-x_P}{x_{E,W}-x_P}$ and $\lambda_{n,s} = \frac{y_{n,s}-y_P}{y_{N,S}-y_P}$, Equation (2.71) can be simplified as

$$\rho c \int\limits_{d\Gamma} \overline{T}(\mathbf{u}\cdot\hat{\mathbf{n}})d\Gamma = a_E^C \overline{T}_E + a_W^C \overline{T}_W + a_N^C \overline{T}_N + a_S^C \overline{T}_S - a_P^C \overline{T}_P - a_M^C \overline{T}_P \qquad (2.72)$$

with

$$a_E^C = \lambda_e(\rho c u_x)_e \Delta y, \ a_W^C = -\lambda_w(\rho c u_x)_w \Delta y, \ a_N^C = \lambda_n(\rho c u_y)_n \Delta x$$

$$a_S^C = -\lambda_s(\rho c u_y)_n \Delta x, \ a_P^C = \left(a_E^C + a_W^C + a_N^C + a_S^C\right)$$

$$a_M^C = [(\rho c u_x)_w - (\rho c u_x)_e]\Delta y + [(\rho c u_y)_s - (\rho c u_y)_n]\Delta x$$

The values of the velocities at the control volume faces can be obtained through suitable interpolation from the known nodal values. In Equation (2.72), the reason for separating the coefficient a_M^C is that it represents an approximation of the integral $\rho c \int_{d\Gamma} (\mathbf{u} \cdot \hat{\mathbf{n}}) d\Gamma = \rho c \int_{dV} \nabla \cdot \mathbf{u} \, dV$, which is identically zero if the given velocity field is incompressible (i.e., $\nabla \cdot \mathbf{u} = 0$). In such cases, $a_M^C = 0$. Equation (2.72) is also second-order accurate and is equivalent to the central differencing scheme (space centered) in finite difference methods.

2.4.3.2 Upwind Scheme

The linear interpolation scheme produces numerical instability if $Pe \gg 1$. In order to alleviate this, an upwind differencing scheme is used similar to the one used in "finite differences." In the present study, a first-order upwind difference method will be used to evaluate the convective term for simplicity. Such a scheme is only first-order accurate and introduces artificial diffusion in the problem. Hence, for more accurate solutions, higher-order upwind schemes like quadratic upwind interpolation for convective kinematics (QUICK) can be utilized. A good review of such schemes can be seen in References [36,55].

Consider Equation (2.71) in compact form:

$$\rho c \int_{d\Gamma} \overline{T}(\mathbf{u} \cdot \hat{\mathbf{n}}) \, d\Gamma = F_e \overline{T}_e - F_w \overline{T}_w + F_n \overline{T}_n - F_s \overline{T}_s \tag{2.73}$$

where

$$F_e = (\rho c u_x)_e \Delta y, \quad F_w = (\rho c u_x)_w \Delta y, \quad F_n = (\rho c u_y)_n \Delta x, \quad F_s = (\rho c u_y)_n \Delta x$$

In the upwind scheme, depending on the signs of $F_{e,w,n,s}$, the upstream nodal values of $\overline{T}$ are picked as the representative values for the faces of the control volume. For example, for the east face (e),

$$\overline{T}_e = \begin{cases} \overline{T}_P, & F_e > 0 \\ \overline{T}_E, & F_e < 0 \end{cases} \tag{2.74}$$

Similar expressions hold true for other faces. Following the above convention, Equation (2.73) can be simplified as follows:

$$\rho c \int_{d\Gamma} \overline{T}(\mathbf{u} \cdot \hat{\mathbf{n}}) d\Gamma = -\left(a_E^C \overline{T}_E + a_W^C \overline{T}_W + a_N^C \overline{T}_N + a_S^C \overline{T}_S - a_P^C \overline{T}_P - a_M^C \overline{T}_P \right) \tag{2.75}$$

where

$$a_E^C = \max(-F_e, 0), \quad a_W^C = \max(F_w, 0), \quad a_N^C = \max(-F_n, 0), \quad a_S^C = \max(F_s, 0)$$

$$a_P^C = \left(a_E^C + a_W^C + a_N^C + a_S^C \right) \quad \text{and} \quad a_M^C = F_w - F_e + F_s - F_n$$

Here, max denotes the maximum of two quantities within parentheses. As noted earlier, $a_M^C = 0$ if the velocity field $\mathbf{u}$ is such that $\nabla \cdot \mathbf{u} = 0$.

For the convection term, either Equation (2.72) or Equation (2.75) can be used depending on the magnitude of the Peclet number.

The sink term (term 3 of the RHS in Equation 2.67) can be evaluated to second-order accuracy as follows:

$$\omega_b \rho_b c_b \int_{dV} \bar{T}\, dV = \omega_b \rho_b c_b \bar{T}_P \Delta x \Delta y = a_P^S \bar{T}_P \tag{2.76}$$

where $a_P^S = \omega_b \rho_b c_b \Delta x \Delta y$. The product $\Delta x \Delta y$ is the volume of the control volume under consideration.

The volumetric source term (term 4 of the RHS in Equation 2.67) is evaluated at point P of the control volume, namely,

$$\int_{dV} Q'''\, dV = Q_P''' \Delta x \Delta y = q_P \tag{2.77}$$

Finally, the transient term (the left-hand side, or LHS, of Equation 2.67) is also evaluated at point P as follows:

$$\rho c \int_{dV} \frac{\partial \bar{T}}{\partial t}\, dV = \rho c \dot{\bar{T}}_P \Delta x \Delta y = m_P^T \dot{\bar{T}}_P \tag{2.78}$$

where the overdot denotes the time derivative and $m_P^T = \rho c \Delta x \Delta y$.

Using Equations (2.70), (2.72) (or 2.75), (2.76), (2.77), and (2.78) in Equation (2.67), and combining coefficients of the same node and rearranging, results in the following algebraic equation for the node P:

$$m_P^T \dot{\bar{T}}_P + a_P \bar{T}_P = a_E \bar{T}_E + a_W \bar{T}_W + a_N \bar{T}_N + a_S \bar{T}_S + q_P \tag{2.79}$$

where $a_P = a_P^D + a_P^C + a_M^C + a_P^S$ and $a_{E,W,N,S} = a_{E,W,N,S}^D + a_{E,W,N,S}^C$. In terms of the indices (i, j) (for point P in 2D), Equation (2.79) can be rewritten as

$$m_{i,j}^T \dot{\bar{T}}_{i,j} + [a_{i,j}\bar{T}_{i,j} - (a_{i+1,j}\bar{T}_{i+1,j} + a_{i-1,j}\bar{T}_{i-1,j} + a_{i,j+1}\bar{T}_{i,j+1} + a_{i,j-1}\bar{T}_{i,j-1})] = q_{i,j} \tag{2.80}$$

In compact linear algebraic form, Equations (2.79) and (2.80) are equivalent to

$$\mathbf{M}\dot{\bar{\mathbf{T}}} + \mathbf{K}\bar{\mathbf{T}} = \mathbf{Q} \tag{2.81}$$

where the entries of matrices $\mathbf{M}_{ij}$, $\mathbf{K}_{ij}$, and $\mathbf{Q}_i$ are the coefficients of $\bar{T}$ given in Equation (2.80).

Equation (2.81) is a transient equation for $\overline{T}$ and resembles Equation (2.45). In order to evolve the equation in time, a time discretization of $\overline{T}$ is performed. The main time discretization methods are the same as presented in Equation (2.46) and will not be repeated here.

2.4.3.3 Treatment of Boundary Conditions

In the finite volume scheme, with control volume faces coinciding with the boundary (see Figure 2.6), there is no need for special treatment of the boundary grid points. All the available boundary data such as the temperature or heat flux are directly substituted at the boundary face. For example, with regard to Figure 2.6, if the temperature is specified at point W, $\overline{T}_W$ is replaced with the given temperature in Equation (2.79). If the heat flux is specified as in Equation (2.36b), the given data are directly substituted in Equation (2.68) for the flux on face w (for the term $k_{eff}(\frac{\partial \overline{T}}{\partial x})_w$).

2.4.3.4 Multimaterial Contact

As mentioned above, the finite volumes are constructed such that their faces coincide with the interface boundaries. Both flux and temperature continuity is satisfied in such a scheme without any additional treatment. If the interface acts as a source or sink of heat flux, it can always be added into Equation (2.68).

Indik and Indik [56] used a finite volume discretization similar to the method described above for the three-dimensional Pennes equation to compute steady-state temperatures from ferromagnetic seed heating. They solved the discretized equation (steady-state form of Equation 2.81 with $\mathbf{M} = 0$) using multigrid methods [37] and concluded that their multigrid technique was much faster than other iterative techniques like the conjugate gradient, the successive overrelaxation technique, and the preconditioned conjugate gradient methods. Chua et al. [57] developed an analytical model to study the rate of cell destruction within a liver tumor subjected to a freeze-thaw cryosurgical process. Temperature transients were obtained using a finite volume method on a 3D Pennes equation. Their numerical results showed good comparison with experimental studies. Details of the model and the boundary conditions used are given in Reference [57]. In a later study, Chua et al. [58] used a 3D Pennes equation to predict the temperature distribution in a human eye when subjected to a laser source of heating. The resulting set of equations (Equation 2.81) was transformed into a tridiagonal matrix format and solved. The temperature at every time step in their scheme was updated using an underrelaxation scheme given as

$$T_P^{n+1} = T_P^n + \varepsilon\left(T_P^{n+1}\right)_{\text{computed}}$$

where $\left(T_P^{n+1}\right)_{\text{computed}}$ is the iterative solution obtained by solving Equation (2.81) and ε is the relaxation factor with $0 < \varepsilon < 1$. Becker and Kuznetsov [59,60] used an implicit finite volume scheme to numerically model thermal effects during the skin electroporation process representing the composite layers of the skin. Their domain for thermal analysis consisted of a blood vessel, composite tissue, and an electrode plate. The composite tissue domain thermal analysis was carried out using a Pennes equation. The heat transfer in the blood vessel domain was described by the convection–diffusion energy equation with the blood velocity assumed to be unidirectional with a parabolic profile along the axis. A variable rectangular grid was used in their finite volume simulation. The convective term in the blood vessel heat transfer equation was modeled using an upwind scheme similar to the one explained in Section 2.4.3.2 under convective term discretization. Reinders et al. [61] studied the safety and efficacy of endometrial thermal balloon ablation treatment for menorrhagia. The heat transfer in a simplified geometry of a uterus was studied using a 1D Pennes equation that included the dependence of blood perfusion rate on balloon pressure, hyperthermia, and the extent of vascular damage. A stable Crank-Nicolson finite volume discretization was used to numerically solve the above problem in a symmetrical spherical coordinate system.

2.5 COUPLED TISSUE–VASCULAR MODELS

In the foregoing discussion of vascular models, the temperature in the blood sub-volume was estimated from simplifications to the temperature distribution in the tissue. Conversely, the continuum models for the tissue were made tractable by simplifying assumptions with regard to the blood. However, neither the vascular models such as the equilibration length analysis nor any of the continuum models can yield simultaneous detail on the local details of the temperature in the blood vessels *and* the adjacent tissue. When such information is needed, there is little alternative to solving a fully coupled formulation that links the blood and tissue subvolumes.

The overall structure of coupled models will include all or most of the elements shown in Figure 2.7. First, the boundary conditions as defined by the vascular architecture and the flow field must be known to create a well-posed problem. Next a means of solving the problem is used to predict the blood and tissue temperatures. And finally, the temperatures must be interpreted in the context of the assumptions introduced to pose and solve the problem. In some cases, the post-processor will yield details along particular vascular paths. In others, a statistical summary of expected temperatures might result.

The first step, a sufficiently detailed description of the vascular architecture and flow characteristics, will typically require a combination of real anatomical data about the larger vessels and some assumptions about the more numerous smaller vessels. The largest blood vessels can be readily located by current imaging technologies or standard anatomical databases. Unfortunately, the smaller thermally significant vessels are too small to be routinely located with

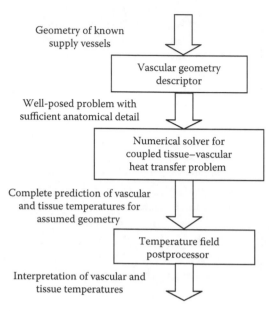

Figure 2.7   Coupled tissue–vascular models.

routine imaging technologies and display considerable variability between individuals. Several approaches are used for modeling the smaller vessels. They are as follows:

1. The smaller vessels are combined into the tissue subvolume, with their convective effects neglected. A few large blood vessels form boundaries on the combined tissue–small blood vessel subvolume.
2. The smaller blood vessels and tissue are modeled with a continuum formulation, such as one of those introduced earlier that can approximately account for the thermal effects of blood perfusion.
3. The vascular architecture and flow in the smaller vessels are idealized based on careful anatomical observations that sometimes make use of fine-scale castings of the entire vascular tree.
4. The missing, fine details of the vasculature are reconstructed by computer algorithms that mimic the growth processes of the blood vessels that would be expected in this anatomical site. An ideal algorithm should be able to produce vessels with the proper number, length, diameter, orientation, connectivity, and branching patterns. An algorithm that predicts the flow in each blood vessel must be linked to the geometrical algorithm.

The selection of the approach will depend on the availability of data and computing power as well as the needs of the modeler for accurate detail. The following paragraphs review an example of a coupled tissue–vascular model. Alternative algorithms for each element of the model exist.

2.5.1 Vascular Geometry–Generating Algorithm

Baish [16] used the following algorithm suggested by Gottlieb [62] to generate the centerline of the vascular geometry:

1. Begin with an incomplete vascular tree.
2. Assume a grid of tissue cells (typically larger than true biological cells) in the volume surrounding the existing tree.
3. Calculate the distance from each cell to the nearest point on the existing tree.
4. If a cell is farther than a set threshold distance, then add a new vessel from the nearest point on the existing tree to the cell.
5. Increase the size of the tissue and the tree by the same factor.
6. Introduce new cells so that each cell remains at constant volume.
7. Repeat steps 3 through 6 until the newly added vessels are as numerous as desired.

Numerous variations on the basic algorithm are feasible by varying the rate of growth per generation, the threshold distance, and the methods of creating tissue cells. Various alternatives to this vessel growth algorithm are available [63–66].

After creating a skeleton of vascular centerlines, the next step is to define a radius for each segment of blood vessel (Figure 2.8). Variations on Murray's law [67,68] can be applied such that each vessel junction satisfies

$$r_{bi}^n = \sum_j r_{bj}^n \tag{2.82}$$

where j is the daughter of vessel i. Murray recommended $n = 3$. Empirical studies suggest other exponents such as $n = 2.7$ [69].

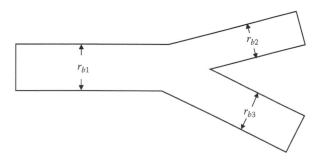

Figure 2.8   Vessels 2 and 3 join to form vessel 1 such that $r_{b1}^n = r_{b2}^n + r_{b3}^n$.

2.5.2 Solution of Coupled Tissue–Vascular Models

The basic formulation of the coupled tissue–vascular heat transfer problem differs little among the various numerical approaches available. Typically, the heat transfer into discrete segments of blood vessel is related to the blood temperature in each segment and the temperature in the adjacent tissue. The differences among the available methods arise primarily in how the temperature in the tissue subvolume is calculated.

In steady state, the temperature in the tissue subvolume satisfies

$$\nabla \cdot k_t \nabla T_t(\mathbf{r}) + q_t'''(\mathbf{r}) = 0 \tag{2.83}$$

and while blood temperature along the axes of each vessel satisfies

$$\dot{m}_i c_b \frac{d\bar{T}_b(s)}{ds} = q_b'(s) \tag{2.84}$$

the rate of heat flow into the vessel is given by

$$q_b'(s) = \pi N u k_b (T_w(s) - \bar{T}_b(s)) \tag{2.85}$$

where $T_w(s)$ is the temperature on the vessel wall at s.

The vessels are divided into N short segments along their axes, where the heat flow is assumed to be constant over the length of the segment at the rate q_{bi}' so that the temperature of the blood exiting the segment is given by the following (and see Figure 2.9):

$$\bar{T}_{bi}(s_i) = \bar{T}_{bi}(0) + \frac{q_{bi}' s_i}{\dot{m}_i c_b} \tag{2.86}$$

Mass is conserved at vascular junctions:

$$\dot{m}_i = \sum_j \dot{m}_j \tag{2.87}$$

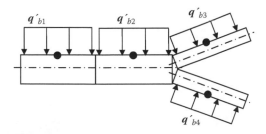

Figure 2.9   Discretization of blood vessels into short segments with heat flows to be determined from the interaction between tissue and blood subvolumes.

And the vascular temperature is continuous at the bifurcation of arteries (subscript a):

$$\bar{T}_{ai}(L_i) = \bar{T}_{aj}(0) \tag{2.88}$$

while mixing occurs at the junction of veins (subscript v):

$$\dot{m}_i \bar{T}_{vi}(L_i) = \sum_j \dot{m}_j \bar{T}_{vj}(0) \tag{2.89}$$

The temperature in the tissue subvolume may be calculated by the finite element, finite difference, or finite volume methods reviewed in Section 2.4. Alternatively, Baish [16] employs a Green's function approach that links the tissue and vessel subvolumes as follows:

$$T(\mathbf{r}) - T_{ref} = \int_{r' \in V_{tissue}} q_t'''(\mathbf{r}')G(\mathbf{r},\mathbf{r}')dV - \int_{r' \in V_{tree}} q_b'(\mathbf{r}')G(\mathbf{r},\mathbf{r}')ds \tag{2.90}$$

where the Green's function is an infinite, homogeneous medium and is given by

$$G(\mathbf{r},\mathbf{r}') = \frac{1}{4\pi k_t |\mathbf{r} - \mathbf{r}'|} \tag{2.91}$$

(In finite media, the approach can be modified by suitable changes to the Green's function or by integrating the effects of additional source terms over the exterior bounding surfaces. See, for example, Reference [70] for a mathematically similar treatment of oxygen transport.)

The goal is then to find the values of all N q_{bi}''s such that temperature and flux conditions between the blood and tissue subvolumes match near the midpoint of each blood vessel segment. That is,

$$q_{bi}' = \pi N u k_b \left(T\left(\mathbf{r}\left(\frac{L_i}{2} \right) \right) - \bar{T}_{bi}\left(\frac{L_i}{2} \right) \right) \tag{2.92}$$

where $T(\mathbf{r}(L_i/2)$ is the temperature in the tissue subvolume at a point on the vessel segment wall near its midpoint and $\bar{T}_{bi}(L_i/2)$ is the blood temperature at the vessel midpoint. The temperature at any point in the tissue subvolume is given by

$$T(\mathbf{r}) - T_{ref} = \int_{r' \in V_{tissue}} q_t'''(\mathbf{r}')G(\mathbf{r},\mathbf{r}')dV - \sum_{i=1}^{N} T_i(\mathbf{r}) \tag{2.93}$$

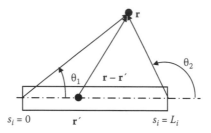

Figure 2.10   Geometry near a blood vessel segment.

where $T_i(\mathbf{r})$ is the local contribution to the tissue temperature from a given vessel segment. Assuming that the flux from each segment is approximately uniform along its length, we can use

$$T_i(\mathbf{r}) = \frac{q'_{bi}}{4\pi k_t} \int\limits_{s_i}^{L_i} \frac{1}{|\mathbf{r}-\mathbf{r}'|} ds_i = \frac{q'_{bi}}{4\pi k_t} \log\left[\frac{\tan\left(\frac{\theta_2}{2}\right)}{\tan\left(\frac{\theta_1}{2}\right)} \right]$$

(2.94)

where the angles are defined in Figure 2.10.

A heat generation field $q'''_t(\mathbf{r}')$ must be specified. For example, a constant volumetric heat source in a spherical volume of radius tissue yields a convenient analytical expression for the first volume integral in Equation (2.93):

$$\int\limits_{\mathbf{r}' \in V_{tissue}} q'''_t(\mathbf{r}') G(\mathbf{r}, \mathbf{r}') dV = \begin{cases} \dfrac{q'''_t}{6k_t}\left(3r^2_{tissue} - r^2\right), & r < r_{tissue} \\[4mm] \dfrac{q'''_t r^2_{tissue}}{3k_t r}, & r \geq r_{tissue} \end{cases}$$

(2.95)

In the absence of such a simple heat generation pattern, routine numerical methods may be employed to evaluate the volume integral.

Temperatures must be specified at the inlets to all vessel segments that do not simply collect blood from upstream. For the major supply arteries, the inlet temperatures must be specified. The case of the smallest veins is somewhat more complex depending on the source of anatomical data. If the veins collect blood from specific terminal arteries or capillaries, no special accommodations need to be made since the blood subvolume is fully connected. There may, however, be cases in which the actual connection between the terminal arteries and the smallest collecting veins is not fully specified. Here, we may introduce various closure schemes. Since the terminal arteries and smallest veins are likely to be thermally equilibrated with the surrounding tissue, it is reasonable to assign the small-vein inlet temperature to be either the local tissue temperature or the temperature exiting the nearest terminal artery. Both assumptions are sound provided that the vascular architecture is modeled to a sufficiently fine scale.

2.5.3 Solution

The Green's function approach given in Section 2.5.2 yields a system of N linear algebraic equations with N unknowns, the values of the q'_{bi}'s. After the q'_{bi}'s have been obtained, the temperature at any point or set of points in the tissue can then be readily calculated from Equation (2.93). The coefficients in the matrices depend on the geometrical configuration of the vessels, the flow rates in the vessels, and the values of the heat generation integral. Since the temperature near each vessel depends on the heat flow into every other vessel, the matrices are fully populated, creating a computationally intensive solution per unknown. Even though each interaction between two distant vessel segments is weak due to the $1/r$ decay with distance, the total interaction arising from all distant vessels cannot be neglected.

In contrast, finite element, finite difference, and finite volume methods yield large, but sparse, matrices because each unknown depends only on local interactions with near neighbors. The advantage of the Green's function approach is that the number of unknowns may be less than in the other methods because only the vascular subvolume is discretized into line segments, whereas finite element, finite difference, and finite volume methods require discretization of both tissue and blood subvolumes.

Baish [16] improved the efficiency of the Green's function method by using a relaxation technique that began by finding an approximate solution using only the largest arteries and their nearest neighbors. The solution then was improved by adding smaller, more numerous vessels.

2.5.4 Statistical Interpretation

Models using computer-simulated vasculature [16,64,71–75] can be used to predict various statistics about the effects of blood perfusion. While the vessel-generating algorithms employed above may not reproduce all details of the actual vasculature, such algorithms can be quite good at mimicking the numbers and sizes of the vessels typically present. By sampling the temperatures throughout one or more instantiations of a vascular growth algorithm, we may obtain stable statistics on the tissue and blood temperatures. For example, a local mean temperature $\overline{T}_t(\mathbf{r})$ can be estimated by the following average over n randomly chosen points within a desired sample volume δV:

$$\overline{T}_t(\mathbf{r}) \approx \frac{1}{n} \sum_{i=1}^{n} T_t(\mathbf{r}_i) \tag{2.96}$$

This method has an advantage over the use of the continuum models in that the size of the sampling volume can be precisely selected and controlled rather than be dictated by the vague assumptions underlying some continuum formulations. Sampling can also yield a histogram of local temperatures giving insights into local maximum, minimum, and variance. The range of temperatures may be significant because small volumes that go undertreated or dangerously overtreated may affect the clinical outcome. For example, local hyperthermia is based on the equivalent time at 43°C. Small fractions of the tumor volume that are heated to lower temperatures will likely experience rapid tumor regrowth. In addition, the range of temperatures expected in a small sampling volume can change the way an implanted temperature sensor might be interpreted. Incidental placement of a sensor near a blood vessel that produces a local temperature disturbance might produce a misleading view of the local average temperature.

The effectiveness of the heat sink can be determined by comparing the local average temperature to that predicted by the Pennes equation. This yields effectiveness values. In the absence of large-scale gradients, we have

$$\varepsilon(\mathbf{r}) = \frac{\omega_b \rho_b c_b}{q'''} [\bar{T}(\mathbf{r}) - T_a(0)] \qquad (2.97)$$

In Baish [16], the value of ε was found to depend on whether the arteries and veins were in close proximity with each other. Vessels spaced approximately one vessel diameter apart gave $\varepsilon \approx 0.8$, which is in the same range as found by Weinbaum et al. [14] and Brinck and Werner [17] by other methods.

All the above models for vasculature consider the flow of blood to be steady and Newtonian, with the vessel walls assumed to be rigid. However, blood flow is essentially pulsatile in nature, with the frequency characterized by the Womersley number. A blood vessel wall expands and contracts with changes in pressure. Pulsatility effects on heat transfer are most significant with large arteries [76,77]. Hence, the viscoelasticity of the blood vessel walls must be accounted for in such cases. With the branching of blood vessels, the reflection and transmission of pressure waves need to be considered. Blood flow is also non-Newtonian due to two reasons: (1) blood rheology is shear thinning, and (2) blood viscosity depends on the diameter of the vessel in blood vessels less than 500 µm in diameter. The inhomogeneous nature of blood starts to have an effect on the apparent viscosity at such a small diameter (the Fahraeus-Lindqvist effect). An introductory discussion of all these features is available in Chapter 17 of Ayyaswamy's book [78], and some of the commonly used models to approximate the shear-thinning rheology such as the Casson model can be found in Mukundakrishnan et al. [79].

2.6 CONCLUSIONS

The numerical methods employed to model heat transfer in perfused tissues generally differ little from the standard methods available in numerous general texts provided that continuum methods are used. In fact, commercial packages such as COMSOL® (Burlington, Massachusetts) can often be directly used with little or no modification. For example, the heat sink term used to model the blood flow in the Pennes equation is readily introduced into parameter windows without need for special macros or externally linked computer code. Only when a fully coupled model of the tissue and vasculature is desirable do the methods become sufficiently specialized that user-developed code may be required.

REFERENCES

1. Diller, K. R., Modeling of bioheat transfer processes at high and low temperatures, in *Bioengineering heat transfer: advances in heat transfer*, Y. I. Cho, Editor. 1992, Boston: Academic Press. pp. 157–357.
2. Rabin, Y. and A. Shitzer, Numerical solution of the multidimensional freezing problem during cryosurgery. *Journal of Biomechanical Engineering*, 1998. 120: pp. 32–37.
3. Zhang, J. et al., Numerical simulation of heat transfer in prostate cancer cryosurgery. *Journal of Biomechanical Engineering*, 2005. 127: pp. 279–294.
4. Incropera, F. P. et al., *Fundamentals of heat and mass transfer*, 4th ed. 2007, New York: John Wiley and Sons.
5. Chato, J. C., Heat transfer to blood vessels. *Journal of Biomechanical Engineering*, 1980. 102: pp. 110–118.
6. Victor, S. A. and V. L. Shah, Steady state heat transfer to blood flowing in the entrance region of a tube. *International Journal of Heat and Mass Transfer*, 1976. 19: pp. 777–783.
7. Barozzi, G. S. and A. Dumas, Convective heat transfer coefficients in the circulation. *Journal of Biomechanical Engineering*, 1991. 113(3): pp. 308–313.
8. Baish, J. W., P. S. Ayyaswamy, and K. R. Foster, Small-scale temperature fluctuations in perfused tissue during local hyperthermia. *Journal of Biomechanical Engineering*, 1986. 108: pp. 246–250.
9. Baish, J. W., P. S. Ayyaswamy, and K. R. Foster, Heat transport mechanisms in vascular tissues: a model comparison. *Journal of Biomechanical Engineering*, 1986. 108: pp. 324–331.
10. Klemick, S. G., M. A. Jog, and P. S. Ayyaswamy, Numerical evaluation of heat clearance properties of a radiatively heated biological tissue by adaptive grid scheme. *Numerical Heat Transfer, Part A*, 1997. 31(5): pp. 451–467.
11. Weinbaum, S., L. M. Jiji, and D. E. Lemons, Theory and experiment for the effect of vascular microstructure on surface tissue heat transfer—Part I: Anatomical foundation and model conceptualization. *Journal of Biomechanical Engineering*, 1984. 106: pp. 321–330.
12. Chen, M. M. and K. R. Holmes, Microvascular contributions in tissue heat transfer. *Annals of the New York Academy of Sciences*, 1980. 325: pp. 137–150.
13. Weinbaum, S. and L. M. Jiji, A new simplified bioheat equation for the effect of blood flow on local average tissue temperature. *Journal of Biomechanical Engineering*, 1985. 107: pp. 131–139.
14. Weinbaum, S. et al., A new fundamental bioheat equation for muscle tissue: Part I. Blood perfusion term. *Journal of Biomechanical Engineering*, 1997. 119(3): pp. 278–288.
15. Pennes, H. H., Analysis of tissue and arterial blood temperatures in the resting forearm. *Journal of Applied Physiology*, 1948. 1: pp. 93–122.
16. Baish, J. W., Formulation of a statistical model of heat transfer in perfused tissue. *Journal of Biomechanical Engineering*, 1994. 116(4): pp. 521–527.

17. Brinck, H. and J. Werner, Efficiency function: improvement of classical bioheat approach. *Journal of Applied Physiology*, 1994. 77: pp. 1617–1622.
18. Wulff, W., The energy conservation equation for living tissue. *IEEE Transactions on Biomedical Engineering*, 1974. 21: pp. 494–495.
19. Bazett, H. C. and B. McGlone, Temperature gradients in tissues in man. *American Journal of Physiology*, 1927. 82: pp. 415–428.
20. Arkin, H., K. R. Holmes, and M. M. Chen, A sensitivity analysis of the thermal pulse decay method for measurement of local tissue conductivity and blood perfusion. *Journal of Biomechanical Engineering*, 1986. 108: pp. 54–58.
21. Anderson, G. T. and J. W. Valvano, A small artery heat transfer model for self-heated thermistor measurements of perfusion in kidney cortex. *Journal of Biomechanical Engineering*, 1994. 116(1): pp. 71–78.
22. Anderson, G. T., J. W. Valvano, and R. R. Santos, Self-heated thermistor measurements of perfusion. *IEEE Transactions on Biomedical Engineering*, 1992. 39(9): pp. 877–885.
23. Chen, M. M. and K. R. Holmes, Thermal pulse-decay method for simultaneous measurement of thermal conductivity and local blood perfusion rate of living tissues, in *1980 Advances in bioengineering*, V. C. Mow, Editor. 1980, New York: ASME. pp. 113–115.
24. Johnson, W. R., A. H. Abdelmessigh, and J. Grayson, Blood perfusion measurements by the analysis of heat thermocouple probe's temperature transients. *Journal of Biomechanical Engineering*, 1979. 101: pp. 58–65.
25. Valvano, J. W., J. T. Allen, and H. F. Bowman, The simultaneous measurement of thermal conductivity, thermal diffusivity, and perfusion in small volumes of tissue. *Journal of Biomechanical Engineering*, 1984. 106(3): pp. 192–197.
26. Arkin, H., K. R. Holmes, and M. M. Chen, Theory on thermal probe arrays for the distinction between the convective and perfusive modalities of heat transfer in living tissues. *Journal of Biomechanical Engineering*, 1987. 109: pp. 346–352.
27. Arkin, H., K. R. Holmes, and M. M. Chen, A technique for measuring the thermal conductivity and evaluating the "apparent conductivity" concept in biomaterials. *Journal of Biomechanical Engineering*, 1989. 111: pp. 276–282.
28. Arkin, H. et al., Thermal pulse decay method for simultaneous measurement of local tissue blood perfusion: A theoretical analysis. *Journal of Biomechanical Engineering*, 1986. 108: pp. 208–214.
29. Baish, J. W., Heat transport by countercurrent blood vessels in the presence of an arbitrary temperature gradient. *Journal of Biomechanical Engineering*, 1990. 112(2): pp. 207–211.
30. Roemer, R. B. and A. W. Dutton, A generic tissue convective energy balance equation: Part I. Theory and derivation. *Journal of Biomechanical Engineering*, 1998. 120(3): pp. 395–404.
31. Bowman, H. F., E. G. Cravalho, and M. Woods, Theory, measurement and applications of thermal properties of biomaterials. *Annual Review of Biophysics and Bioengineering*, 1975. 4: pp. 43–80.
32. Chato, J. C., Fundamentals of bioheat transfer, in *Thermal dosimetry and treatment planning*, M. Gautherie, Editor. 1990, New York: Springer-Verlag. pp. 1–56.
33. Erdmann, B., J. Lang, and M. Seebass, Optimization of temperature distributions for regional hyperthermia based on a nonlinear heat transfer model. *Annals of the New York Academy of Sciences*, 1998. 858: pp. 36–46.
34. Lang, J., B. Erdmann, and M. Seebass, Impact of nonlinear heat transfer on temperature control in regional hyperthermia. *IEEE Transactions on Biomedical Engineering*, 1999. 46(9): pp. 1129–1138.

35. Mitchell, J. W. et al., Thermal response of human legs during cooling. *Journal of Applied Physiology*, 1970. 29: pp. 859–866.

36. Ferziger, J. H. and M. Peric, *Computational methods for fluid dynamics*. 3rd rev. ed. 2002, Berlin: Springer.

37. Briggs, W. L., V. E. Henson, and S. F. McCormick, *A multigrid tutorial*. 2nd ed. 2000, Philadelphia: Society for Industrial and Applied Mathematics.

38. Saad, Y., *Iterative methods for sparse linear systems*. 2nd ed. 2003, Philadelphia: Society for Industrial and Applied Mathematics.

39. Kotte, A., N. van Wieringen, and J. J. W. Lagendijk, Modelling tissue heating with ferromagnetic seeds. *Physics in Medicine and Biology*, 1998. 43(1): pp. 105–120.

40. Van Leeuwen, G. M. J. et al., Numerical modeling of temperature distributions within the neonatal head. *Pediatric Research*, 2000. 48(3): pp. 351–356.

41. Dai, W. Z., H. F. Yu, and R. Nassar, A fourth-order compact finite-difference scheme for solving a 1-D Pennes' bioheat transfer equation in a triple-layered skin structure. *Numerical Heat Transfer Part B: Fundamentals*, 2004. 46(5): pp. 447–461.

42. Lele, S. K., Compact finite difference schemes with spectral-like resolution. *Journal of Computational Physics*, 1992. 103(1): pp. 16–42.

43. Dai, W. Z. et al., Optimal temperature distribution in a three dimensional triple-layered skin structure with embedded vasculature. *Journal of Applied Physics*, 2006. 99(10): pp. 809–834.

44. Karaa, S., J. Zhang, and F. Q. Yang, A numerical study of a 3D bioheat transfer problem with different spatial heating. *Mathematics and Computers in Simulation*, 2005. 68(4): pp. 375–388.

45. Deng, Z. S. and J. Liu, Numerical study of the effects of large blood vessels on three-dimensional tissue temperature profiles during cryosurgery. *Numerical Heat Transfer Part A: Applications*, 2006. 49(1): pp. 47–67.

46. Lewis, R. W., *The finite element method in heat transfer analysis*. 1996, Chichester, UK: Wiley.

47. Reddy, J. N. and D. K. Gartling, *The finite element method in heat transfer and fluid dynamics*. 2nd ed. 2001, Boca Raton, FL: CRC Press.

48. Zienkiewicz, O. C., R. L. Taylor, and J. Z. Zhu, *The finite element method: its basis and fundamentals*. 6th ed. 2005, Amsterdam: Elsevier Butterworth-Heinemann.

49. Johnson, C. and C. Johnson, *Numerical solution of partial differential equations by the finite element method*. 1987, Cambridge: Cambridge University Press.

50. Torvi, D. A. and J. D. Dale, A finite-element model of skin subjected to a flash fire. *Journal of Biomechanical Engineering: Transactions of the ASME*, 1994. 116(3): pp. 250–255.

51. Chatterjee, I. and R. E. Adams, Finite-element thermal modeling of the human-body under hyperthermia treatment for cancer. *International Journal of Computer Applications in Technology*, 1994. 7(3–6): pp. 151–159.

52. Dennis, B. H. et al., Finite-element simulation of cooling of realistic 3D human head and neck. *Journal of Biomechanical Engineering: Transactions of the ASME*, 2003. 125(6): pp. 832–840.

53. Wainwright, P. R., The relationship of temperature rise to specific absorption rate and current in the human leg for exposure to electromagnetic radiation in the high frequency band. *Physics in Medicine and Biology*, 2003. 48(19): pp. 3143–3155.

54. Wainwright, P. R., Localized specific absorption rate calculations in a realistic phantom leg at 1–30 MHz using a finite element method. *Physics in Medicine and Biology*, 1999. 44(4): pp. 1041–1052.

55. Patankar, S. V., *Numerical heat transfer and fluid flow*. Series in computational methods in mechanics and thermal sciences. 1980, New York: McGraw-Hill.

56. Indik, R. A. and J. H. Indik, A new computer method to quickly and accurately compute steady-state temperatures from ferromagnetic seed heating. *Medical Physics*, 1994. 21(7): pp. 1135–1144.

57. Chua, K. J., S. K. Chou, and J. C. Ho, An analytical study on the thermal effects of cryosurgery on selective cell destruction. *Journal of Biomechanics*, 2007. 40(1): pp. 100–116.

58. Chua, K. J. et al., On the study of the temperature distribution within a human eye subjected to a laser source. *International Communications in Heat and Mass Transfer*, 2005. 32(8): pp. 1057–1065.

59. Becker, S. M. and A. V. Kuznetsov, Numerical assessment of thermal response associated with in vivo skin electroporation: the importance of the composite skin model. *Journal of Biomechanical Engineering: Transactions of the ASME*, 2007. 129(3): pp. 330–340.

60. Becker, S. M. and A. V. Kuznetsov, Thermal damage reduction associated with in vivo skin electroporation: a numerical investigation justifying aggressive pre-cooling. *International Journal of Heat and Mass Transfer*, 2007. 50(1–2): pp. 105–116.

61. Reinders, D. M., S. A. Baldwin, and J. L. Bert, Endometrial thermal balloon ablation using a high temperature, pulsed system: a mathematical model. *Journal of Biomechanical Engineering: Transactions of the ASME*, 2003. 125(6): pp. 841–851.

62. Gottlieb, M. E., Modelling blood vessels: a deterministic method with fractal structure based on physiological rules. *Proceedings of the Twelfth International Conference of the IEEE EMBS*, 1990. 12: p. 1386.

63. Nelson, T. R. and D. K. Manchester, Modeling of lung morphogenesis using fractal geometries. *IEEE Transactions on Medical Imaging*, 1988. 7(4): pp. 321–327.

64. Kotte, A. N. T. J., G. M. J. Van Leeuwen, and J. J. W. Lagendijk, Modelling the thermal impact of a discrete vessel tree. *Physics in Medicine and Biology*, 1999. 44: pp. 57–74.

65. Van Leeuwen, G. M. J., A. N. T. J. Kotte, and J. J. W. Lagendijk, A flexible algorithm for construction of 3D vessel networks for use in thermal modeling. *IEEE Transactions on Biomedical Engineering*, 1998. 45: pp. 596–605.

66. Van Leeuwen, G. M. J. et al., Temperature simulations in tissue with a realistic computer generated vessel network. *Physics in Medicine and Biology*, 2000. 45: pp. 1035–1049.

67. Murray, C. D., The physiological principle of minimum work. I. The vascular system and the cost of blood volume. *Proceedings of the National Academy of Sciences USA*, 1926. 12: pp. 207–214.

68. Murray, C. D., The physiological principle of minimum work. II. Oxygen exchange in capillaries. *Proceedings of the National Academy of Sciences USA*, 1926. 12: pp. 299–304.

69. Suwa, N. and T. Takahashi, *Morphological and morphometric analysis of circulation in hypertension and ischemic kidney*. 1971, Munich: Urban and Schwarzenberg.

70. Secomb, T. W. et al., Analysis of oxygen transport to tumor tissue by microvascular networks. *International Journal of Radiation Oncology Biology Physics*, 1993. 25(3): pp. 481–489.

71. Raaymakers, B. W., A. N. T. J. Kotte, and J. J. W. Lagendijk, How to apply a discrete vessel model in thermal simulations when only incomplete vessel data are available. *Physics in Medicine and Biology*, 2000. 45: pp. 3385–3401.

72. Huang, H. W., Z. P. Chen, and R. B. Roemer, A counter current vascular network model of heat transfer in tissues. *Journal of Biomechanical Engineering*, 1996. 118(1): pp. 120–129.

73. Shrivastava, D. and R. B. Roemer, Readdressing the issue of thermal significant blood vessels using a countercurrent vessel network. *Journal of Biomechanical Engineering*, 2006. 128: pp. 210–216.

74. Kotte, A. et al., A description of discrete vessel segments in thermal modelling of tissues. *Physics in Medicine and Biology*, 1996. 41: pp. 865–884.

75. Kotte, A. N. T. J., N. van Wieringen, and J. J. K. Lagendijk, Modelling tissue heating with ferromagnetic seeds. *Physics in Medicine and Biology*, 1998. 43: pp. 105–120.

76. Craciunescu, O. I. and S. Clegg, Pulsatile blood flow effects on temperature distribution and heat transfer in rigid blood vessels. *Journal of Biomechanical Engineering*, 2001. 123: pp. 500–505.

77. Horng, T.-L. et al., Effects of pulsatile blood flow in large vessels on thermal dose distribution during thermal therapy. *Medical Physics*, 2007. 34(4): p. 1312.

78. Ayyaswamy, P. S., Introduction to biofluid mechanics, in *Fluid mechanics*, 4th ed., I. Cohen and P. K. Kundu, Editors. 2007, San Diego, CA: Elsevier Academic. pp. 765–840.

79. Mukundakrishnan, K., P. S. Ayyaswamy, and D. M. Eckmann, Finite-sized gas bubble motion in a blood vessel: Non-Newtonian effects. *Physical Review E.*, 2008. 78: p. 036303.

3

Numerical Methods for Solving Bioheat Transfer Equations in Complex Situations

J. Liu and Z.-S. Deng

CONTENTS

3.1 INTRODUCTION

Bioheat transfer simulations have significant applications in a wide variety of clinical, basic, and environmental sciences [1,2]. Especially, understanding the heat transfer in biological tissues involving either the raising or lowering of temperature is a necessity for many therapeutic practices such as cancer hyperthermia [3], burn injury [2,4], brain hypothermia resuscitation [5], disease diagnostics [6], thermal comfort analysis [7], cryosurgery [8,9], and cryopreservation [10,11]. Up to now, much attention has been generally paid to the simulations of hyperthermia and cryosurgery. This is because such endeavors are more often urgently requested in tumor treatment.

In a hyperthermia process, whose primary objective is to raise the temperature of the diseased tissue to a therapeutic value, typically above 43°C, and then thermally destroy it, various apparatuses such as the microwave [12], ultrasound [13], and laser [14] have been used to deposit heat for treating the tumor in the deep biological body. Temperature prediction would be used to find an optimum way to either induce or prevent such thermal damage to the target tissues. In contrast to the principle of hyperthermia therapy, cryosurgery realizes its clinical object by a controlled destruction of tissues through deep freezing and thawing [15]. Applications of this treatment are quite wide in clinics owing to its outstanding virtues such as being quick, clean, and relatively painless, providing good homeostasis, and resulting in minimal scaring. An accurate understanding of the extent of the irregular shape of the frozen region, the direction of ice growth, and the temperature distribution within the iceballs during the freezing process is a basic requirement for the successful operation of a cryosurgery.

Until now, the classical Pennes equation has been commonly accepted as the best practical approach for modeling bioheat transfer in view of its simplicity and excellent validity [1,2]. This is because most of the other models either still lack sound experimental grounding or just appear too complex for mathematical solution. Although the real anatomical geometry of a biological body can be incorporated, the Pennes equation remains the most useful model for characterizing the heat transport process in clinical hyperthermia or cryosurgery. Therefore, this review will be focused on illustrating several typical numerical methods for solving this classical model. However, different from the previous treatments in mathematics, complex situations such as independent solutions of temperature at a single site of interest, irregularities of the calculation domain, a moving boundary due to freezing phase change, the nonlinearity of the bioheat transfer model and boundary conditions due to the temperature-dependent thermal properties or the phase change process involved, the existence of discrete large blood vessels, and the coupling multifields transport due to laser heating and the like will be especially addressed.

For brevity, here only cases for space-dependent thermal properties will be adopted for illustration purposes. A generalized form for the Pennes equation can then be written as

$$\rho c \frac{\partial T(\mathbf{X},t)}{\partial t} = \nabla \cdot k(\mathbf{X}) \nabla [T(\mathbf{X},t)] + w_b(\mathbf{X}) \rho_b c_b [T_a - T(\mathbf{X},t)]$$

$$+ Q_m(\mathbf{X},t) + Q_r(\mathbf{X},t), \quad \mathbf{X} \in \Omega$$

(3.1)

where ρ and c are the density and the specific heat of the tissue, respectively; ρ_b and c_b denote the density and specific heat of the blood, respectively; $\mathbf{X}$ contains the Cartesian coordinates x, y, and z; Ω denotes the analyzed spatial domain, respectively; $k(\mathbf{X})$ is the space-dependent thermal conductivity; and $w_b(\mathbf{X})$ is the space-dependent blood perfusion, which can generally be measured through a thermal clearance method. The value of blood perfusion represents the blood flow rate per unit tissue volume and is mainly from microcirculation, including the capillary network plus small arterioles and venules. T_a is the blood temperature in the arteries supplying the tissue and is often treated as a constant at 37°C; $T(\mathbf{X},t)$ is the tissue temperature; $Q_m(\mathbf{X},t)$ is the metabolic heat generation; and $Q_r(\mathbf{X},t)$ is the distributed volumetric heat source due to externally applied spatial heating.

From the historical viewpoint, we can find that the development of the bioheat transfer art and science can, in fact, be termed as one to modify and improve the Pennes model [1]. Among the efforts, the blood perfusion term in Pennes's equation has been substantially studied, which led to several conceptually innovative bioheat transfer models such as Wulff's continuum model [16], the Chen-Holmes model addressing both the flow and perfusion properties of blood [17], and the Weinbaum-Jiji three-layer model to characterize the heat transfer in the peripheral tissues [18].

The bioheat transfer equation and its extended forms can be directly used to characterize the thermal process of the biological bodies subject to various external or interior factors such as convective interaction with a heated or cooled fluid, radiation by fire or laser, contact with a heating or freezing apparatus, electromagnetic trauma, or a combination among them. Such issues can be treated using different boundary conditions as well as spatial heating or freezing patterns [19–21].

Generally, the geometric shape, dimensions, thermal properties, and physiological characteristics for tissues, as well as the arterial blood temperature, can usually be used as the input to the Pennes equation for a parametric study. According to the situation most commonly encountered in clinics, external energy applicators such as cryoprobes or hyperthermia needles can be inserted into target tissues for localized ablation of tumor tissues (Figure 3.1a). For a

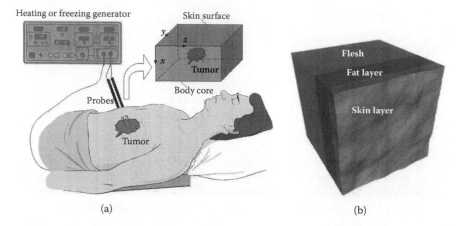

Figure 3.1   (a) Schematic illustration of a clinical hyperthermia or cryosurgery configuration, and (b) a simplified three-layer (skin, fat, and flesh layers) model geometry for the calculation of a region of interest in the human body.

feasible simulation purpose, a relatively complex yet mathematically solvable biological body geometry can generally be stratified as three layers: the skin, fat, and flesh layers (Figure 3.1b). In each layer, the thermal parameters can be treated as constant, or temperature and space dependent, or just different from each other in the areas of interest. Clearly, if desired, more anatomical geometry can still be incorporated into this model. According to a specific need in clinics, the bioheat transfer model can be modified by taking more factors into concern.

So far, many different numerical approaches such as the finite difference method (FDM), the finite element method (FEM), and the boundary element method (BEM) have been developed for solving bioheat transfer problems in various coordinates. Among these, the FDM is convenient for compiling the computer code but is not convenient for resolving the coordinates of the complex biological shape. As an alternative, the FEM has good adaptability to the complex shape. Compared with these, the BEM has the unique virtue of providing a complete problem solution in terms of boundary values only, with substantial savings in computer time and data preparation [22]. Overall, the geometry and properties of biological bodies vary drastically, which make the simulations of bioheat transfer rather complex. The numerical calculations are, therefore, often required. Unlike an analytical solution for mainly tackling relatively simple cases, the numerical simulation would provide more information for optimizing the best treatment protocol in clinics.

This review is dedicated to presenting an overview on several of the latest numerical methods as developed in the authors' laboratory for solving complex bioheat transfer problems. For illustration purposes, their typical applications from low to high temperatures will be discussed.

3.2 MONTE CARLO METHOD FOR SOLVING THE BIOHEAT EQUATION IN HYPERTHERMIA

3.2.1 Probability Model for Monte Carlo Algorithm

Monte Carlo (MC), a branch of experimental mathematics, is a method of directly simulating mathematical relations by random processes [23]. It has been applied to solve many heat conduction problems in nonliving materials with various boundary conditions and has received a great deal of attention in recent years [24–29]. In contrast to the classical FDM, FEM, and BEM, the efficiency of the Monte Carlo method (MCM) depends weakly on the dimensions and geometric details of the problem. Thus, compiling a computer code in this way appears relatively simple.

A particularly attractive feature of the MCM still lies in that the solution at a desired point (which can be arbitrary) can be obtained independently from the solutions of the other points within the domain [28], which will be an asset when temperatures are needed at only some isolated sites. This is rather unusual, since for all the other conventional methods of solving the discretized equations, the temperatures at all mesh points must be computed simultaneously [23]. It is because the MCM allows many independent statistical experiments to be carried out simultaneously that no communication between central processors is needed when using multiprocessor computing systems. The MC algorithms can be easily parallelized, which may greatly speed up the solution [30,31]. In fact, during a hyperthermic treatment of tumors, only temperatures at some desired tissues are of especial interest and must be known [32,33]. Therefore, the MCM turns out to be an excellent tool for tackling such bioheat transfer issues [34]. This will save a large amount of computational time if people wish only to know the thermal information at a certain specific area. It better satisfies the real-time requirement in clinics.

For other numerical methods such as FDM [35,36], FEM [13,37], and BEM [22], special care is needed to handle the complicated geometry and properties of the biological bodies. It is, therefore, difficult to compile a computer code to solve some complex cases with a three-dimensional (3D) domain or variable thermophysical properties. Compared with the above approaches, the MCM successfully avoids this disadvantage since it depends weakly on the dimensions and geometry of the problem. The recent work by Deng and Liu [34] demonstrated that the MCM can be used to solve nearly every kind of bioheat transfer problem with transient or space-dependent boundary conditions, blood perfusion, metabolic rate, and volumetric heat source for tissue.

The core ingredient of the MCM to solve differential equations is the random walk. The following part will illustrate the basic development and the subsequent application of the MCM for the solution of bioheat transfer problems. And the fixed random walk, with the step size and the pathways that are fixed in advance, is used.

Equation (3.1) can be rewritten as

$$\frac{\partial T(\mathbf{X},t)}{\partial t} = \alpha \nabla^2 T(\mathbf{X},t) - \frac{\omega_b \rho_b c_b}{\rho c} T(\mathbf{X},t) + Q(\mathbf{X},t), \quad \mathbf{X} \in \Omega \tag{3.2}$$

where $Q(\mathbf{X},t) = [Q_m(\mathbf{X},t) + Q_r(\mathbf{X},t) + \rho_b c_b \omega_b T_a]/\rho c$, and $\alpha = k/\rho c$ is the thermal diffusivity of tissue. The initial condition for Equation (3.2) can be defined as

$$T(\mathbf{X},t) = T_0(\mathbf{X}), \quad t = 0 \tag{3.3}$$

In an MC solution, there are three major types of walls encountered; these are classified as absorbing, reflecting, and partially absorbing barriers [23]. They correspond to the Dirichlet, Neumann, and mixed boundary conditions [38], respectively, and read as

$$T(\mathbf{X},t) = g(\mathbf{X},t), \quad \mathbf{X} \in \Gamma_1 \tag{3.4}$$

$$-k\frac{\partial T(\mathbf{X},t)}{\partial \mathbf{n}} = q(\mathbf{X},t), \quad \mathbf{X} \in \Gamma_2 \tag{3.5}$$

$$-k\frac{\partial T(\mathbf{X},t)}{\partial \mathbf{n}} = h_f(\mathbf{X})\left[T(\mathbf{X},t) - T_f(\mathbf{X},t)\right], \quad \mathbf{X} \in \Gamma_3 \tag{3.6}$$

where Γ_1 is the Dirichlet boundary, Γ_2 the Neumann boundary, and Γ_3 the mixed boundary; $\mathbf{n}$ is the unit normal vector on the boundaries.

Applying explicit discretization formulation to Equation (3.2), and using the following relation to express the linear term $T(\mathbf{X},t)$ in the right of Equation (3.2),

$$T(\mathbf{X},t) = \beta T(\mathbf{X},t + \Delta t) + (1 - \beta)T(\mathbf{X},t) \tag{3.7}$$

where β is the relaxation factor, and $0 \leq \beta \leq 1$, one can obtain Equation (3.8):

$$T(\mathbf{X},t+\Delta t) = \frac{1 - W(1-\beta)\Delta t - m \cdot Fo}{1 + W\beta\Delta t} T(\mathbf{X},t) + \sum_{i=1}^{m/2} \frac{Fo}{1 + W\beta\Delta t} T(\mathbf{X} + \Delta \mathbf{X}_i, t)$$

$$+ \sum_{i=1}^{m/2} \frac{Fo}{1 + W\beta\Delta t} T(\mathbf{X} - \Delta \mathbf{X}_i, t) + \frac{Q(\mathbf{X},t)\Delta t}{1 + W\beta\Delta t} \tag{3.8}$$

where Δt is the time increment; $W = \frac{\rho_b c_b \omega_b}{\rho c}$; $Fo = \frac{\alpha \cdot \Delta t}{\Delta x^2} = \frac{k \cdot \Delta t}{\rho c \Delta x^2}$ is the Fourier number; $m = 2, 4, 6$ corresponds to the cases of one, two, and three dimensions respectively; and

$$\Delta \mathbf{X}_1 = (\Delta x, 0, 0), \quad \Delta \mathbf{X}_2 = (0, \Delta y, 0), \quad \Delta \mathbf{X}_3 = (0, 0, \Delta z) \tag{3.9}$$

For convenience of presentation, it is assumed that

$$\Delta x = \Delta y = \Delta z \tag{3.10}$$

Since

$$\frac{1-W(1-\beta)\Delta t - m \cdot Fo}{1+W\beta\Delta t} + \sum_{i=1}^{m/2}\frac{Fo}{1+W\beta\Delta t} + \sum_{i=1}^{m/2}\frac{Fo}{1+W\beta\Delta t} = \frac{1-W(1-\beta)\Delta t}{1+W\beta\Delta t} \neq 1 \tag{3.11}$$

Equation (3.8) does not satisfy a probability model required by the classical MC method. In order to solve this problem, a common factor can be extracted from Equation (3.8) as follows (see Reference [34]):

$$T(\mathbf{X},t+\Delta t) = \frac{1-W(1-\beta)\Delta t}{1+W\beta\Delta t}\left\{\frac{Q(\mathbf{X},t)\Delta t}{1-W(1-\beta)\Delta t} + \frac{1-W(1-\beta)\Delta t - m \cdot Fo}{1-W(1-\beta)\Delta t}T(\mathbf{X},t)\right.$$

$$\left.+\sum_{i=1}^{m/2}\frac{Fo}{1-W(1-\beta)\Delta t}T(\mathbf{X}+\Delta\mathbf{X}_i,t) + \sum_{i=1}^{m/2}\frac{Fo}{1-W(1-\beta)\Delta t}T(\mathbf{X}-\Delta\mathbf{X}_i,t)\right\} \tag{3.12}$$

The laws of probability do not allow negative probability; one can therefore have

$$1-W(1-\beta)\Delta t - m \cdot Fo > 0 \tag{3.13}$$

Besides the explicit form of finite difference representation, the implicit formulation can also serve as the basis for the MC solution using the fixed random walk formulation, and it can be used to test the effect of various differential schemes on the computational results. The implicit probability model for Equation (3.2) is given as follows:

$$T(\mathbf{X},t+\Delta t) = \frac{1+m \cdot Fo - W(1-\beta)\Delta t}{1+m \cdot Fo + W\beta\Delta t}\left\{\frac{1-W(1-\beta)\Delta t}{1+m \cdot Fo - W(1-\beta)\Delta t}T(\mathbf{X},t)\right.$$

$$+\sum_{i=1}^{m/2}\frac{Fo}{1+m \cdot Fo - W(1-\beta)\Delta t}T(\mathbf{X}+\Delta\mathbf{X}_i,t+\Delta t)$$

$$+\sum_{i=1}^{m/2}\frac{Fo}{1+m \cdot Fo - W(1-\beta)\Delta t}T(\mathbf{X}-\Delta\mathbf{X}_i,t+\Delta t)$$

$$\left.+\frac{Q(\mathbf{X},t)\Delta t}{1+m \cdot Fo - W(1-\beta)\Delta t}\right\} \tag{3.14}$$

Similarly, the probability must satisfy:

$$1 - W(1-\beta)\Delta t > 0 \tag{3.15}$$

The discretized initial and boundary conditions can be written as

$$T(\mathbf{X}_P, t_P) = T_0(\mathbf{X}_P), \quad P \in \{P | P \in \Omega, t = 0\} \tag{3.16}$$

$$T(\mathbf{X}_P, t_P) = g(\mathbf{X}_P, t_P), \quad P \in \Gamma_1 \tag{3.17}$$

$$T(\mathbf{X}_P, t_P) = T(\mathbf{X}_I, t_I) - \frac{q(\mathbf{X}_P, t_P) \cdot \Delta x}{k}, \quad P \in \Gamma_2 \tag{3.18}$$

$$T(\mathbf{X}_P, t_P) = \frac{1}{1 + \mathrm{Bi}(\mathbf{X})} T(\mathbf{X}_I, t_I) + \frac{\mathrm{Bi}(\mathbf{X})}{1 + \mathrm{Bi}(\mathbf{X})} T_f(\mathbf{X}_P, t_P), \quad P \in \Gamma_3 \tag{3.19}$$

where $\mathrm{Bi}(\mathbf{X}) = h_f(\mathbf{X}) \cdot \Delta x / k$. For convenience of presentation, it is often assumed that the unit normal vector on the boundary $\mathbf{n}$ is parallel to one of the three coordinate axes, and point I is the interior node next to point P along the direction of the normal vector.

3.2.2 The Fixed Random Walk Formulation

The random walk refers to a statistical procedure that is used to solve the difference form of various partial or ordinary differential equations. The description of a random walk is well documented in the literature [23]. In the fixed random walk method, a domain grid is constructed, and a random walker is dispatched from the desired interior node, as shown in Figure 3.2 [34]. The MC procedure determines the path of the random walk, which terminates when the Dirichlet boundary is reached. The computation of the temperature at any interior mesh point (x_0, y_0, z_0), as indicated in Figure 3.2, can be illustrated as follows. Here, the case where the boundary temperature is prescribed will be considered. A random-walking

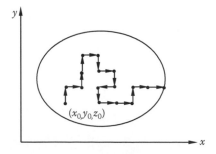

Figure 3.2 Schematic of the path of a random walk that a particle travels.

particle is dispatched from the point (x_0, y_0, z_0) whose temperature is desired. It steps randomly within the domain grid in terms of the probabilities, stated by the finite difference representation of the governing equation, until it arrives at a boundary node. Then the temperature at that boundary node is tallied. Similarly, a second particle is dispatched and reaches a boundary node whose temperature is also tallied. This procedure is repeated for the third, the fourth, and so on to the nth particle dispatched from the same point (x_0, y_0, z_0), and the corresponding boundary temperatures are recorded. Then, the sum of these temperatures divided by n, the total number of particles, is the approximate temperature of the point (x_0, y_0, z_0) where the particles start. In the limit as n approaches infinity, the temperature thus calculated would approach the real solution.

The above is a brief description of the fixed random walk procedure. In real bioheat processes, this procedure is much more complex, which can be found from the finite difference form of Equation (3.2) (i.e., Equations 3.12 and 3.14). The relation given in Equation (3.12) indicates that the temperature at a point within an isotropic region is related to the temperature of its neighboring points. Note that Equation (3.12), under the condition required by Equation (3.13), guaranteed that (1) each probability for the particle to walk along any direction is positive, and (2) the sum of all probabilities is equal to 1, that is,

$$\frac{1 - W(1-\beta)\Delta t - m \cdot Fo}{1 - W(1-\beta)\Delta t} + \sum_{i=1}^{m/2} \frac{Fo}{1 - W(1-\beta)\Delta t} + \sum_{i=1}^{m/2} \frac{Fo}{1 - W(1-\beta)\Delta t} = 1 \qquad (3.20)$$

Now, Equation (3.12) can be given a probabilistic interpretation: if a random walker is momentarily at a point $(\mathbf{X}, t + \Delta t)$, a fraction of the random walk equal to $[1 - W(1-\beta)\Delta t - m \cdot Fo]/[1 - W(1-\beta)\Delta t]$ steps to $T(\mathbf{X}, t)$, a fraction of the random walk equal to $Fo/[1 - W(1-\beta)\Delta t]$ steps to $T(\mathbf{X} + \Delta\mathbf{X}_1, t)$, a fraction of the random walk equal to $Fo/[1 - W(1-\beta)\Delta t]$ steps to $T(\mathbf{X} - \Delta\mathbf{X}_1, t)$, and so on. Once the random walker has completed its first step, the procedures should be taken in a similar manner for other steps until the boundary of the region or the time $t = 0$ is encountered [34]. That is to say, the procedure for every random walk, beginning at interior mesh point $P_0 \in \Omega$ for $t > 0$ and terminating at initial point $P \in \{P | P \in \Omega, t = 0\}$ or boundary point $P \in \Gamma, \Gamma = \Gamma_1 + \Gamma_2 + \Gamma_3$, follows:

$$\gamma_P : P_0 \to P_1 \to P_2 \to \cdots \to P_i \to P_{i+1} \to \cdots \to P_{k-1} \to P \in \Omega_p \qquad (3.21)$$

where γ_P denotes a random walk, and $\Omega_p = \{P | P \in \Omega, t = 0\} + \Gamma$. Then, one can define a stochastic variable ξ as

$$\xi = u(\gamma_p)$$

$$= \sum_{i=0}^{k-1} \left[\left(\frac{1 - W(1-\beta)\Delta t}{1 + W\beta\Delta t} \right)^{i+1} \cdot \frac{Q(\mathbf{X}_{p_i}, t_{p_i})\Delta t}{1 - W(1-\beta)\Delta t} \right] + \left(\frac{1 - W(1-\beta)\Delta t}{1 + W\beta\Delta t} \right)^{k} \cdot f(P) \qquad (3.22)$$

Therefore, at the end of each random walk, the corresponding stochastic variable ξ is tallied.

In the same way, Equation (3.14) can also be given a probabilistic interpretation: if a random walker begins at point $(\mathbf{X}, t+\Delta t)$, the probability of stepping to any one of the points $(\mathbf{X}, t)$, $T(\mathbf{X}+\Delta\mathbf{X}_i, t+\Delta t)$, and $T(\mathbf{X}-\Delta\mathbf{X}_i, t+\Delta t)$ (where, $i=1,\ldots,m$) is the respective value of $\frac{1-W(1-\beta)\Delta t}{1+m \cdot Fo - W(1-\beta)\Delta t}$, $\frac{Fo}{1+m \cdot Fo - W(1-\beta)\Delta t}$, and $\frac{Fo}{1+m \cdot Fo - W(1-\beta)\Delta t}$. Once the random walker has completed its first step, the procedure is repeated for the second step, the third step, and so on until the boundary of the region or $t=0$ is encountered. This random walk procedure follows:

$$\gamma_P : P_0 \rightarrow P_1 \rightarrow P_2 \rightarrow \cdots \rightarrow P_i \rightarrow P_{i+1} \rightarrow \cdots \rightarrow P_{k-1} \rightarrow P \in \Omega_p \qquad (3.23)$$

The stochastic variable ξ can be defined as

$$\xi = u(\gamma_p)$$

$$= \sum_{i=0}^{k-1} \left[\left(\frac{1+m \cdot Fo - W(1-\beta)\Delta t}{1+m \cdot Fo + W\beta\Delta t} \right)^{i+1} \cdot \frac{Q(\mathbf{X}_{p_i}, t_{p_i})\Delta t}{1+m \cdot Fo - W(1-\beta)\Delta t} \right] \qquad (3.24)$$

$$+ \left(\frac{1+m \cdot Fo - W(1-\beta)\Delta t}{1+m \cdot Fo + W\beta\Delta t} \right)^{k} \cdot f(P)$$

Similarly, at the end of a random walk, the corresponding stochastic variable ξ is tallied.

In Equations (3.22) and (3.24), the value of $f(P)$ depends on the point type encountered at the end of a random walk. When the random walker ends at a mesh point in the initial state (i.e., $P \in \{P | P \in \Omega, t=0\}$), the initial temperature is scored, and $f(P)$ in Equations (3.22) and (3.24) takes the value of $T_0(\mathbf{X}_P)$. Whenever the random walk arrives at a boundary point, the random walk may be absorbed or sent to a neighboring node depending on a preassigned statistical chance. The stochastic variable ξ described above can be applied for any one of the three major types of boundary conditions only by assigning the value of $f(P)$ as follows [34]:

1. For the surface temperature boundary condition (i.e., $P \in \Gamma_1$), the surface temperature is scored; accordingly, $f(P) = g(\mathbf{X}_P, t_P)$.
2. For prescribed surface heat flux instead of surface temperature (i.e., $P \in \Gamma_2$), the computation can be accomplished by a similar scheme. As indicated in Equation (3.18), the random walk will not terminate at this wall; instead, it may go back to the interior node next to point P. From the combination of Equation (3.12) or (3.14) and Equation (3.18), it can be made out that when a random walk is at a surface location with prescribed heat flux, the probability of returning to the interior domain

is $\frac{2Fo}{1-W(1-\beta)\Delta t}$ for the case of explicit discretization formulation and $\frac{2Fo}{1+m\cdot Fo-W(1-\beta)\Delta t}$ for the case of implicit formulation, while the probability of stepping along other axes remains as given by Equations (3.12) and (3.14). Whether the random walk steps to a different location or remains at the same boundary point, the tally is the value of $-\frac{2Fo\cdot q(\mathbf{X}_P,t_P)\cdot\Delta x/k}{1-W(1-\beta)\Delta t}$ for the case of explicit discretization formulation and $-\frac{2Fo\cdot q(\mathbf{X}_P,t_P)\cdot\Delta x/k}{1+m\cdot Fo-W(1-\beta)\Delta t}$ for the case of implicit formulation. Clearly, the random walk will proceed to wander from point to point along the nodes of the grid until the initial temperature is scored or another surface with the prescribed temperature is encountered.

3. For the convection boundary condition (i.e., $P \in \Gamma_3$), it is clearly indicated in Equation (3.19) that the random walk will not necessarily terminate at this wall, and it may go back to the interior node next to point P. At this point, for the case of explicit discretization formulation, the probabilities of returning to the interior domain and terminating the random walk are respectively $\frac{2Fo}{1-W(1-\beta)\Delta t}\cdot\frac{1}{1+\mathrm{Bi}(\mathbf{X})}$ and $\frac{2Fo}{1-W(1-\beta)\Delta t}\cdot\frac{\mathrm{Bi}(\mathbf{X})}{1+\mathrm{Bi}(\mathbf{X})}$; and for the case of implicit formulation, the probabilities are respectively $\frac{2Fo}{1+m\cdot Fo-W(1-\beta)\Delta t}\cdot\frac{1}{1+\mathrm{Bi}(\mathbf{X})}$ and $\frac{2Fo}{1+m\cdot Fo-W(1-\beta)\Delta t}\cdot\frac{\mathrm{Bi}(\mathbf{X})}{1+\mathrm{Bi}(\mathbf{X})}$. The probability of stepping along other axes remains as given by Equations (3.12) and (3.14). When the random walk terminates at the same boundary point, the tally is the value of $T_f(\mathbf{X}_P,t_P)$ for both cases of explicit and implicit discretization formulation.

The computation of the temperature at any interior mesh point $(\mathbf{X}_0,t_0)$ can now be described. An MC procedure begins by starting N random walks at the same time, referred to as the sample size. For the ith random walk, the tally is expressed as ξ_i, where $i = 1, 2, \ldots, N$. Each random walk procedure is similar to the above-mentioned ones. Following the completion of the nth random walk, the MC estimate for $T(\mathbf{X}_0,t_0)$ can be written as

$$T(\mathbf{X}_0,t_0) = \frac{1}{N}\sum_{i=1}^{N}\xi_i \qquad (3.25)$$

Figures 3.3a,b,c depict temperature distributions at different profiles in tissues with tumors by using a different random walk number N [34]. The temperature near the tumor site is obviously higher than at other locations. Comparing Figure 3.3a with Figure 3.3b, it is clearly shown that the larger the random walk number, the higher the computational accuracy. Otherwise, the temperature curve appears relatively coarse. In Figures 3.3a,c, the same random walk numbers are adopted. However, the accuracy of results in Figure 3.3c for the boundary temperature is poor, while that in Figure 3.3a is fine. This indicates that the computing accuracy at the boundary points is worse than in the interior points.

Although the generation of true random numbers is possible, the process is currently impossible for computer-based numerical computation. It is thus usually routine for one to use a sequence of pseudorandom numbers that can be generated with ease by a random number generator. Before generation, the generator must be initialized by selecting a seed, or, in other words, a starting value. Different seed numbers will generate different sequences of pseudorandom numbers and will inevitably influence the computational results. Figure 3.4

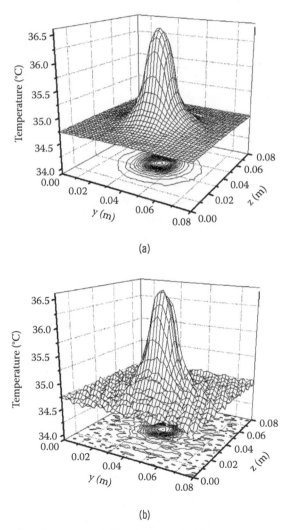

(a)

(b)

Figure 3.3   Temperature distributions at different profiles ($t = 2000$ seconds), in which (a) $x = 0.014$ m, $N = 10000$; (b) $x = 0.014$ m, $N = 1000$; and (c) $x = 0$, $N = 10000$. (*Note:* x denotes the tissue depth from the skin surface, while y and z are along the surface.)

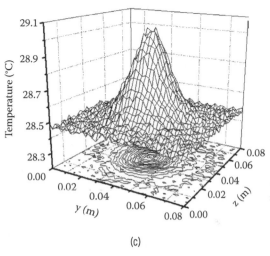

(c)

Figure 3.3   (Continued)

illustrates the influence of seed number on temperature distribution, in which the seed numbers are selected stochastically [34]. Clearly, the influence of different seed numbers on MC results is not obvious, which is beneficial for practical application purposes.

In an MC solution, it can be demonstrated that if the random number approaches infinity (i.e., $N \to \infty$), the temperature thus determined will approach the exact solution. The influence of random walk number on the MC results is depicted in Figure 3.5. As a comparison, Figure 3.5a,b gives temperature values spanning two different random walk scopes [34]. It is clear from Figure 3.5 that the computed temperature at a certain point will no longer change with the increase

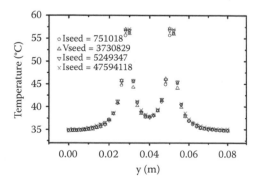

Figure 3.4   Influence of seed number on the temperature distribution ($x = 0.014$, $z = 0.04$, $t = 2000$ seconds).

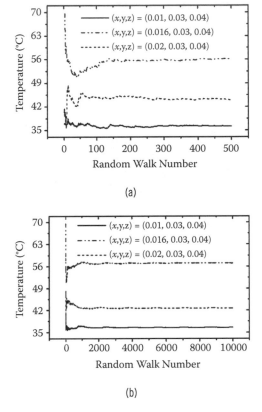

(a)

(b)

Figure 3.5 Influence of random walk number on the temperature distribution ($t = 2000$ seconds), in which (a) $N = 1,2,3,\cdots,500$ and (b) $N = 1,2,3,\cdots,10000$.

of random walk number, and the computed temperature is approximately equal to this value when $N \geq 2000$.

3.2.3 The Absorption Formulation

For explicit and implicit formulations, as given in Equations (3.12) and (3.14), respectively, the random walk will proceed to wander from point to point along the nodes of the grid until the initial temperature is tallied or another surface with the prescribed temperature is encountered. This increases the duration of a random walk, thereby prolonging the computation time. To avoid time consumption, the absorption formulation is introduced as an alternative [34].

As stated above, Equation (3.8) does not satisfy the probability model of the MC method. But, if one is taking $1 - \frac{1-W(1-\beta)\Delta t}{1+W\beta\Delta t}$ as the absorption probability at the present point (namely, when a particle is momentarily at a point P_i), the probability of terminating at point P_i is $1 - \frac{1-W(1-\beta)\Delta t}{1+W\beta\Delta t}$). Equation (3.8) can also be given a probabilistic interpretation: if a random walker starts at point $(\mathbf{X}, t + \Delta t)$, the probabilities of stepping to any one of the points $(\mathbf{X}, t), T(\mathbf{X} + \Delta\mathbf{X}_i, t + \Delta t)$, and $T(\mathbf{X} - \Delta\mathbf{X}_i, t + \Delta t)$ (where $i = 1, \cdots, m$) are $\frac{1-W(1-\beta)\Delta t - m \cdot Fo}{1+W\beta\Delta t}$, $\frac{Fo}{1+W\beta\Delta t}$, and $\frac{Fo}{1+W\beta\Delta t}$, respectively, and the probability of terminating a random walk at the present point is $1 - \frac{1-W(1-\beta)\Delta t}{1+W\beta\Delta t}$. Once the random walker has completed its first step, the procedure is repeated for the second step. In a similar manner, a third step is taken and so on until the random walk is absorbed at a certain point or until the boundary of the region or $t = 0$ is encountered. If the random walk procedure is absorbed at point P_s, that is,

$$\gamma_P : P_0 \to P_1 \to P_2 \to \cdots \to P_i \to P_{i+1} \to \cdots \to P_k \tag{3.26}$$

the stochastic variable ξ can be defined as

$$\xi = u(\gamma_p) = \sum_{t=0}^{k} \frac{Q(\mathbf{X}_{p_i}, t_{p_i})\Delta t}{1 + W\beta\Delta t} \tag{3.27}$$

If the random walk procedure is terminated either at a certain boundary point or when $t = 0$ is encountered, one has

$$\gamma_P : P_0 \to P_1 \to P_2 \to \cdots \to P_i \to P_{i+1} \to \cdots \to P_{k-1} \to P \in \Omega_p \tag{3.28}$$

where $\Omega_p = \{P | P \in \Omega, t = 0\} + \Gamma$. Then the corresponding stochastic variable ξ can be defined as follows:

$$\xi = u(\gamma_p) = \sum_{i=0}^{k-1} \frac{Q(\mathbf{X}_{p_i}, t_{p_i})\Delta t}{1 + W\beta\Delta t} + f(P) \tag{3.29}$$

In Equation (3.29), the value assigned to $f(P)$ is the same as that of the algorithms developed above. At the end of each random walk, the corresponding stochastic variable ξ is tallied. Similarly, the MC estimation for temperature at a specific point can also be obtained by Equation (3.25).

The explicit and implicit formulations are time-consuming. Comparison of the computation time among different formulations shows that [34] at boundary points, the absorption formulation seems to have higher computational accuracy than that of the explicit and implicit formulations. Besides, the absorption formulation is the most timesaving among the three formulations, and the explicit formulation takes the second place (Figure 3.6).

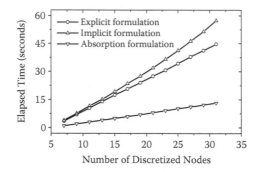

Figure 3.6 Comparison of computation time for different algorithms ($t = 2000$ seconds).

3.2.4 Monte Carlo Algorithm for Nonlinear Boundary Conditions

As is well known, the generalized boundary condition for the heat transfer occurring at the skin surface is generally composed of three parts (i.e., convection, radiation, and evaporation), which can be written as follows [6]:

$$-k\frac{\partial T}{\partial n}\bigg|_{skin} = h_f(T_s - T_f) + \sigma\varepsilon(T_s^4 - T_f^4) + Q_e \tag{3.30}$$

where h_f is the convection heat transfer coefficient; T_s and T_f are the skin and surrounding air temperatures, respectively; ε is the skin emissivity; σ is the Stefan-Boltzmann constant; and Q_e is the evaporative heat losses due to sweat secretion (refer to References [7,19]).

$$Q_e = Q_{dif} + Q_{rsw} \tag{3.31}$$

$$Q_{dif} = 3.054(0.256T_s - 3.37 - P_a) \quad \text{W/m}^2 \tag{3.32}$$

$$Q_{rsw} = 16.7h_f W_{rsw}(0.256T_s - 3.37 - P_a) \quad \text{W/m}^2 \tag{3.33}$$

where Q_{dif} is the heat loss by evaporation of implicit sweat secretion when the skin is dry, Q_{rsw} is the heat loss by evaporation of explicit sweat secretion, W_{rsw} is the skin humidity ($0 \leq W_{rsw} \leq 1$ and $W_{rsw} = 0,1$, respectively, mean that the skin is dry and that it is entirely wet), P_a is the vapor pressure in ambient air, and $P_a = \phi_a P_a^*$ (where ϕ_a is the relative humidity of surrounding air, and P_a^* is the saturated vapor pressure at surrounding air temperature). In Equations (3.32) and (3.33), the unit of pressure is kPa. Substituting Equations (3.31) through (3.33) into Equation (3.30) yields

$$-k\frac{\partial T}{\partial n}\bigg|_{skin} = h_f(T_s - T_f) + \sigma\varepsilon\left(T_s^4 - T_f^4\right) + (3.054 + 16.7h_f W_{rsw})(0.256T_s - 3.37 - P_a) \tag{3.34}$$

This nonlinear boundary condition, due to occurrence of the $\sigma \varepsilon T_s^4$ term, can be solved through an iteration as follows: (1) if the random walk ends at the skin surface, assume the predicted skin temperature $T_{s,a}$; (2) rewrite Equation (3.34) as the normal format:

$$-k \frac{\partial T}{\partial n}\bigg|_{skin} = h_f[T_s - f(Q)] \tag{3.35}$$

in Equation (3.35),

$$f(Q) = T_f - \frac{\sigma \varepsilon (T_{s,a}^4 - T_f^4) + (3.054 + 16.7 h_f W_{rsw})(0.256 T_{s,a} - 3.37 - P_a)}{h_f} \tag{3.36}$$

(3) then calculate the skin temperature T_s based on $T_{s,a}$; (4) determine the corrected skin temperature by

$$T_c = T_{s,a} + \gamma(T_s - T_{s,a}) \tag{3.37}$$

where $0 < \gamma \leq 1$ is the relaxation factor; (5) check convergence: if $|T_c - T_{s,a}|/|T_c| > \varepsilon_{max}$ (where ε_{max} is a prescribed maximum acceptable error), set $T_{s,a} = T_c$, and continue the iterative operations from step 2 until the relative error becomes less than or equal to ε_{max}; and (6) if convergence is achieved, end the iteration and record $T_s = T_c$.

Solving the generalized bioheat transfer model and the boundary and initial conditions, the relative contribution of each thermal factor to skin temperature distribution can be clarified. Presented in Figure 3.7 is a comparison of the heat fluxes due to radiation, convection, and evaporation, and the sum of all these heat fluxes at the skin surface [6]. It indicates that the heat loss due to evaporation when the skin is dry is much less than that while the skin is partially wet. Consequently, the nonhomogeneous skin humidity can also result in an abnormal temperature

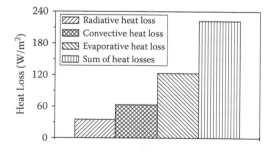

Figure 3.7 The heat fluxes at the skin's surface due to radiation, convection, and evaporation ($\varepsilon = 0.9$, $W_{rsw} = 0.2$, $\phi = 40\%$, $T = 25°C$, and $h_f = 10$ W/m²°C).

distribution at the skin surface. These complexities should be considered in some specific clinical practices such as thermal diagnosis. But for a general analysis, evaporative and radiative heat transfer at the skin surface is often omitted or just attributed to the apparent convective heat transfer term for simplicity.

3.3 SIMULATION OF BIOHEAT TRANSFER PROCESS IN CRYOMEDICAL ENGINEERING

3.3.1 Particularities in Modeling of Bioheat Transfer with Phase Change

A major difference between simulation of cryosurgery and that in hyperthermia is that a phase change process occurs in the former case. The main difficulties encountered in cryosurgical simulation are the unknown extent of the irregular shape of the frozen region, the direction of ice growth, and the temperature distribution within the iceballs during the freezing process. Such moving boundary problems are highly nonlinear. For example, the thermal conductivities for the frozen tissue and blood-perfused region were generally different (e.g., a smaller value for tissue was often taken as $k_l = 0.5$ W/m°C, while in the frozen region a larger value as $k_s = 2$ W/m°C was usually used) [39]. Besides, blood perfusion and metabolic heat generation exist in the unfrozen region, while they disappear after being frozen. Considering that biological tissues in cryosurgery experience a wide range of temperature change, their thermal properties (including specific heat and thermal conductivity) are expected to change significantly over the freezing and thawing process. That is to say, the thermal parameters of the biological tissues are generally temperature dependent. Several previous studies have shown that inclusion of temperature dependence has a significant effect on phase change predictions [40,41]. As an alternative, constant assumptions were also often adopted or only space-dependent parameters were taken into concern. For example, to avoid the expensive and intensive numerical iteration, a multisegmental constant thermal conductivity has been used to approximate the temperature-dependent case [41]. Further, due to the nonideal solution property, the phase change temperature for the biological tissues usually occurs in a rather wide range, say, between -1°C (upper limit) and -8°C (lower limit), not just fixed at 0°C, as is assumed in most of the calculations.

In the long-term development of cryosurgery technology, a number of numerical models to solve the phase change problems of biological tissues have been proposed. Generally, the existing numerical schemes can be divided into two basic approaches [42]: one is based on the front-tracking technique, while another is based on the nonfront-tracking technique including enthalpy formulation and the effective heat capacity method.

For a front-tracking technique, the heat transfer equations for both frozen and blood-perfused regions should be separately described and solved simultaneously. In the unfrozen tissue, the classical Pennes heat transfer model was often used, that is,

$$\rho_u C_u \frac{\partial T_u(\mathbf{X},t)}{\partial t} = \nabla \cdot k_u \nabla[T_u(\mathbf{X},t)] + w_b C_b[T_a - T_u(\mathbf{X},t)] + Q_m, \quad \mathbf{X} \in \Omega_u(t) \qquad (3.38)$$

where subscript u indicates the unfrozen phase.

For the frozen region, due to the absence of blood perfusion and metabolic activities, the heat balance is given by

$$\rho_f C_f \frac{\partial T_f(\mathbf{X},t)}{\partial t} = \nabla \cdot k_f \nabla[T_f(\mathbf{X},t)], \quad \mathbf{X} \in \Omega_f(t) \qquad (3.39)$$

where subscript f indicates frozen tissue.

For ideal biological tissues, the temperature continuum and energy balance conditions at the moving solid–liquid interface are given as follows (assuming that the density of tissue ρ is the same constant for both liquid and solid phases):

$$T_f(\mathbf{X},t) = T_u(\mathbf{X},t) = T_m, \quad \mathbf{X} \in \Gamma_{m.i.} \qquad (3.40)$$

$$k_f \frac{\partial T_f(\mathbf{X},t)}{\partial n} - k_u \frac{\partial T_u(\mathbf{X},t)}{\partial n} = Q_l V_n, \quad \mathbf{X} \in \Gamma_{m.i.} \qquad (3.41)$$

where n denotes the unit outward normal; Q_l and T_m are, respectively, the latent heat and freezing point of tissue; $\Gamma_{m.i.}$ is the moving boundary (i.e., the moving interface resulted by phase change); and V_n is the normal velocity of the moving interface.

Due to the high nonlinearity of Equations (3.38) through (3.41), complex iteration at the moving boundary is inevitable by directly discretizing these governing equations. In some recently emerging cryosurgery assisted by hyperthermia, more complicated situations will be encountered. For example, due to alternate strong freezing and heating, many different phase change interfaces will be produced in the tissue domain (Figure 3.8). Especially for multiple probes application, such a three-dimensional calculation would become extremely difficult.

In order to avoid the iteration at the moving boundary, the effective heat capacity method is also often adopted [42]. Since first proposed by Bonacina et al. [43], the effective heat capacity method has been used by many investigators to solve phase change problems. The advantage of this method lies in that a fixed grid can be used for the numerical computation, and that the nonlinearity at the moving boundary can thus be avoided. The essence of the effective heat capacity

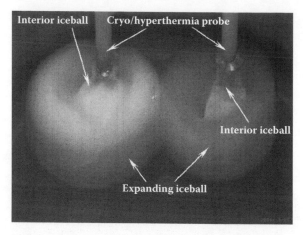

Figure 3.8 The iceballs produced in phantom gel by two probes during a freeze–heat cycle using a combined cryosurgery and hyperthermia system.

method is to approximate the latent heat by a generalized effective heat capacity over a small temperature range near the freezing point. Following this strategy, the numerical solution can be carried out on a fixed grid throughout the calculation process, which is much easier to implement.

To apply the effective heat capacity method, Equations (3.38) and (3.39) must be substituted by a uniform energy equation, which can be constructed as follows [42]:

$$\Lambda \frac{\partial T(\mathbf{X},t)}{\partial t} = \nabla \cdot k(T)\nabla[T(\mathbf{X},t)] + w_b(T)C_b[T_a - T(\mathbf{X},t)] + Q_m(T), \quad \mathbf{X} \in \Omega \qquad (3.42)$$

In Equation (3.42), it is assumed that the phase change occurs over a very small temperature range near the freezing point, that is, $(T_m - \Delta T, T_m + \Delta T)$, in which ΔT is an extremely small quantity, and

$$\Lambda = \rho C(T) + \frac{Q_l \delta(T - T_m)}{2\Delta T} \qquad (3.43)$$

$$\rho C(T) = \begin{cases} \rho C_f, & T < T_m \\ \rho C_u, & T > T_m \end{cases} \qquad (3.44)$$

$$\delta(T - T_m) = \begin{cases} 0, & T < T_m - \Delta T \\ 1, & T_m - \Delta T \le T \le T_m + \Delta T \\ 0, & T > T_m + \Delta T \end{cases} \qquad (3.45)$$

$$k(T) = \begin{cases} k_f, & T < T_m \\ k_u, & T > T_m \end{cases} \tag{3.46}$$

$$Q_m(T) = \begin{cases} 0, & T < T_m \\ Q_m, & T > T_m \end{cases} \tag{3.47}$$

$$w_b(T) = \begin{cases} 0, & T < T_m \\ w_b, & T > T_m \end{cases} \tag{3.48}$$

Since the phase change of real biological tissue does not take place at a specific temperature but within a temperature range, it is reasonable to substitute a large effective heat capacity over a temperature range (T_{ml}, T_{mu}) for the latent heat, where T_{ml} and T_{mu} are, respectively, the lower and upper phase transition temperatures of tissue. Introducing the effective heat capacity $\tilde{C}$, the effective thermal conductivity $\tilde{k}(T)$, the effective metabolic heat generation $\tilde{Q}_m$, and the effective blood perfusion $\tilde{\omega}_b(T)$, respectively, as follows (assuming that k_u, k_f, C_u, and C_f are all constant):

$$\tilde{C}(T) = \begin{cases} \rho C_f, & T < T_{ml} \\ \dfrac{Q_l}{(T_{mu} - T_{ml})} + \dfrac{\rho C_f + \rho C_u}{2}, & T_{ml} \leq T \leq T_{mu} \\ \rho C_u, & T > T_{mu} \end{cases} \tag{3.49}$$

$$\tilde{k}(T) = \begin{cases} k_f, & T < T_{ml} \\ (k_f + k_u)/2, & T_{ml} \leq T \leq T_{mu} \\ k_u, & T > T_{mu} \end{cases} \tag{3.50}$$

$$\tilde{Q}_m(T) = \begin{cases} 0, & T < T_{ml} \\ 0, & T_{ml} \leq T \leq T_{mu} \\ Q_m, & T > T_{mu} \end{cases} \tag{3.51}$$

$$\tilde{w}_b(T) = \begin{cases} 0, & T < T_{ml} \\ 0, & T_{ml} \leq T \leq T_{mu} \\ w_b, & T > T_{mu} \end{cases} \tag{3.52}$$

Then Equation (3.42) can be written as

$$\tilde{C}\frac{\partial T}{\partial t}=\nabla\cdot\tilde{k}\nabla T+\tilde{w}_b C_b(T_a-T)+\tilde{Q}_m, \quad \mathbf{X}\in\Omega \tag{3.53}$$

Consequently, the complex nonlinear phase change problems are simplified as nonhomogeneous ones, which can be easily dealt with by a general numerical method.

3.3.2 Finite Difference Formulation

Numerical calculations on the effective heat capacity phase change problems of biological tissues can be done by the FDM, FEM, finite volume method, or BEM. Presented below is an algorithm developed using the FDM. Applying the explicit finite difference formulation to Equation (3.53), and using Equation (3.7) to express the linear term $T(\mathbf{X},t)$ on the right-hand side of Equation (3.53), Equation (3.53) can be discretized as follows:

$$T(\mathbf{X},t+\Delta t)=\frac{1-W(1-\beta)\Delta t-m\cdot Fo}{1+W\beta\Delta t}T(\mathbf{X},t)+\sum_{i=1}^{m/2}\frac{Fo}{1+W\beta\Delta t}T(\mathbf{X}+\Delta\mathbf{X}_i,t)$$

$$+\sum_{i=1}^{m/2}\frac{Fo}{1+W\beta\Delta t}T(\mathbf{X}-\Delta\mathbf{X}_i,t)+\frac{[\tilde{Q}_m+\tilde{\omega}_b C_b T_a]\Delta t}{1+W\beta\Delta t} \tag{3.54}$$

where Δt is the time increment; $W=\frac{\tilde{\omega}_b C_b}{\tilde{C}}$; $Fo=\frac{\tilde{k}\cdot\Delta t}{\tilde{C}\cdot\Delta x^2}$ is the Fourier number; $m=2$, 4, 6 corresponds to the cases of one, two, and three dimensions, respectively; and the definitions of $\Delta\mathbf{X}_1, \Delta\mathbf{X}_2$, and $\Delta\mathbf{X}_3$ can be found in Equation (3.9).

Applying the boundary conditions at time $t+\Delta t$ and substituting the calculated results at the previous time t, the unknown T at time $t+\Delta t$ can be solved from Equation (3.54). After the temperature distributions at time $t+\Delta t$ have been solved, the anterior and posterior moving boundaries can be determined by the isotherms of T_{mu} and T_{ml}, respectively.

The FDM algorithm can also be flexibly used for solving more bioheat transfer cases. Deng and Liu had extended this algorithm for treating more complex problems that involve both freezing and heating processes [42]. Readers are referred there for more detail.

3.3.3 Dual-Reciprocity Boundary Element Method (BEM) Formulation

The BEM has advantages over others due to its requiring only discretization on the boundaries of the domain. However, the traditional BEM for problems involving nonlinearities may still yield difficulty in deriving the so-called fundamental solution and thus require additional domain discretization [22]. There exist severe restrictions in the traditional BEM for solving the bioheat transfer equation (BHTE). One is that the fundamental solution to the BHTE with a non-homogeneous blood perfusion term is hard to obtain. Besides, in most cases, it is inconvenient to alter the program by incorporating a new fundamental solution when the user wishes to study a slightly different bioheat equation. Furthermore, the nonhomogeneous term accounting for the spatial heating needs to be included in the ordinary BEM formulation by means of domain integrals, which makes the technique time-consuming and loses the attraction of its "boundary-only" character. In this regard, the dual-reciprocity boundary element method (DRBEM) can avoid the above restrictions [44].

DRBEM is a transformation originating from but superior to the traditional BEM. The basic idea is to employ a fundamental solution corresponding to a simpler equation and to treat the remaining terms as well as other nonhomogeneous terms in the original equation through a procedure that involves a series expansion using global approximation functions and the application of reciprocity principles [44,45].

Deng and Liu had extended the DRBEM to solve multidimensional phase change problems of biological tissues during cryosurgery [44]. In order to avoid the complex iteration at the moving boundary, the effective heat capacity was also adopted to simplify the governing equations. As a result, the complex non-linear phase change problems can be simplified as nonhomogeneous ones, which can be easily dealt with by DRBEM.

Without losing generality, one can write the transient bioheat transfer equation under two-dimensional coordinates and constant properties as

$$\nabla^2 T = a\frac{\partial T}{\partial t} + bT + c = d_f$$

(3.55)

where $a = \frac{\tilde{c}}{k}$, $b = \frac{\tilde{\omega}_b C_b}{k}$, and $c = -\frac{\dot{Q}_m + \tilde{\omega}_b C_b T_a}{k}$. The solution of Equation (3.55) can be expressed as the sum of the solution of Laplace's equation and a particular solution $\hat{T}$ for the Poisson equation $\nabla^2\hat{T} = d_f$. Due to the difficulties of finding a solution $\hat{T}$, the dual-reciprocity method uses a series of particular solutions $\hat{T}_j$ instead of a single function $\hat{T}$. The number of $\hat{T}_j$ used is equal to the total number of nodes in the problem, that is, $N+L$, where N and L are the numbers of boundary nodes and internal nodes, respectively. Applying the dual-reciprocity technique [44,45], the following relations can be deduced:

$$d_f \approx \sum_{j=1}^{N+L} \gamma_j(t)f_j(r)$$

(3.56)

$$\nabla^2\hat{T}_j = f_j$$

(3.57)

where γ_j is the time-dependent coefficient to be determined, r is the distance from the point j to any other point under consideration, f_j is the function of distance r, and $f = 1 + r$ is used here. Then, Equation (3.55) can be rewritten as

$$\nabla^2 T = \sum_{j=1}^{N+L} \left(\gamma_j \cdot \nabla^2 \hat{T}_j \right) \tag{3.58}$$

Applying the conventional BEM based on the use of the fundamental solution of Laplace's equation and integrating by parts the Laplacian terms, the integral equation for each source node i can be deduced:

$$C_i T_i + \int_\Gamma q^* T d\Gamma - \int_\Gamma T^* q d\Gamma = \sum_{j=1}^{N+L} \gamma_j \left(\int_\Gamma q^* \hat{T}_j d\Gamma - \int_\Gamma T^* \hat{q}_j d\Gamma + C_i \hat{T}_{ij} \right) \tag{3.59}$$

where C_i is a constant that only depends on the geometry at the node i; T^* is the fundamental solution of the Poisson equation; q^* is the normal derivative of T^* along the boundary; $\hat{q}_j$ is defined as $\hat{q}_j = \partial \hat{T}_j / \partial n$; and n is the unit outward normal to $\Gamma_{m.i.}$. For the two-dimensional problem considered here, T^*, q^*, $\hat{T}$, and $\hat{q}$ can be derived as [44,45]

$$T^* = \frac{1}{2\pi} \ln\left(\frac{1}{r}\right) \quad q^* = -\frac{1}{2\pi r}\left(r_x \frac{\partial x}{\partial n} + r_y \frac{\partial y}{\partial n} \right) \tag{3.60}$$

$$\hat{T}_j = \frac{r^2}{4} + \frac{r^3}{9} \quad \hat{q} = \left(\frac{1}{2} + \frac{r}{3} \right)\left(r_x \frac{\partial x}{\partial n} + r_y \frac{\partial y}{\partial n} \right) \tag{3.61}$$

Introducing the interpolation functions and integrating over each boundary element, Equation (3.59) can be written in terms of nodal values as

$$C_i T_i + \sum_{k=1}^{N} H_{ik} T_k - \sum_{k=1}^{N} G_{ik} q_k = \sum_{j=1}^{N+L} \gamma_j \left(\sum_{k=1}^{N} H_{ik} \hat{T}_{kj} - \sum_{k=1}^{N} G_{ik} \hat{q}_{kj} + C_i \hat{T}_{ij} \right) \tag{3.62}$$

where G_{ik} and H_{ik} are called as influence coefficients, and can be written as

$$G_{ik} = \int_\Gamma T^* d\Gamma, \quad H_{ik} = \int_\Gamma q^* d\Gamma + \frac{1}{2}\delta_{ij} \tag{3.63}$$

where δ is the Kronecker delta. From Equation (3.56), the coefficient γ in Equation (3.62) can be obtained as $\gamma = \mathbf{F}^{-1}(a\dot{T} + bT + c)$, where $\mathbf{F}$ is the matrix form of the function f, and $\dot{T}$ means $\partial T/\partial t$. Then the matrix form of Equation (3.62) can be derived as

$$\mathbf{HT} - \mathbf{Gq} = (\mathbf{H\hat{T}} - \mathbf{G\hat{q}})\mathbf{F}^{-1}(a\dot{T} + bT + c) \tag{3.64}$$

where $\mathbf{H}$ and $\mathbf{G}$ are the matrices of H_{ik} and G_{ik}, respectively. The matrices $\mathbf{T}$, $\mathbf{q}$, $\hat{\mathbf{T}}$, and $\hat{\mathbf{q}}$ correspond to vectors T_k, q_k, $\hat{T}_{kj}$, and $\hat{q}_{kj}$, respectively. Substituting $\mathbf{S} = (\mathbf{H\hat{T}} - \mathbf{G\hat{q}})\mathbf{F}^{-1}$, $T = (1 - \zeta_T)T^m + \zeta_T T^{m+1}$, $q = (1 - \xi_q)q^m + \xi_q q^{m+1}$, and $\dot{T} = (T^{m+1} - T^m)/\Delta t$ into Equation (3.64) gives

$$\left(\frac{a\mathbf{S}}{\Delta t} + b\zeta_T \mathbf{S} + \zeta_T \mathbf{H} \right) \mathbf{T}^{m+1} - \xi_q \mathbf{Gq}^{m+1} + c\mathbf{S}$$

$$= \left[\frac{a\mathbf{S}}{\Delta t} - b(1 - \zeta_T)\mathbf{S} - (1 - \zeta_T)\mathbf{H} \right] \mathbf{T}^m + \left(1 - \xi_q \right)\mathbf{Gq}^m \tag{3.65}$$

Applying the boundary conditions at time $(m+1)\Delta t$ and substituting the calculated results at the previous time $m\Delta t$, the unknown T and q at time $(m+1)\Delta t$ can be solved from Equation (3.65). After the temperature distributions at time $(m+1)\Delta t$ have been solved, the anterior and posterior moving boundaries can be respectively determined by the isotherms of T_{mu} and T_{ml}.

For the case of temperature-dependent blood perfusion such as $\omega_b = w_1 + w_2 T$ or $\omega_b = w_1 \exp(w_2 T)$ (where w_1, w_2 are constant coefficients), a DRBEM algorithm corresponding to Equation (3.65) can also be derived. The obtained nonlinear transient algorithm due to the occurrence of the complex temperature term can be solved through iteration at each time step [46].

3.3.4 Phase Change Cases Based on 3D Simulation

Figure 3.9 depicts a typical temperature distribution for freezing by three probes with identical insertion depth in the tissue [42]. It was shown that the temperature responses at tissues surrounding the three probe tips are much different from those of the rest of the tissues, and that there appear to be three identical valleys in the temperature distributions during freezing, while three identical peaks were produced at the same positions during heating. Figure 3.10 gives the location and size of the iceball produced by the freezing of three probes. Compared with the case of a single freezing probe, the iceball formed by the freezing of three probes is much larger. It must be noticed that the domain of the iceball should exceed the area of the tumor, which is required for complete destruction of the target. The use of multiple probes permits overlapping the requested frozen and heated areas in the treatment of large tumors, and provides a method of destroying the tissue to the desired size and shape in complex tumor ablation. Meanwhile, excessive freezing might also cause irreversible injury to

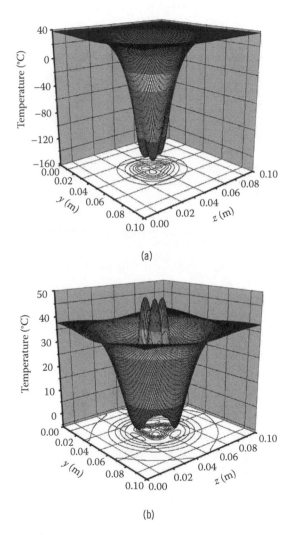

Figure 3.9   Tissue temperature distributions at a cross section of $x = 0.027$ m for the case of three probes with identical insertion depth, in which the heating–cooling procedure includes 1200 seconds of freezing immediately followed by 800 seconds of heating, in which (a) $t = 1200$ seconds, (b) $t = 1400$ seconds, and (c) $t = 2000$ seconds.

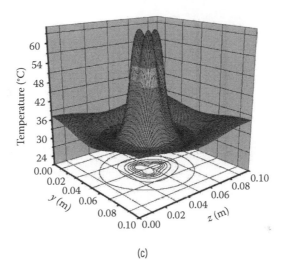

(c)

Figure 3.9   (Continued)

the neighboring healthy tissues. Therefore, numerical calculations can provide a very informative prediction on the tissue temperature responses and thus help optimize the treatment parameters before tumor operation. For the method to optimize the cryosurgical protocols, readers are referred to References [47,48] for more detail.

In cryosurgery, the major injury factors have been found to be the lowest temperature achieved, the dwell time at this lowest temperature, the number of

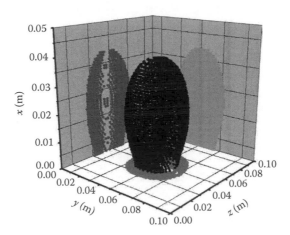

Figure 3.10   The location and size of an iceball after freezing for $t = 1200$ seconds.

freeze-thaw cycles, and the temperature change rate. Although no single factor is solely responsible for the tissue damage, it is argued that tissue will likely be injured if its temperature is reduced below a certain critical value, which temperature is likely to be tissue dependent and generally ranges from –20°C to –70°C [42,43]. The isotherm based on the critical temperature is commonly adopted by cryosurgeons as a boundary of lethality for tissue. Locating this isotherm within an iceball during clinical application is problematic due to the limitations of magnetic resonance imaging (MRI), ultrasound, and computerized tomography (CT) imaging modalities.

For similar reasons, the tissue will likely be burned if its temperature is increased over a certain critical value. Such maps of three-dimensional isotherms for tissue can also be obtained through numerical simulation. The moving boundaries at different times inside the prefrozen iceball during heating by three probes are depicted in Figure 3.11 as illustration (the moving boundaries include the lower and upper boundaries, which are determined by the isotherm of $T_{ml} = -8°C$ and $T_{mu} = -1°C$, respectively). It clearly presents the evolutions of the moving boundaries of phase change. Bearing such information in mind, a cryosurgeon can at least acquire somewhat quantitative information for planning the specific extent and size of the freezing lesion and its growth rate. In addition, it is even possible to output in advance the results of the numerical computations as a series of critical isotherm, which can be used as a surgical atlas. This would allow the cryosurgeon to obtain a quantitative feeling for the lesion size and its growth rate.

In another study [49], numerical investigations were also performed for the cases using multiple probes for combined cryosurgical and hyperthermic treatment. Figure 3.12 shows part of the results, in which three probes with different configurations are applied (the freeze-heat cycle includes 10 minutes of freezing followed by 10 minutes of heating). Results in Figures 3.12a,b indicate that either insufficient or excessive freezing and heating have resulted with the corresponding probe configurations. It is shown in Figure 3.12c that the lethal area produced by freezing has totally encompassed the whole tumor with similar shape, and that parameters of probes under the third configuration have been optimal. The current results also suggest that the combined cryosurgery and hyperthermia system has strong freezing and heating capabilities, and that it may improve the treatment effect by providing double chances to possibly kill tissues.

For the cases where freezing and heating probes can be moved along their initial insertion path, numerical algorithms can also be modified for treating such complex moving freezing and heating problems [50]. The results will not be repeated here for the sake of brevity.

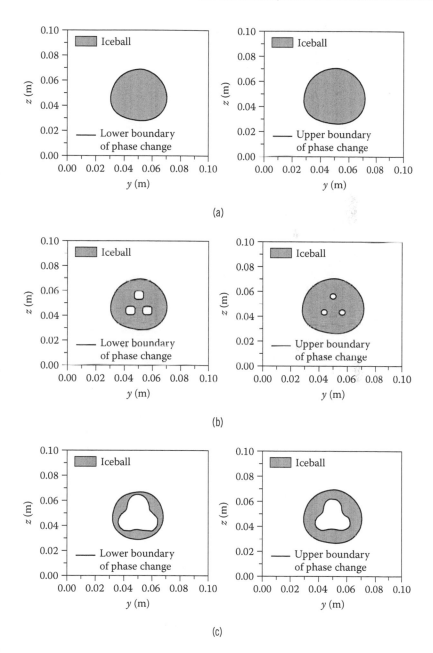

Figure 3.11 The moving boundaries at different times inside the prefrozen iceball during heating by three probes ($x = 0.027$ m), in which (a) $t = 1210$ seconds, (b) $t = 1220$ seconds, and (c) $t = 1300$ seconds.

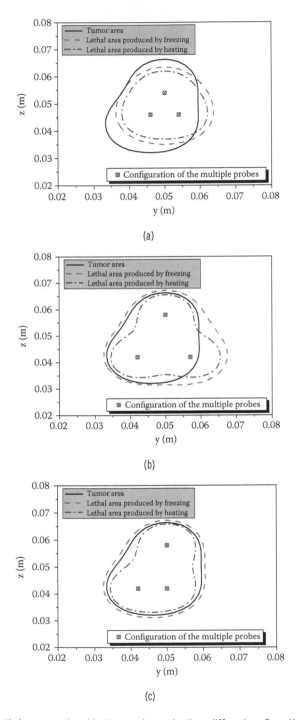

Figure 3.12   Lethal areas produced by three probes under three different configurations.

3.4 SIMULATION OF LIGHT–HEAT–FLOW TRANSPORT IN TISSUES SUBJECT TO LASER-INDUCED THERMOTHERAPY

3.4.1 Particularities of Laser Heating of Tissues

During a laser-induced thermotherapy (LITT), a minimally invasive treatment of malign and benign tumors, the laser light can be delivered either through external laser beam irradiated on skin surface or by placing a special scattering laser applicator on pathological tissues. The light absorbed usually leads to an increased local temperature between 45°C and 100°C, which will result in hyperthermic, coagulative, and necrosis effects in living tissues. For safety consideration in clinics, it is essential to ensure necrosis of the total tumor cells within the desired volume of treatment while minimizing the thermal damage of healthy tissues surrounding the tumor [51].

A great many efforts had been made to calculate the 3D transient temperature distribution in laser-irradiated biological tissues [52–62]. However, most of them either did not take blood perfusion into consideration or are still not applicable to the situation where convective heat loss by blood flow in vessels is significant. In some studies [63], the distortion of temperature distribution by the presence of a blood vessel was considered. But the vessel there was simply represented by several node points of increased absorption and decreased scattering. An important and rather complex phenomenon (i.e., the convective cooling by the flowing blood) was not addressed until recently.

Depicted in Figure 3.13 is a typical geometrical configuration showing laser-fiber irradiation in tissues embedded with two large countercurrent blood vessels [51]. This is often encountered in tumor treatment by LITT. The tissue domain surrounding the large blood vessels can be prescribed in a rectangular geometry, and a 3D Cartesian coordinate system can be adopted to solve the coupled tissue–vessel energy equations. In simulating the optothermal response of tissues embedded with large blood vessels during LITT, two kinds of problems need to be addressed. The first one is how the light distribution is affected by the presence of large blood vessels. The other one is how to model the localized cooling effect of the large blood vessels [51,53,59,60].

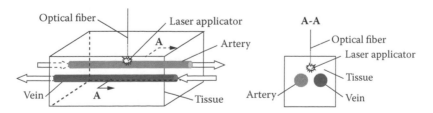

Figure 3.13 Physical model for the case of inserted fiber irradiation on tissues embedded with two large countercurrent blood vessels.

Interfaces between a blood vessel and its surrounding tissue may lead to reflection or absorption of the coming laser light. This consists of the toughest issue for modeling and calculation. Most of the existing efforts just treat this by considering a collective model. Up to date, little attention has been paid to the effect of a single large blood vessel to the tissue temperature prediction during laser-induced thermotherapy. To tackle this important issue in clinics, Zhou and Liu developed a numerical algorithm to simultaneously solve 3D light and heat transport in several typical tissue domains with either one single blood vessel or two countercurrent blood vessels [51]. The 3D heat transfer and blood flow models were established to characterize the temperature transients over the whole area. The coupled equations for heat and blood flow in multiple regions were solved using the blocking-off method. Particularly, the Monte Carlo method was modified to calculate the light transport inside the tissues as well as the blood vessel. Theoretical algorithms to deal with the complex interfaces between the tissues and vessels and the tissue–air interface were developed. The heat generation pattern due to absorption of laser light was thus obtained by the Monte Carlo simulation and then adopted in the heat and flow transport equations to predict the 3D temperature transients over the whole domain. It indicated that without considering large blood vessels inside the tissues, a very different temperature response would be induced when subject to the same laser heating. Such information is expected to be critical for accurate treatment planning in clinics.

3.4.2 Monte Carlo Formulation for Light Propagation

The most widely used equation describing the light propagation in biological tissue is the radiative transfer equation [64]:

$$s \cdot \nabla I(\vec{r}, \hat{s}) + \mu_t I(\vec{r}, \hat{s}) = \frac{\mu_t}{4\pi} \int_{4\pi} p(\hat{s}, \hat{s}') I(\vec{r}, \hat{s}') d\omega' \qquad (3.66)$$

where $\vec{r}$ is the position vector; $\hat{s}$ and $\hat{s}'$ are direction vectors; $I(\vec{r}, \hat{s})$ is the intensity of laser light at position $\vec{r}$ in the $\hat{s}$ direction; ω' is the solid angle; μ_t is the total attenuation coefficient defined as the sum of the absorption coefficient μ_a (defined as the probability of photon absorption per unit infinitesimal path length) and the scattering coefficient μ_s (defined as the probability of photon scattering per unit infinitesimal path length); and $p(\hat{s}, \hat{s}')$ is the phase function, which is a probability density function giving the probability of photon scattering from an initial propagation direction $\hat{s}'$ to a final direction $\hat{s}$. By assuming tissue is homogeneous and isotropic, the scattering from $\hat{s}'$ to $\hat{s}$ can be simply represented by the angle between $\hat{s}'$ and $\hat{s}$. So the phase function $p(\hat{s}, \hat{s}')$ can be written as $p(\hat{s}, \hat{s}') = p(\hat{s} \bullet \hat{s}')p(\phi) = p(\cos\theta)p(\phi)$, where θ is the deflection angle between $\hat{s}$ and $\hat{s}'$ (the symbol $\bullet$ denotes the scalar product of two vectors) and ϕ is the azimuthal angle about $\hat{s}$. $p(\cos\theta)$ describes the forward-scattering nature of light in tissue, and $p(\phi)$ is uniformly distributed over the 2π radians at $\hat{s}$.

The analytical solutions for the radiative transfer equation can be obtained only for very simple cases [65], for example one-dimensional geometry. Monte Carlo simulation [66] of photon propagation offers a flexible yet rigorous approach toward photon transport in turbid materials like biological tissues, and it can easily deal with two- or three-dimensional problems with complex geometry. For these reasons, it has been widely used to simulate light transport in tissues for various applications [67–72].

The Monte Carlo method simulates the "random walk" of photons in a medium that contains absorption and scattering. The rules of photon propagation are expressed as probability distributions for the step sizes of photon movement between two photon–tissue interaction sites, for the angles of deflection when a scattering event occurs, and for the probability of transmittance or reflectance at interfaces (such as the tissue–air interface and tissue–blood interface). A random number generator is used to sample discrete events from which the probability distributions for the laser–tissue interaction coefficients and phase function can be derived. In a Monte Carlo simulation, the photons are generally treated as neutral particles, and thus the polarization and wave phenomenon are neglected. The photon transport behavior depends on the following four quantities: (1) the absorption coefficient μ_a, (2) the scattering coefficient μ_s, (3) the phase function, and (4) the refractive index of tissue relative to the surrounding medium n.

For the calculation domain as illustrated in Figure 3.13, two coordinate systems are used in the Monte Carlo simulation at the same time [51]:

1. A Cartesian coordinate system is used to trace photon movements and to score internal photon absorption as a function of x, y, and z, which are the three components of the Cartesian coordinate system. Since solving the energy equations is directly based on the internal photon absorption, the Cartesian coordinate system used for the Monte Carlo simulation is the same as that for heat transfer calculation, and the same discrete grid system can thus be used for its numerical computations.
2. A moving spherical coordinate system, whose z-axis is dynamically aligned with the photon propagation direction, is used for sampling of the propagation direction change of a photon. In this spherical coordinate system, the deflection angle θ and the azimuthal angle ϕ due to scattering can be first sampled. Then, the photon direction is updated in terms of the direction cosines in the Cartesian coordinate system.

For tracing photons in tissues embedded with large blood vessels with the Monte Carlo simulation, each incident photon is initially assigned a weight W equal to unity. Weighting the photons in this way can result in better statistical precision for a given number of input photons [51]. During the photon transport history, it can hit two kinds of interfaces: the tissue–air interface or tissue–blood interface. The photon will experience transmittance or reflection at these interfaces due to the mismatched refractive indices. At the tissue–air interface, part of the photon weight is transmitted out into the air, and the remaining part of the photon weight is reflected back to the tissue. Only the reflection part needs to be continuously traced. This situation is somewhat different at the tissue–blood interface, where the entire photon weight is either reflected or transmitted. At each interaction point, the photon is assumed to deposit a fraction of its current weight as absorbed energy, and the remaining fraction is continuously traced as scattering energy. Once the photon has taken a step, some attenuation of the photon weight due to absorption by the tissues, $W\mu_a/\mu_t$, must occur. Therefore, the accumulated deposited photon weight in grid element (x, y, z) would be the sum of the old one and the new part absorbed by the tissues (i.e., $Q(x, y, z) + W\mu_a/\mu_t$). The new weight should then be updated as follows by the right-hand side term:

$$Q(x, y, z) \leftarrow Q(x, y, z) + W\mu_a/\mu_t \tag{3.67}$$

where μ_a is the absorption coefficient, and μ_t is the total attenuation coefficient.

Considering the photon scattering, the new photon weight, W, is calculated as follows by the right-hand side term:

$$W \leftarrow W\mu_s/\mu_t \tag{3.68}$$

where W represents the photon weight, and μ_s is the scattering coefficient.

If the photon weight, W, has been sufficiently decremented such that it falls below a threshold value (e.g., $W_{threshold} = 0.0001$), then further propagation of the photon yields little information. However, proper termination must be executed to ensure conservation of energy without skewing the distribution of the photon deposition. A technique called "roulette" is used to terminate the photon when $W < W_{threshold}$. The roulette technique gives the photon one chance in m (e.g., $m = 10$) of surviving with a weight of mW. Otherwise, the photon weight is reduced to zero, and the photon is terminated [51]. A random number, ξ, in the interval $[0,1]$ is generated by the computer. The photon weight is updated according to the following decision:

$$\begin{aligned} & \textit{if } \xi \leq 1/m \text{ then } W \leftarrow mW \\ & \textit{if } \xi > 1/m \quad \text{then } W = 0 \end{aligned} \tag{3.69}$$

where ξ denotes a random number in the interval $[0,1]$; and $W \leftarrow mW$ means to update W by mW. This method conserves energy and eventually terminates photons in an unbiased manner.

The rate of heat generation at any node $[i,j,k]$ due to the absorption of light in tissue, $q_l[i, j, k]$, can then be calculated by the following relation:

$$q_l[i, j, k] = \{Q[i, j, k]P\}/NV(i, j, k) \tag{3.70}$$

where N is the total photon number, $V(i, j, k)$ is the element volume of node $[i,j,k]$, P is the total laser power delivered at the surface or by the laser applicator, and $Q[i, j, k]$ is the accumulated deposited photon weight in grid element $[i,j,k]$ in units of photon weight.

The local accumulated diffuse reflectance can be converted to a useful form by the following expression:

$$R[i, j] = \{R[i, j]\}/[N \cdot \Delta x_i \cdot \Delta y_i] \tag{3.71}$$

where i and j represent the index number of grid points in the x-axis and y-axis, respectively; and Δx_i and Δy_i are the lengths of the grid point $[i,j]$ in the x-axis and y-axis, respectively.

The total reflectance R_t is given by

$$R_t = R_{sp} + \int_0^{x_l} \int_0^{y_l} R[i, j] dx dy \tag{3.72}$$

where R_{sp} is the specular reflectance, which can be described by $R_{sp} = (n_1 - n_2)^2/(n_1 - n_2)^2$ (n_1 and n_2 are refractive indices of medium 1 and 2, respectively). The total transmittance T_t is calculated in a similar manner like that of the total diffuse reflectance R_t.

It should be mentioned that the existence of large blood vessels significantly increases the difficulty for modeling and calculating the photon transport throughout the tissues [51]. Since the optical properties for the tissues and the blood are different, it is necessary to know in which region the photon is located before sampling of step size. For example, if the photon is currently located in the tissue region, the optical properties of tissues should be used to determine the step size s. If the photon is located in the blood region, then the optical properties of blood should be adopted. When the photon is located just at the tissue–blood interface, the situation becomes a little more complicated. As shown in Figure 3.14,

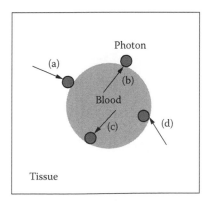

Figure 3.14   Four possibilities when photons reach the tissue–blood interface.

one cannot decide which properties should be used simply according to the position of the photon because there are four possibilities while the photon is located at the tissue–blood interface. The photon position combined with its history information (i.e., from where the photons reach the tissue–blood interface and whether or not the photon has transmitted the interface) can provide complete information for one to choose proper optical properties. For example, when the photon comes from the tissue region and no transmittance occurs (Figure 3.14a) or when the photon comes from the blood region but there is transmittance (Figure 3.14b), the optical properties of tissue should be used to determine the step size using $s = -\ln \xi / \mu_t$. Similarly, for the cases of Figures 3.14c,d, the optical properties of blood should be used to calculate the step size.

Zhou and Liu established a relatively complete method of developing the Monte Carlo algorithm for launching and moving the photon, sampling the photon step size, detecting boundary crossing, reflecting and transmittance at interfaces, and photon absorption and scattering during laser heating on tissues embedded with large blood vessels [51]. Since the derivations are rather lengthy, they will not be repeated here. Readers are referred to Reference [51] for more detail. It is worth mentioning, however, that the basic route as described above is also useful for tackling the light transport in moving boundary problems during thawing of frozen tissues due to laser heating [73]. In that case, reflection, transmittance at the moving solid–liquid interface should be carefully addressed.

3.4.3 Conjugated Model for Heat and Flow Transport

In most of the blood vessel thermal analysis, heat transfer coefficients or Nusselt numbers were used to calculate heat transfer to large blood vessels. However, the conjugate heat transfer problem can be solved numerically by modeling the heat transfer in tissues and blood regions as a whole [51]. If one assumes that the velocity profile at the entrance of the blood vessel is known, only the energy equation is required to be solved in the fluid. The energy equations used to describe the transient temperature distributions in the blood and tissue domains take the same form:

$$\rho c \frac{\partial T}{\partial t} + \rho c v \frac{\partial T}{\partial y} = k \left(\frac{\partial^2 T}{\partial x^2} + \frac{\partial^2 T}{\partial y^2} + \frac{\partial^2 T}{\partial z^2} \right) + q_l \tag{3.73}$$

where v denotes the velocity component in the y-axis (here, the blood flow is treated as fully developed; therefore, the velocity components in the x-axis and z-axis, say u and w, are zero), ρ denotes density, c is specific heat, T is temperature, t is time, k is thermal conductivity, and q_l is the rate of heat generation due to the absorption of laser light.

In writing Equation (3.73), the blood perfusion and metabolic heat production in the tissue region are neglected. The blood vessel wall is approximated as very thin compared to the whole computational domain and has no thermal resistance. When using Equation (3.73) to solve the temperature distributions in the tissues and blood region as a whole, the velocity distribution in the blood vessel region is assigned to be parabolic, whereas the velocity in the tissue region is assigned to be zero. The conductive terms can be discretized by central difference, while the convection term is discretized according to an upwind difference scheme. The resulting system of discretization equations was solved using the alternate direction implicit (ADI) method. Different specific heat and density are used for different regions (tissue or blood). The conductivity discontinuities at the vessel–tissue interface are dealt with by using the harmonic mean of the vessel and tissue conductivities [74].

For the case of a countercurrent vessel pair, the blood velocity in the artery is taken as positive and the blood velocity in the vein is negative to account for the countercurrent flowing in these two kinds of vessels. For simplification, the capillary bleed-off between the countercurrent vessels can be neglected. As mentioned earlier, large blood vessels can carry away a significant amount of the energy deposited during the laser heating. This results in a local cooling of the tumor cells immediately adjacent to the vessel and leads to possible recurrence of the tumor. Vessel occlusion (i.e., reducing the velocity of the blood before it arrives at the tumor region) is often adopted to reduce the localized cooling of large blood vessels. When the velocity of the blood flowing through the vessel is reduced drastically, the buoyancy-driven flow due to density variations may become significant. Clearly, Equation (3.73) cannot model the above-mentioned buoyancy-driven flow phenomenon. In fact, for certain physiological reasons (e.g., vessel bifurcation), nonzero velocity components may exist along the x-axis and z-axis. To account for these phenomena, one must use 3D Navier-Stokes equations to calculate the flow field.

Strictly speaking, blood is a multiphase system consisting of plasma and blood cells. The diameter of the main blood cell (i.e., a red blood cell) is in the order of several microns (~7.6 μm) under normal conditions [75], which is much smaller than the diameter of the blood vessel. Therefore, it is reasonable to assume that blood flow behavior can be described by a continuum model, such as Navier-Stokes equations. Other assumptions still include the following: (1) blood is considered a Newtonian fluid with a constant coefficient of viscosity (this is valid when the shear rates are relatively high, e.g., greater than 100 sec^{-1}) [76], (2) the vessel elasticity and thus the pressure wave effect are neglected, (3) the flow through the blood vessel is assumed to be laminar and incompressible, (4) the viscous dissipation is neglected, (5) the density is constant except in the expression of the buoyancy term (Boussinesq approximation), and (6) no slip boundary condition at the blood–vessel interface is assumed.

With these assumptions, the governing equations in a rectangular coordinate system (shown in Figure 3.13) can be given, in their unsteady form, by the following [77]:

Continuity Equation:

$$\frac{\partial u}{\partial x} + \frac{\partial v}{\partial y} + \frac{\partial w}{\partial z} = 0 \qquad (3.74)$$

Momentum Equations (Navier-Stokes equations):

$$\rho\left(\frac{\partial u}{\partial t} + u\frac{\partial u}{\partial x} + v\frac{\partial u}{\partial y} + w\frac{\partial u}{\partial z}\right) = -\frac{\partial p}{\partial x} + \mu\left(\frac{\partial^2 u}{\partial x^2} + \frac{\partial^2 u}{\partial y^2} + \frac{\partial^2 u}{\partial z^2}\right) \qquad (3.75)$$

$$\rho\left(\frac{\partial v}{\partial t} + u\frac{\partial v}{\partial x} + v\frac{\partial v}{\partial y} + w\frac{\partial v}{\partial z}\right) = -\frac{\partial p}{\partial y} + \mu\left(\frac{\partial^2 v}{\partial x^2} + \frac{\partial^2 v}{\partial y^2} + \frac{\partial^2 v}{\partial z^2}\right) \qquad (3.76)$$

$$\rho\left(\frac{\partial w}{\partial t} + u\frac{\partial w}{\partial x} + v\frac{\partial w}{\partial y} + w\frac{\partial w}{\partial z}\right) = -\frac{\partial p}{\partial z} + \mu\left(\frac{\partial^2 w}{\partial x^2} + \frac{\partial^2 w}{\partial y^2} + \frac{\partial^2 w}{\partial z^2}\right) + \rho g \qquad (3.77)$$

Energy Equation:

$$\rho c\left(\frac{\partial T}{\partial t} + u\frac{\partial T}{\partial x} + v\frac{\partial T}{\partial y} + w\frac{\partial T}{\partial z}\right) = k\left(\frac{\partial^2 T}{\partial x^2} + \frac{\partial^2 T}{\partial y^2} + \frac{\partial^2 T}{\partial z^2}\right) + q_1 \qquad (3.78)$$

where g denotes acceleration of gravity, and μ is dynamic viscosity. In the buoyancy term, the density ρ is approximated by $\rho = \rho_{ref}[1 - \beta(T - T_{ref})]$, where ρ_{ref} is the reference density, T_{ref} is the reference temperature, and β is the coefficient of thermal expansion.

The above governing equations can be calculated using the SIMPLE algorithm for velocity–pressure coupling, based on primitive variables and a staggered-grid scheme [74].

The tissue and the flowing blood are treated as a whole in the conjugated method. In doing this, a technique called the "blocking-off method" [74] can be used, in which the velocity in the solid region (i.e., tissue region) is assigned to be zero and the viscosity of the solid region is set as infinitely large (e.g., 10^{35}). The resulting viscosity discontinuities at the vessel–tissue interface can be easily dealt with by using the harmonic mean of the vessel and the tissue viscosity.

3.4.4 Typical Results on Laser–Tissue–Vessel Interactions

All the calculations demonstrate that for tissues with different vessel configurations, a very different temperature field will be induced for the same laser power heating on the biological body (Figure 3.15). Without carefully addressing the tissue vasculature, serious errors on predicting the thermal information will be obtained, and clinical problems such as thermal injury on healthy tissues due to overheating or poor treatment on tumors due to inefficient irradiation will be aroused [51].

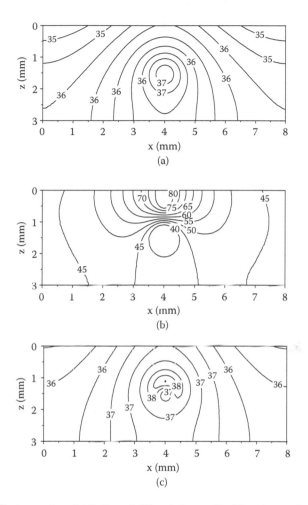

Figure 3.15   The temperature distributions at different planes with different y coordinate values corresponding to Figure 3.13: (a) 12.3 mm upstream the midplane, (b) midplane, and (c) 12.3 mm downstream the midplane.

   When the blood flow velocity in the vessel is very low as a result of certain physiological reasons (such as vessel occlusion), studies reveal that the buoyancy-driven flow due to density variations induced by laser irradiation may be significant [51]. Figure 3.16 presents a typical calculated result when the velocity at the vessel entrance is decreased to about 0.052 m/s. It is noted that the velocity distribution at the midplane and the exit differs from the entrance velocity distribution. In addition, the temperature distribution when natural convection plays a role is also different from the temperature when not considering natural convection. Further calculations still indicate that when the vessel diameter becomes small enough, the effect of the natural convection will be significantly

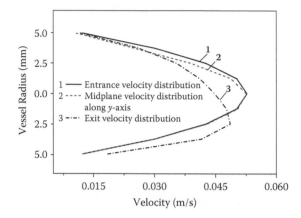

Figure 3.16 The natural convection effect becomes significant when the velocity at the tube entrance is low.

weakened even if the blood flow velocity is low. This is purely the result of a low Rayleigh number flow. For the case of a very small vessel diameter, the biological heat transfer analysis can be significantly simplified since consideration of the effect of buoyancy-driven flow may not be very necessary. However, even in this case, the effect of a single vessel could still be remarkable as a result of other reasons such as laser light reflection and transmission at the vessel–tissue interface.

Clearly, different geometrical parameters for the blood vessels such as configuration, position, diameter, length, the centerline-to-centerline distance between two countercurrent vessels, and the thermal parameters for the blood flow such as velocity and entrance temperature, the optical scattering and absorption coefficients at tissue–blood interfaces, and so on all contribute to the final output of the tissue temperatures.

Overall, the Monte Carlo method as well as the blocking-off method are very useful for simultaneously solving the light transport and the coupled heat and flow equations. As is well known, humans and animals have very complex anatomical structures over their whole body. Although examples as explained above have included the most typical vasculatures (i.e., tissues with one single blood vessel or two countercurrent vessels), it can never be said to be an exhaustive one. Further efforts can be made on more complex light and heat transfer due to the adoption of real vasculature.

3.5 CONCLUSIONS

In this review, simulations of temperature distributions in the biological tissues under several complex high- or low-temperature situations were explained. The methods as described would find significant value in treatment planning for tumor hyperthermia and cryosurgery. However, tremendous efforts are still needed to gain a more comprehensive understanding of the freezing and heating effect. For example, uncertainties exist in the modeling, especially with respect to the influence of blood perfusion, which may lead to either lower or higher temperatures than expected. Besides, the effect of the large blood vessels or spatially distributed vasculature should also be investigated in more detail. In addition, the local thermal properties of the cancerous or healthy tissue, albeit of smaller influence, should be accurately measured for the specifically tested subject through experiment. Knowledge regarding heat transfer in many organs such as the liver, the brain, and the kidneys and in the prostate gland is still rather limited. For these reasons, there still exist many obstacles in developing a general bioheat transfer model and numerical simulation algorithm. Therefore, externally applied three-dimensional heating and/or freezing patterns by the medical apparatus should always be carefully justified, especially when the research is clinic oriented.

To improve the efficiency of treatment planning, further research can be done by incorporating the following factors: (1) the real shape of a biological body, (2) accurate thermophysical properties, (3) the circulatory process inside the body, (4) the real heating and freezing patterns and boundary conditions at the skin surface, and (5) model justification via experimental measurements. In terms of the design of new medical systems, the goals are to increase the ablation efficiency on target diseased tissues, reduce the injury to the surrounding healthy tissues, and optimize a more targeted heat and/or cold delivery method. Obviously, development of a powerful computational technique would significantly support this endeavor.

Undoubtedly, with the rapid progress of computational technology, accurate simulations with respect to time and space are possible in the three-dimensional domain. Simulations would help improve the performance of a surgical process in many aspects such as (1) the design and optimization of specific freezing and/or heating therapies, (2) the identification of a desirable heat and/or cold delivery system for obtaining an optimal thermal dose, (3) the development and evaluation of new surgical systems, (4) a guide for controlling the complex heat transport for therapeutic processes, and (5) an investigation of the tissue–medical device interactions.

It is expected that a clinically useful software for bioheat transfer simulation will finally become true in the coming future.

ACKNOWLEDGMENT

Part of the research performed in this review has been supported by the National Natural Science Foundation of China under several grants: 50325622, 50576104, 50776097, and 50436030.

REFERENCES

1. J. Liu, Bioheat Transfer Model, in *Wiley Encyclopedia of Biomedical Engineering* (ed. M. Akay), pp. 1–11, New York: John Wiley & Sons, 2006.
2. K. R. Diller, Modeling of Bioheat Transfer Processes at High and Low Temperatures, in *Advances in Heat Transfer*, vol. 22 (ed. Y. I. Cho), pp. 157–357, Boston: Academic Press, 1992.
3. R. B. Roemer, Engineering Aspects of Hyperthermia Therapy, *Annual Review of Biomedical Engineering*, vol. 1, pp. 347–376, 1999.
4. Y. G. Lv, J. Liu, and J. Zhang, Theoretical Evaluation on Burn Injury of Human Respiratory Tract Due to Inhalation of Hot Gas at the Early Stage of Fires, *Burns*, vol. 32, pp. 436–446, 2006.
5. J. Liu, Cooling Strategies and Transport Theories for Brain Hypothermia Resuscitation, *Frontiers of Energy and Power Engineering in China*, vol. 1, pp. 32–57, 2007.
6. Z. S. Deng and J. Liu, Mathematical Modeling of Temperature Mapping over Skin Surface and Its Implementation in Thermal Disease Diagnostics, *Computers in Biology and Medicine*, vol. 34, pp. 495–521, 2004.
7. P. O. Fanger, *Thermal Comfort*, pp. 19–67, New York: McGraw-Hill, 1970.
8. B. Rubinsky, Cryosurgery, *Annual Review of Biomedical Engineering*, vol. 2, pp. 157–187, 2000.
9. J. C. Bischof, Quantitative Measurement and Prediction of Biophysical Response during Freezing in Tissues, *Annual Review of Biomedical Engineering*, vol. 2, pp. 257–288, 2000.
10. T. C. Hua and H. S. Ren, *Cryogenic Biomedical Technology* (in Chinese), Beijing: Science Press, 1993.
11. J. O. M. Karlsson and M. Toner, Long-Term Storage of Tissues by Cryopreservation: Critical Issues, *Biomaterials*, vol. 17, pp. 243–256, 1996.
12. G. T. Martin, M. G. Haddad, E. G. Cravalho, and H. F. Bowman, Thermal Model for the Local Microwave Hyperthermia Treatment of Benign Prostatic Hyperplasia, *IEEE Transactions on Biomedical Engineering*, vol. 39, pp. 836–844, 1992.
13. P. M. Meaney, R. L. Clarke, G. R. Ter Haar, and I. H. Rivens, 3-D Finite-Element Model for Computation of Temperature Profiles and Regions of Thermal Damage during Focused Ultrasound Surgery Exposures, *Ultrasound in Medicine and Biology*, vol. 24, pp. 1489–1499, 1998.
14. B. M. Kim, S. L. Jacques, S. Rastegar, S. Thomsen, and M. Motamedi, Nonlinear Finite-Element Analysis of the Role of Dynamic Changes in Blood Perfusion and Optical Properties in Laser Coagulation of Tissue, *IEEE Journal of Selected Topics in Quantum Electronics*, vol. 2, pp. 922–932, 1996.
15. A. A. Gage and J. Baust, Mechanism of Tissue Injury in Cryosurgery, *Cryobiology*, vol. 37, pp. 171–186, 1998.
16. W. Wulff, The Energy Conservation Equation for Living Tissues, *IEEE Transactions in Biomedical Engineering*, vol. 21, pp. 494–497, 1974.
17. M. M. Chen and K. R. Holmes, Microvascular Contributions in Tissue Heat Transfer, *Annals of the New York Academy of Sciences*, vol. 335, pp. 137–150, 1980.
18. S. Weinbaum and L. M. Jiji, A New Simplified Bioheat Equation for the Effect of Blood Flow on Local Average Tissue Temperature, *ASME Journal of Biomechanical Engineering*, vol. 107, pp. 131–139, 1985.
19. J. Liu and C. Wang, *Bioheat Transfer* (in Chinese), pp. 285–290, Beijing: Science Press, 1997.
20. J. Liu, *Principle of Cryo-Biomedical Engineering* (in Chinese), Beijing: Science Press, 2007.

21. J. P. Abraham, E. M. Sparrow, and S. Ramadhyani, Numerical Simulation of a BPH Thermal Therapy: A Case Study Involving TUMT, *ASME Journal of Biomechanical Engineering*, vol. 129, pp. 548–557, 2007.

22. C. L. Chan, Boundary Element Method Analysis for the Bioheat Transfer Equation, *ASME Journal of Biomechanical Engineering*, vol. 114, pp. 358–365, 1992.

23. A. Haji-Sheikh, Monte Carlo Methods, in *Handbook of Numerical Heat Transfer* (ed. W. J. Minkowycz, E. M. Sparrow, G. E. Schneider, and R. H. Pletcher), Chap. 16, p. 673–722, New York: John Wiley & Sons, 1988.

24. A. Haji-Sheikh and E. M. Sparrow, The Solution of Heat Conduction Problems by Probability Methods, *ASME Journal of Heat Transfer*, vol. 89, pp. 121–131, 1967.

25. G. E. Zinsmeister and J. A. Sawyerr, Method for Improving the Efficiency of Monte Carlo Calculation of Heat Conduction Problems, *ASME Journal of Heat Transfer*, vol. 96, pp. 246–248, 1974.

26. L. C. Burmeister, Monte Carlo Procedure for Straight Convecting Boundaries, *International Journal of Heat and Mass Transfer*, vol. 28, pp. 717–720, 1985.

27. M. H. N. Naraghi and S. C. Tsai, A Boundary-Dispatch Monte Carlo (Exodus) Method for Analysis of Conductive Heat Transfer Problems, *Numerical Heat Transfer, Part B*, vol. 24, pp. 475–487, 1993.

28. F. Kowsary and M. Arabi, Monte Carlo Solution of Anisotropic Heat Conduction, *International Communications in Heat and Mass Transfer*, vol. 26, pp. 1163–1173, 1999.

29. M. Grigoriu, A Monte Carlo Solution of Heat Conduction and Poisson Equations, *ASME Journal of Heat Transfer*, vol. 122, pp. 40–45, 2000.

30. K. K. Sabelfeld, *Monte Carlo Methods in Boundary Value Problems*, Berlin: Springer-Verlag, 1991.

31. A. Haji-Sheikh and F. P. Buckingham, Multidimensional Inverse Heat Conduction Using the Monte Carlo Method, *ASME Journal of Heat Transfer*, vol. 115, pp. 26–33, 1993.

32. G. M. Hahn, *Hyperthermia and Cancer*, New York: Plenum Press, 1982.

33. S. Mizushina, H. Ohba, K. Abe, S. Mizoshiri, and T. Sugiura, Recent Trends in Medical Microwave Radiometry, *IEICE Transactions on Communications*, vol. E78-B, pp. 789–798, 1995.

34. Z. S. Deng and J. Liu, Monte Carlo Method to Solve Multi-Dimensional Bioheat Transfer Problem, *Numerical Heat Transfer, Part B: Fundamentals*, vol. 42, pp. 543–567, 2002.

35. C. F. Babbs, N. E. Fearnot, J. A. Marchosky, C. J. Moran, J. T. Jones, and T. D. Plantenga, Theoretical Basis for Controlling Minimal Tumor Temperature during Interstitial Conductive Heat Therapy, *IEEE Transactions on Biomedical Engineering*, vol. 37, pp. 662–672, 1990.

36. S. A. Haider, T. C. Cetas, and R. B. Roemer, Temperature Distribution in Tissues from a Regular Array of Hot Source Implants, *IEEE Transactions on Biomedical Engineering*, vol. 40, pp. 408–417, 1993.

37. D. A. Torvi and J. D. Dale, Finite Element Model of Skin Subjected to a Flash Fire, *ASME Journal of Biomechanical Engineering*, vol. 116, pp. 250–255, 1994.

38. S. Chantasiriwan, Determination of Sensitivity Coefficients in Linear Heat Conduction Problems by Random-Walk Method, *Numerical Heat Transfer, Part B*, vol. 34, pp. 103–120, 1998.

39. Y. Rabin and A. Shitzer, Exact Solution to One-Dimensional Inverse-Stefan Problem in Nonideal Biological Tissues, *ASME Journal of Heat Transfer*, vol. 117, pp. 425–431, 1995.

40. B. Han, A. Iftekhar, and J. C. Bischof, Improved Cryosurgery by Use of Thermophysical and Inflammatory Adjuvants, *Technology in Cancer Research and Treatment*, vol. 3, pp. 103–111, 2004.

41. Z. S. Deng and J. Liu, Numerical Simulation of Selective Freezing of Target Biological Tissues Following Injection of Solutions with Specific Thermal Properties, *Cryobiology*, vol. 50, pp. 183–192, 2005.

42. Z. S. Deng and J. Liu, Numerical Simulation on 3-D Freezing and Heating Problems for the Combined Cryosurgery and Hyperthermia Therapy, *Numerical Heat Transfer, Part A: Applications*, vol. 46, pp. 587–611, 2004.

43. C. Bonacina, G. Comini, A. Fasano, and M. Primicero, Numerical Solution of Phase-Change Problems, *International Journal of Heat Mass Transfer*, vol. 16, pp. 1825–1832, 1973.

44. Z. S. Deng and J. Liu, Modeling of Multidimensional Freezing Problem during Cryosurgery by the Dual Reciprocity Boundary Element Method, *Engineering Analysis with Boundary Elements*, vol. 28, pp. 97–108, 2004.

45. D. Nardini and C. A. Brebbia, A New Approach to Free Vibration Analysis Using Boundary Elements, in *Boundary Elements in Engineering* (ed. C. A. Brebbia), Berlin: Springer-Verlag, 1982.

46. J. Liu and L. X. Xu, Boundary Information Based Diagnostics on the Thermal States of Biological Bodies, *International Journal of Heat and Mass Transfer*, vol. 43, pp. 2827–2839, 2000.

47. R. G. Keanini and B. Rubinsky, Optimization of Multiprobe Cryosurgery, *ASME Journal of Heat Transfer*, vol. 114, pp. 796–801, 1992.

48. J. G. Baust and A. A. Gage, Progress toward Optimization of Cryosurgery, *Technology in Cancer Research and Treatment*, vol. 3, pp. 95–101, 2004.

49. Z. S. Deng and J. Liu, Conformal Cryosurgical Treatment of Tumor by Use of Nanoparticles: Feasibility Study, *1st Annual IEEE International Conference on Nano/Molecular Medicine and Engineering*, Macau, China, August 6–9, 2007.

50. J. F. Yan, Z. S. Deng, and J. Liu, New Modality for Maximizing Cryosurgical Killing Scope while Minimizing Mechanical Incision Trauma Using Combined Freezing-Heating System, *ASME Journal of Medical Device*, vol. 1, pp. 264–271, 2007.

51. J. H. Zhou and J. Liu, Numerical Study on 3-D Light and Heat Transport in Biological Tissues Embedded with Large Blood Vessels during Laser-Induced Thermotherapy, *Numerical Heat Transfer, Part A: Applications*, vol. 45, pp. 415–449, 2004.

52. T. Halldorsson and J. Langerholc, Thermodynamic Analysis of Laser Irradiation of Biological Tissue, *Applied Optics*, vol. 17, pp. 3948–3958, 1978.

53. A. J. Welch, E. H. Wissler, and L. A. Priebe, Significance of Blood Flow in Calculations of Temperature in Laser Irradiated Tissue, *IEEE Transactions on Biomedical Engineering*, vol. 27, pp. 164–166, 1980.

54. L. I. Grossweiner, A. M. Al-Karmi, P. W. Johnson, and K. R. Brader, Modeling of Tissue Heating with a Pulsed Nd:YAG Laser, *Lasers in Surgery and Medicine*, vol. 10, pp. 295–302, 1990.

55. A. Sagi, A. Shitzer, A. Katzir, and S. Akselrod, Heating of Biological Tissue by Laser Irradiation: Theoretical Model, *Optical Engineering*, vol. 31, pp. 1417–1423, 1992.

56. T. N. Glenn, S. Rastegar, and S. L. Jacques, Finite Element Analysis of Temperature Controlled Coagulation in Laser-Irradiated Tissue, *IEEE Transactions on Biomedical Engineering*, vol. 43, pp. 79–86, 1996.

57. L. B. Director, S. E. Frid, V. Y. Mendeleev, and S. N. Scovorod'ko, Computer Simulation of Heat and Mass Transfer in Tissue during High-Intensity Long-Range Laser Irradiation, *Annals of the New York Academy of Sciences*, vol. 858, pp. 56–65, 1998.

58. R. K. Shah, B. Nemati, L. V. Wang, and S. M. Shapshay, Optical-Thermal Simulation of Tonsillar Tissue Irradiation, *Lasers in Surgery and Medicine*, vol. 28, pp. 313–319, 2001.

59. J. H. Torres, M. Motamedi, J. A. Pearce, and A. J. Welch, Experimental Evaluation of Mathematical Models for Predicting the Thermal Response of Tissue to Laser Irradiation, *Applied Optics*, vol. 32, pp. 597–606, 1993.

60. C. M. Beacco, S. R. Mordon, and J. M. Brunetaud, Development and Experimental in Vivo Validation of Mathematical Modeling of Laser Coagulation, *Lasers in Surgery and Medicine*, vol. 14, pp. 362–373, 1994.

61. Y. Yamada, T. Tien, and M. Ohta, Theoretical Analysis of Temperature Variation of Biological Tissues Irradiated by Light, *ASME/JSME Thermal Engineering Conference*, vol. 4, pp. 575–581, 1995.

62. C. Sturesson and S. Andersson-Engels, A Mathematical Model for Predicting the Temperature Distribution in Laser-Induced Hyperthermia: Experimental Evaluation and Applications, *Physics in Medicine and Biology*, vol. 40, pp. 2037–2052, 1995.

63. L. Cummins and M. Nauenberg, Thermal Effects of Laser Radiation in Biological Tissue, *Biophysical Journal*, vol. 42, pp. 99–102, 1983.

64. A. Ishimaru, *Wave Propagation and Scattering in Random Media*, New York: Academic Press, 1978.

65. M. F. Modest, *Radiative Heat Transfer*, 2nd ed., New York: Academic Press, 2003.

66. O. I. Craciunescu, T. V. Samulski, J. R. MacFall, and S. T. Clegg, Perturbations in Hyperthermia Temperature Distributions Associated with Counter-Current Flow: Numerical Simulations and Empirical Verification, *IEEE Transactions on Biomedical Engineering*, vol. 47, pp. 435–443, 2000.

67. S. T. Flock, M. S. Patterson, B. C. Wilson, and D. R. Wyman, Monte Carlo Modeling of Light Propagation in Highly Scattering Tissues—I: Model Predications and Comparison with Diffusion Theory, *IEEE Transactions on Biomedical Engineering*, vol. 36, pp. 1162–1168, 1989.

68. M. Keijzer, S. L. Jacques, S. A. Prahl, and A. J. Welch, Light Distributions in Artery Tissue: Monte Carlo Simulations for Finite-Diameter Laser Beams, *Lasers in Surgery and Medicine*, vol. 9, pp. 148–154, 1989.

69. Y. Hasegawa, Y. Yamada, M. Tamura, and Y. Nomura, Monte Carlo Simulation of Light Transmission through Living Tissues, *Applied Optics*, vol. 30, pp. 4515–4520, 1991.

70. L. Wang, S. L. Jacques, and L. Zheng, MCML—Monte Carlo Modeling of Light Transport in Multi-Layered Tissues, *Computer Methods and Programs in Biomedicine*, vol. 47, pp. 131–146, 1995.

71. S. L. Jacques and L. Wang, Monte Carlo Modeling of Light Transport in Tissues, in *Optical-Thermal Response of Laser-Irradiated Tissue* (ed. A. J. Welch and M. J. C. van Gemert), New York: Plenum Press, 1995.

72. A. J. Welch and C. M. Gardner, Monte Carlo Model for Determination of the Role of Heat Generation in Laser-Irradiated Tissue, *ASME Journal of Biomechanical Engineering*, vol. 119, pp. 489–495, 1997.

73. J. H. Zhou, J. Liu, and A. B. Yu, Study on the Thawing Process of Biological Tissue Induced by Laser Irradiation, *ASME Journal of Biomechanical Engineering*, vol. 127, pp. 416–431, 2005.

74. S. V. Patankar, *Numerical Heat Transfer and Fluid Flow*, New York: Hemisphere Publishing, 1980.
75. D. W. Liepsch, Flow in Tubes and Arteries: A Comparison, *Biorheology*, vol. 23, pp. 395–433, 1986.
76. Y. C. Fung, *Biomechanics: Mechanical Properties of Living Tissues*, 2nd ed., New York: Springer-Verlag, 1993.
77. W. M. Kays and M. E. Crawford, *Convective Heat and Mass Transfer*, New York: McGraw-Hill, 1980.

4

Discrete Vasculature (DIVA) Model Simulating the Thermal Impact of Individual Blood Vessels for In Vivo Heat Transfer

B. W. Raaymakers, A. N. T. J. Kotte,
and J. J. W. Lagendijk

CONTENTS

4.1 INTRODUCTION

To calculate the *in vivo* temperature distribution, the impact of blood flow on heat transfer has to be accounted for [1]. Basically, two approaches can be utilized when taking the blood flow into account in a thermal model: the continuum models and the discrete vessel models. The first approach models the thermal impact of all blood vessels with a single, global parameter; the latter models the thermal impact of each vessel individually.

4.1.1 Bioheat Model

The continuum model used most often is the Pennes bioheat or heat sink model [2]. The thermal impact of blood is described by introducing an energy drain that is proportional with the volumetric perfusion level and the elevation of the local tissue temperature over that of the arterial temperature. Equation (4.1) states the local heat balance for this model.

$$\rho_{tis} c_{tis} \frac{\partial T}{\partial t} = \nabla \cdot \left(k_{tis} \nabla T \right) - c_b W_b (T - T_{art}) + P \tag{4.1}$$

In this equation, $T(K)$ is the local tissue temperature; $t(s)$ is the time; ρ_{tis}, c_{tis}, and k_{tis} are the tissue density (kg m^{-3}), specific heat (JK^{-1} kg^{-1}), and thermal conductivity (WK^{-1} m^{-1}), respectively; c_b is the specific heat of blood; W_b is the volumetric perfusion rate (kg m^{-3} s^{-1}); T_{art} is the arterial blood temperature (K); and P is the sum of the absorbed power and metabolic heat production (Wm^{-3}). The term on the left-hand side of the equal sign describes the temperature evolution in time, and the first term on the right-hand side of the equal sign describes the heat transfer due to conduction. The second term is the actual contribution of blood to the heat transfer; the assumption is (1) that arterial blood with temperature T_{art} is heated in the capillaries to the local tissue temperature T without any preheating, and (2) that the venous blood leaves the tissue without any thermal interaction.

4.1.2 Equilibration Length

In reality, blood can be preheated by the interaction with heated surrounding tissue. The rate at which blood is heated can be quantified by the equilibration length [3], characterizing the distance it takes for the blood in a vessel to thermally equilibrate to the temperature of the surrounding tissue. Main arteries typically have a high flow and large equilibration length in the order of meters [4]. Capillaries have low flow and very short equilibration lengths (0.1 μm) [4], and are therefore always in thermal equilibrium with the surrounding tissue. Chen and Holmes [3] and Van Leeuwen et al. [5] showed that vessels with a diameter between 0.2 and 0.5 mm take care of a major part of the thermal equilibration of the arterial blood and are therefore thermally significant.

4.1.3 Thermal Modeling of Individual Vessels

In a discrete vessel model, the thermal impact of many vessel segments within a complex vasculature has to be taken into account individually. The impact on the temperature distribution can be calculated analytically only for very basic vessel configurations, as has been done for single vessels by Chen and Holmes [3], Crezee and Lagendijk [4], and Huang et al. [6] and for countercurrent vessels by Weinbaum et al. [7], Baish et al. [8], Wissler [9], M. Zhu et al. [10,11], and L. Zhu and Weinbaum [12]. Therefore, modeling the thermal impact of a complex, detailed, discrete vasculature has to be done numerically.

The first numerical models could only cope with straight vessels [13–16] or branching networks with only perpendicular vessel connections [17]. Mooibroek and Lagendijk [18] presented a more versatile model that could cope with curved vessel networks. However, the vessel description is directly coupled with the resolution of the discretized tissue, which makes describing extensive and curved vasculature cumbersome.

More flexible alternatives are from Stanczyk et al. [19], who present a two-dimensional (2D) model, and from Liu et al. [20] and Dughiero and Curazza [21], who present a model for detailed simulations around bifurcations. However, the most flexible and versatile solution for modeling the thermal impact of individual blood vessels is the discrete vasculature (DIVA) model [22,23], developed at our Department of Radiotherapy in the University Medical Center Utrecht, the Netherlands. DIVA decouples the tissue description and the vessel description and accounts for the heat transfer between tissue and vessel separately. This concept of decoupling the heat transfer in tissue and vessels has also been implemented by Blanchard et al. [24].

4.2 DISCRETE VASCULATURE (DIVA) MODEL

4.2.1 Introduction

DIVA is a versatile model for modeling *in vivo* heat transfer, including the impact of discrete vasculature [22,23]. The original aim of the model was to calculate the temperature distribution during hyperthermia treatments [1], that is, a cancer treatment where the sensitivity of the tumor (e.g., for radiotherapy) is enhanced by elevating the temperature to approximately 44°C. This means the model does not take phase transitions of the tissue into account such as freezing or evaporation.

DIVA takes a voxelized anatomy and a geometric description of the vasculature network as an input. The latter means that the spatial precision of the vessel description is independent of the anatomy grid resolution. The heat flow rate between adjacent voxels is calculated by solving Fourier's law of heat conduction using a finite difference iteration scheme. The interaction between the blood vessels and the tissue is calculated by determining the heat flow rate between a blood vessel and the surrounding tissue voxels. The flow rate is calculated using the tissue temperature samples, a blood temperature sample, and the distance between the center of the vessel and the location of the tissue temperature sample.

4.2.2 Conductive Heat Transfer in Tissue

The anatomy has to be defined as a grid of rectangular voxels. For instance, by using a magnetic resonance imaging (MRI) or computerized tomography (CT) data set, each voxel gets assigned a tissue type identifier and a tissue temperature. In a separate table, the tissue properties and initial temperatures of each tissue type are defined.

For one grid direction, the heat conduction between voxels is illustrated in Figure 4.1. The x-coordinates of the voxel centers have a unique value denoted by $x_i = x_0 + i\Delta x (i \in \mathbb{Z})$. The goal is to calculate the temperatures at unique moments in time, given by $t_n = t_0 + n\Delta t (n \in \mathbb{N})$. The simulated field, T, describing temperatures at a certain time and location is then denoted by $T_i^n = T(t_n, x_i)$ [25]. The finite difference scheme can be derived by considering the thermal interaction between two adjacent voxels using the concept of "thermal resistance" [26].

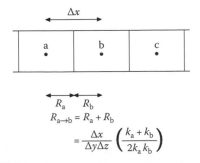

Figure 4.1   The thermal resistance, in one dimension, between the centers of two adjacent voxels is simply given by the sum of the two thermal resistances between a voxel center and the voxel boundary. The contact area is given by $\Delta y \Delta z$, and the thermal conductivity is denoted by k.

Thermal resistance is defined by the thermal conductivities of the voxels involved, as explained below. At the same time, the thermal resistance also couples the temperature difference and the heat flux between these two voxels (Equations 4.2 and 4.3); the total thermal resistance $R_{a \to b}$ between voxels a and b is defined by Equation (4.2):

$$R_{a \to b} = \frac{T_a - T_b}{q_{a \to b}} = \frac{L_{ab}}{kS} \tag{4.2}$$

where T_a and T_b are the temperature of voxels a and b, respectively; $q_{a \to b}$ is the heat flow between voxels a and b; L_{ab} is the distance between the temperature samples in voxels a and b; S is the contact surface between a and b; and k is the appropriate thermal conductivity between voxels a and b.

Consider the interaction between voxels a and b, as depicted in Figure 4.1. The voxel volume is given by $\Delta x \Delta y \Delta z$, while the interaction surface for the two voxels under consideration is $\Delta y \Delta z$. This results in a thermal resistance, as given by Equation (4.3).

$$R_{a \to b} = \frac{\Delta x}{\Delta y \Delta z} \left(\frac{k_a + k_b}{2 k_a k_b} \right) \quad [\text{KW}^{-1}] \tag{4.3}$$

The thermal energy change of voxel b due to the thermal interaction with the neighboring voxels a and c during a time step Δt is given by

$$\rho_{tis} c_{tis} \left(T_b^{n+1} - T_b^n \right) \Delta x \Delta y \Delta z = \left(q_{a \to b} + q_{c \to b} \right) \Delta t = \left(\frac{T_a - T_b}{R_{a \to b}} + \frac{T_c - T_b}{R_{c \to b}} \right) \Delta t \Rightarrow$$

$$\frac{\rho_{tis} c_{tis}}{\Delta t} \left(T_b^{n+1} - T_b^n \right) = \frac{2 k_a k_b}{k_a + k_b} \frac{1}{\Delta x^2} \left(T_a^n - T_b^n \right) + \frac{2 k_b k_c}{k_b + k_c} \frac{1}{\Delta x^2} \left(T_c^n - T_b^n \right) \tag{4.4}$$

This results in the general one-dimensional difference scheme based on the voxel energy balance:

$$\frac{\rho_{tis} c_{tis}}{\Delta t} \left(T_i^{n+1} - T_i^n \right) = \frac{1}{\Delta x^2} \left[\frac{2 k_i k_{i+1}}{k_i + k_{i+1}} T_{i+1}^n + \frac{2 k_i k_{i-1}}{k_i + k_{i-1}} T_{i-1}^n - \left(\frac{2 k_i k_{i+1}}{k_i + k_{i+1}} + \frac{2 k_i k_{i-1}}{k_i + k_{i-1}} \right) T_i^n \right] \tag{4.5}$$

where ρ_{tis} is the tissue density [kg m^{-3}], c_{tis} is the tissue-specific heat capacity [J kg^{-1} K^{-1}], and k is the tissue thermal conductivity [W m^{-1} K^{-1}].

For the three-dimensional (3D) scheme, this equation is applied in all three grid directions, each with its own uniform spacing. The total energy balance for a voxel results from the summation of the interactions with all its nearest neighbors (six in total). This results in a 3D temperature grid.

4.2.3 Convective Heat Transfer by a Single Vessel Segment

The blood temperature profile in a vessel segment (i.e., a part of a vessel without any branches) is registered separately and is updated separately from the tissue temperatures. The impact of the vasculature on the temperature distribution is basically accounted for as follows: one iteration to update the tissue temperatures is followed by one iteration to update the blood temperature profiles in the vessels.

4.2.3.1 Geometric Vessel Description

A vessel segment is a nonbranching part of the vasculature. It is defined by a parameterized curve in 3D with an associated diameter. Points along the segment are addressed by an index, running from 0 to 1, that represents the distance from the beginning point of the segment. This way, the location of each point along the segment can be determined in the tissue grid.

4.2.3.2 Blood Temperature Profile

DIVA uses the "mixing cup" blood temperature [27,28] to serve as the axial blood temperature profile. This mixing cup temperature T_{mix} is a velocity weighted average of the temperature of the blood flowing through an imaginary plane perpendicular to the heart line of the vessel.

The use of the mixing cup temperature for the axial blood temperature profile has the advantage that, together with the mean blood velocity, the specific heat and density for blood, it models the axially convected thermal energy rate. The mixing cup blood temperature profile is discretized, yielding an array of blood temperature samples (or buckets).

The sample or bucket density can be chosen independent of the vessel geometry. The array of buckets can also be addressed with an index, running from 0 to 1. Since the curve representing the geometry of the vessel is also indexed between zero and one, every point of the segment can be addressed with a specific index, and with the same index the temperature of that location can be found. The temperature of a geometric point is calculated by linear interpolation between the neighboring bucket centers.

4.2.3.3 Convective Heat Transfer by Blood Flow

Using the mixing cup temperature as the blood temperature, one value for the blood velocity, $\langle v \rangle$, can be defined. When the blood flows for a certain amount of time, δt, the temperature profile must be shifted over a certain distance, $\delta = \langle v \rangle \delta t$.

The shift can be directed forward, $x \rightarrow x + \delta$, or in the opposite direction, $x \rightarrow x - \delta$. We will only show the first; the second is very similar. A schematic picture of this shift procedure is given in Figure 4.2.

Before shifting the total temperature profile, the temperature of the incoming blood, T_{new}, has to be specified. This is implemented by extending the blood temperature array to the range $(-\infty, 1)$ with the same sample density as the original

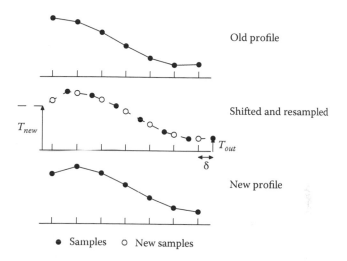

Figure 4.2 Flow implementation; the heat exchange with the environment is not depicted.

array $T_{(-\infty,\,1]}(x)$. All the samples in the range $(-\infty, 0)$ will have the value T_{new}. From this infinite array, the new "shifted" array $T_{[0,1]}^{new}(x)$ can be created by applying

$$T_{[0,1]}^{new}(x) = T_{(-\infty,1]}(x - \delta)$$

for the indexes of all the samples of $T_{[0,1]}^{new}(x)$.

The shift results in new values for the samples of the array and a single value "shifted out." This result will be the average of all the samples with an index > $1 - \delta$.

4.2.3.4 Interaction between Vessel and Tissue

For every bucket, the heat flow to or from the tissue grid is calculated. The heat flow toward a bucket can be estimated from the (blood) temperature of the bucket and the tissue temperatures in its immediate vicinity. To sample the tissue temperature around the bucket, the "estimation set," S_{est}, is devised. This set contains the voxels located just outside the bucket; the heat flow results in a change of the bucket's temperature. The heat flow is accounted for in the tissue grid in the voxels of the "exchange set," S_{exc}. In Figure 4.3, a bucket with its two sets is shown.

The heat flow rate between a bucket and the tissue is determined as follows. Consider a bucket with radius r_{ves} embedded in a coaxial tissue cylinder with radius R; the temperature evolution of the tissue in time is then stated by Equation (4.6) in cylinder coordinates (r, φ, z) [4,27].

$$\rho_{tis} c_{tis} \frac{\partial T_{tis}}{\partial t} = \frac{1}{r} \frac{d}{dr} \left(kr \frac{dT_{tis}}{dr} \right) + P \quad [\text{W m}^{-3}] \tag{4.6}$$

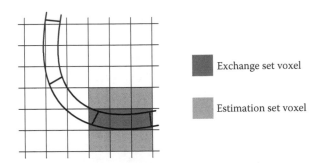

Figure 4.3   Part of a vessel segment projected onto a grid. Three buckets are indicated. For one bucket, the two sets involved in the heat exchange between the bucket and tissue are shown.

For the considered system, the heat flux at the vessel wall $\phi|_{r=r_{ves}}$ can be calculated analytically. In Crezee and Lagendijk [4], the following expression is given for $\phi|_{r=r_{ves}}$ under the assumption of fully developed velocity and temperature profiles:

$$\phi|_{r=r_{ves}} = -\frac{Nu\,k_b}{2r_{ves}}(T_{tis}(r_{ves}) - T_{mix}) \quad [\mathrm{W\,m^{-2}}] \tag{4.7}$$

where Nu is the Nusselt number, k_b the blood thermal conductivity [W m^{-1} K^{-1}], T_{mix} the mixing cup blood temperature [K], and $T_{tis}(r_{ves})$ the temperature of the vessel wall. Note that $\phi|_{r=r_{ves}}$ is directed radially out of the vessel.

Solving Equation (4.6) and using Equation (4.7) to eliminate $T_{tis}(r_{ves})$ result in an expression for the heat flux at the vessel wall.

$$\phi|_{r=r_{ves}} = -\frac{1}{\frac{2k}{Nu\,k_b} + \ln\left(\frac{R}{r_{ves}}\right)} \left\{ \frac{k}{r_{ves}}(T_{tis}(R) - T_{mix}) + Pr_{ves}\left(\frac{1}{4}\left(\left(\frac{R}{r_{ves}}\right)^2 - 1\right) - \frac{1}{2}\ln\left(\frac{R}{r_{ves}}\right) \right) \right\} \tag{4.8}$$

The total heat flow rate Φ into a cylindrical volume of the vessel (bucket) is the integral of the density of the heat flow rate ϕ at the surface of this volume:

$$\Phi = -\int \phi|_{r=r_{ves}} \cdot \hat{n}\, d\Omega \quad [\mathrm{W}] \tag{4.9}$$

where $d\Omega$ is an infinitesimal vessel surface element and $\hat{n}$ is the unit radial vector on this surface element.

This definition of the heat flow rate (Equation 4.9 together with Equation 4.8) will be used to estimate the heat flow rate from or toward a bucket. The estimate is established by sampling the tissue temperature from the estimation set (S_{est}). The

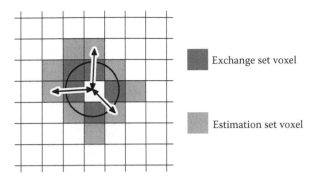

Exchange set voxel

Estimation set voxel

Figure 4.4   Projection of a vessel onto the tissue grid together with the two associated voxel sets. The interaction is calculated based on the distances between the temperature samples and the center of the vessel.

heat flow is accounted for in the exchange set (S_{exc}), that is, the energy exchange of the blood with the tissue is done in these exchange set voxels. In Figure 4.4, the two sets of a bucket are schematically drawn in the tissue grid. Once the two sets, S_{est} and S_{exc}, are created, the heat exchange can be calculated straightforwardly. In the grid holding the discretized temperature field, tissue temperature samples can be found for all the members of S_{est}. For all these samples, the heat flux on the vessel wall is calculated using Equation (4.8). Averaging these fluxes and multiplying with the bucket surface area result in the used heat flow rate of the bucket.

In summary, one iteration for updating the vessel temperature profile is as follows:

1. Calculate the heat flow toward every bucket, using Equation (4.8).
2. The total heat flow is distributed over the members of the exchange set in the drain grid.
3. Update the (mixing cup) temperature of every bucket. Before the temperature sample is updated, a weighted average of the heat flow to this bucket with the flow to the next bucket (in the direction of blood flow) is calculated. This is done to reflect the flowing character of the blood: while the blood is heated by means of the heat flow, it flows from one bucket to the next. This interpolated heat flow is used to calculate the blood temperature change.
4. Shift the profile of the updated vessel temperature samples.

4.2.4 Convective Heat Transfer by a Vessel Network

The vessel segments as presented in Section 4.2.3 can be connected to form a vessel tree. Multiple vessel trees together constitute the total vasculature. Both

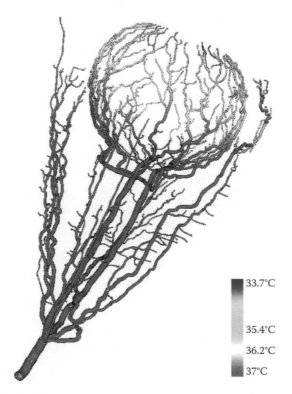

33.7°C

35.4°C

36.2°C

37°C

Figure 4.5 Example of an arterial vasculature as used for thermal calculations in the orbit [29]. The blood temperature in the vessels is gray-scale coded and shows the situation for a normothermic exposure of the eye to the surroundings.

arterial and venous trees can be constructed; arterial and venous trees cannot be connected to each other. A vessel tree starts with a single segment that can have an indefinite number of child segments, each child segment in its turn can have an indefinite number of child segments, and so on. The difference between arteries and veins is the blood flow direction, from the first segment downward for arteries and toward the first segment for veins. An example of an arterial vasculature of the eye is shown in Figure 4.5.

When the blood leaves a segment, there are various possibilities: the blood is exchanged with connected segments (segment–segment interface), the blood is exchanged with tissue (segment–tissue interface), or, since blood conservation between connected segments is not mandatory, part of the blood is exchanged with the connected segment and the other part with tissue (mixed interface). An exchange of blood with tissue is implemented by an additional heat exchange between blood and tissue after the blood leaves the vessel.

4.2.4.1 Segment–Segment Interface

In Figure 4.6, a parent with two child segments is depicted. Here, we will describe the case where the sum of the child flows does equal the parent's flow. The sign of the blood flow, discriminating between arteries and veins, must be the same for parent and child segments: it is not possible to connect an artery to a vein directly.

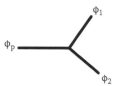

Figure 4.6 Segment p with (two) child segments. The total blood flow φ_p is equal to the total blood flow into the child segments: $\varphi_p = \Sigma_i \varphi_i$.

The temperature of the blood flowing out of an arterial parent segment becomes the blood inflow temperature of all its child segments. The inflow blood temperature for a venous parent segment is obtained from the "flow-averaged outflow blood temperature" of all its child segments. The blood exchange between segments might lead to local deviations in the blood temperature when the bucket length l_b is not the same for parent and child segments.

4.2.4.2 Segment–Tissue Interface

Whenever blood crosses the vessel–tissue boundary at the end of a vessel segment, there must be an interaction with the surrounding tissue. For a venous segment, this means that the blood inflow temperature has to be established from the local tissue temperature, whereas blood leaving an arterial segment will generally have a temperature unequal to the local tissue temperature and need to equilibrate with the tissue temperature. In both cases, the tissue temperature around the end of a vessel segment is sampled from a so-called sink set. This sink set is a collection of tissue voxels that can be chosen freely for each individual segment; this is schematically shown in Figure 4.7. The geometry of the set is not limited to a sphere, nor does it have to enclose the endpoint of the corresponding segment.

The sink set is used to thermally equilibrate the outflowing arterial blood with its associated tissue. All the voxels in this sink set are given an *extra* perfusion equal to the blood flow (from the segment) into the tissue (φ_{tis}) divided by the number of voxels present in the sink set. Then we use the local blood temperature T_{out}^{art} and the local tissue temperature together with the local blood perfusion to calculate the amount of heat that has to be exchanged with every tissue voxel separately. In fact, this strategy is a true implementation

Figure 4.7 Parent segment without any child segments. The total blood flow φ_p has to be equilibrated with the surrounding tissue, so the total sink geometry gets a perfusion of $\varphi_{tis} = \varphi_p$.

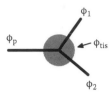

Figure 4.8   Segment with (two) child segments. The total blood flow of the child segments is lower than the flow of the parent segment. The flow associated with the tissue is set to the difference: $\varphi_{tis} = \varphi_p - \Sigma_i \varphi_i$.

of the well-known [2] "bioheat equation." For a venous segment, we use the average temperature of the sink set (for veins, it is called the "sample set") as the inflow temperature of the venous segment.

4.2.4.3 Mixed Interface

The blood flow can be specified for each segment individually, so blood flow conservation at a segment junction is not mandatory. This yields the possibility to model "bleed-off" by side branches too small to be modeled discretely. When the total blood flow into the child segments is not equal to the blood flow in the parent segment, a combination of the two interfaces described above is used. This implies that every segment junction can have a sink and sample set. For example, to model bleed-off for a single straight segment, this segment has to be described as a concatenation of smaller segments with a sink set at every concatenation, similar as in Figure 4.8 but then with one child per connection. The sign of the tissue blood flow φ_{tis} must be the same as that of the associated parent flow φ_p so there are no sample sets in an arterial vessel tree and there are no sink sets in a venous vessel tree.

4.2.5 Choices for Using Sink Sets

The sink sets as presented in Section 4.2.4 provide a means to complete the thermal interaction of blood leaving the arterial vasculature or sampling a realistic inflow temperature for veins. Ideally, these sink sets or sample sets would be the anatomical territory perfused by this vessel [30]. In practice, it is very hard to determine these territories. If so, various options for creating appropriate sink and sample sets are available, where the choice is led by the available data. When the blood flow in (some of) the vessels is known, the choice is different from when the perfusion distribution is known.

4.2.5.1 Blood Flow in the Vessels Is Known

An intuitive way of constructing the sets is to define the volume of the sets as the anatomy volume divided by the number of branches. Then assign spherical volumes with this volume to each terminating blood vessel segment (or to the branch point where no flow conservation is present; for the sake of brevity, we will refer to terminating branches only). For a dense vasculature, this is a fine approximation, since then the sets will be small and rather homogeneously distributed over the anatomy. However, when only a rudimentary vasculature is present for modeling, this approach fails [31]. This is because likely sets will be overlapping (which is allowed), which results in very high local heat sinks in these areas, whereas some anatomy volumes are not covered by sink sets at all and experience no additional heat sink.

To prevent the sets from overlapping, Kotte et al. [23] proposed a strategy in which the anatomy grid is divided into subvolumes, for instance cubes, such that the entire volume is covered. Then the subvolume is assigned as a sink or sample set to the vessel segment that ends there. A subvolume can be assigned to multiple vessels. The subvolumes without a vessel terminating in them are assigned to the closest terminating segment, since obviously these subvolumes must have some kind of perfusion in reality. Now, the sets will differ in volume, which means that for similar blood flow from all terminating vessels, the local perfusion in the sets will vary.

If one does not want to couple the terminating vessels to a specific region, an alternative is to assign the entire anatomy to each terminating vessel. In fact, one then takes the average outflowing blood temperature and equilibrates this to the tissue temperature averaged over the entire anatomy using the [2] heat sink model.

4.2.5.2 Perfusion Distribution Is Known

So far, we have assumed that we know the blood flow of the terminating vessels and have used this to create a perfusion distribution by choosing the appropriate sink sets. However, often the tracks of the vasculature are known, together with the perfusion distribution, while there are no data on the flow inside the vessels. Then, the flow in the vessels can be set such that this meets the known perfusion distribution. Of course, the resulting flow in the vessels is again determined by the choice of the sink sets.

4.2.5.3 Discrepancies between Flow and Perfusion Data

Even more often, there are partial data available on flow in the vessels and a perfusion distribution, and these data do not necessarily agree. Then, the strategy is to use both data; the flow in the vessels is used, but the corresponding perfusion in the sink sets as determined by the vessels is discarded and replaced by a separate absolute perfusion map instead. This means that an additional heat sink is applied using the separate perfusion map and, for each set, the local outflow blood temperature. Again, also in this case it is possible to assign the entire anatomy to each terminating segment, which in fact means that the average outflowing blood temperature is equilibrated to the average tissue temperature using the [2] heat sink model and the separate perfusion map.

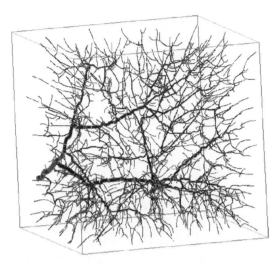

Figure 4.9   An example of an artificially generated vessel tree. In this example, the vessel density of the upper left half of the volume is chosen to be half of the vessel density in the lower right half of the volume.

4.2.6 Artificial Generation of Vessels for Thermal Calculations

In Sections 4.2.4 and 4.2.5, the sink sets were discussed as a means to complete the thermal interaction between the blood and the tissue in the case of an incomplete vessel network. Yet another approach is the artificial generation of additional vasculature in order to complete the thermal interaction between blood and tissue. Van Leeuwen et al. [32] present a method to generate an artificial yet realistic vasculature. The method can generate an entire network from scratch (an example is shown in Figure 4.9), but also the extension of existing networks is possible.

Virtually any geometry can be created. One has to define a number of endpoints and a single starting point for the tree. During construction, a (random) endpoint is chosen to start growing toward the starting point. Then, a next endpoint is chosen to start growing toward the constructed vessel. The "grow" process is defined by parameters that determine the attraction of branches toward the starting point and toward the already-generated vessel segments. Both the attraction toward the root point and the mutual artery–artery, vein–vein, and artery–vein attraction can be controlled; see Van Leeuwen et al. [32] for details. Figure 4.10 shows possible results for a few different parameter settings.

Applying this on a patient-specific basis is hard, simply because apparently the data on some parts of the vasculature are missing. However, creating an artificial network allows the investigation of the generic behavior of a full discrete vasculature; see Section 4.4 for examples.

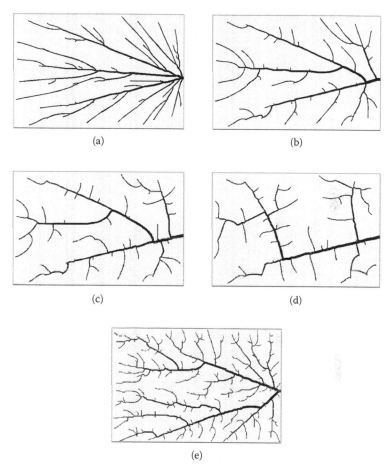

Figure 4.10   Vessel networks in a thin tissue slice constructed with different parameter sets [32]. Clearly, the tendency to grow toward the starting point can be influenced by adapting the parameters. In networks (a) through (d), an increasing mutual attraction and a decreasing attraction by the starting point are chosen for the same set of 54 endpoints. Network (e) shows a denser network with 216 endpoints with parameters similar to those of (c).

4.3 VALIDATION

For simple vessel geometries, DIVA has been validated by comparing simulated temperatures with analytic results [33,34], and very good correspondence is achieved.

DIVA has also been validated experimentally. Raaymakers et al. [35] compared the measured and simulated temperature profiles from an agar-agar phantom heated by three hot-water tubes and well-defined boundary conditions. The hot-water tubes are modeled as blood vessels. Nearly perfect correspondence was found.

Figure 4.11 Photograph of the experimental validation setup. The isolated, perfused bovine tongue is shown, heated by hot-water tubes (hw). The two arteries are connected to a roller pump, and the tongue is flushed with water to apply well-defined boundary conditions.

More difficult to interpret is the comparison of measured and simulated temperature profiles from an isolated, perfused bovine tongue at three different perfusion levels (0, 6, and 24 ml, $(100gr)^{-1}$, min^{-1}) [35]. Again, the tongue is heated by three hot-water tubes, and well-defined boundary conditions are applied (see Figure 4.11), while the arterial vasculature is connected to a roller pump in order to control the perfusion. For the simulations, the tongue, the vasculature down to 0.5 mm diameter, as well as the hot-water tubes were reconstructed from cryotome microslices (see Figure 4.19).

For no flow, the correspondence was good but worse than for the agar–agar phantom, indicating that reconstruction of the bovine tongue was less precise than a simple agar–agar phantom. For the 6 and 24 ml $(100gr)^{-1}$, min^{-1}) cases, three different simulations were done: two simulations include the vasculature, and the third is the heat sink model. The blood flow in the vessels was controlled by a pump, so the blood flow was known; therefore, the intuitive approach for determining the sink sets from Section 4.2.5 and the strategy where the entire anatomy is assigned as a sink set to each terminating vessel were used. The three methods in fact yield similar results: good agreement, but clearly the reconstruction is not precise enough for very high-precision prediction of the temperature profiles at mm resolution. However, the discrete vessel strategies did predict the local distortion around a blood vessel, which was obviously missed by the heat sink model.

The conclusion is that high-resolution prediction of temperature profiles is only feasible in very well-defined anatomies. Prediction of temperature distributions *in vivo* is feasible; however, care should be exercised when interpreting a patient-specific temperature distribution. The precision is hampered by limited data acquisition needed to reconstruct the volume of interest. Additionally, *in vivo* the physiologic reaction to temperature variations [36], especially changes in perfusion, should be accounted for in a model. Incorporating this in the model is relatively easy; see, for instance, Reference [37]. The problem is acquiring the correct physiologic data, which further complicates a reliable prediction of the temperature distribution.

4.4 APPLICATIONS OF DIVA

4.4.1 Patient-Specific Temperature Distribution during Regional Hyperthermia

As discussed in Section 4.3, calculation of the patient-specific temperature distribution, for instance during a hyperthermia treatment, is hard. Recently, Van den Berg et al. [38] made an attempt to calculate the temperature distribution of a prostate during regional hyperthermia [39]. The idea was that with the advent of sophisticated angiography and dynamic contrast-enhanced (DCE) imaging techniques, it has become possible to image small vessels and blood perfusion, bringing the ultimate goal of patient-specific thermal modeling within close reach. In this study, dynamic contrast-enhanced multislice CT-imaging techniques are employed to investigate the feasibility of this concept for regional hyperthermia treatment of the prostate. The results are retrospectively compared with clinical thermometry data of a patient group from an earlier trial [39]. Also, the role of the prostate vasculature in the establishment of the prostate temperature distribution is studied.

Quantitative 3D perfusion maps of the prostate were constructed for five patients using a distributed-parameter tracer kinetics model to analyze dynamic CT data. CT angiography was applied to construct a discrete vessel model of the pelvis; see Figure 4.12. It is currently not possible to visualize the intraprostatic vasculature *in vivo*. Therefore, a discrete vessel model of the prostate vasculature was constructed of a prostate taken from a human corpse, as shown in

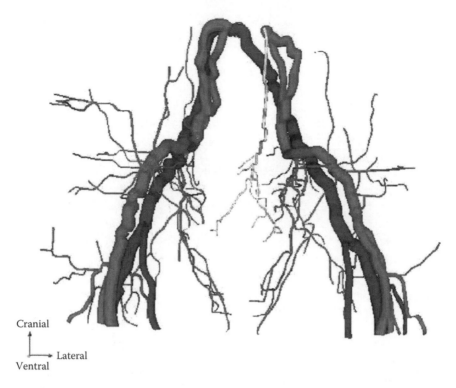

Cranial

Ventral ⌐→ Lateral

Figure 4.12   Overview of the pelvic vasculature. The light gray vessels represent the arterial system, which is dominated by the large iliac arteries and their side branches. The artery that seems to branch off from the root of the left iliac artery, which is the inferior mesenteric artery. It actually branches off from the aorta. The dark gray vessels correlate to the iliac venous system.

Figure 4.13. Three thermal-modeling schemes with increasing inclusion of the patient-specific physiological information were used to simulate the temperature distribution of the prostate during regional hyperthermia.

Prostate perfusion is in general heterogeneous, and T3 prostate carcinomas (T3 refers to tumor evasion beyond the prostate capsule), are often characterized by a strongly elevated tumor perfusion (up to 70 to 80 ml 100gr^{-1} min^{-1}), which was confirmed by the measured quantitative perfusion maps. Taking into account a realistic heterogeneous perfusion distribution with elevated tumor perfusion leads to 1°C to 2°C lower tumor temperatures than thermal simulations based on a homogeneous prostate perfusion. Furthermore, the comparison shows that the simulations with the measured perfusion maps result in consistently lower prostate temperatures than were clinically achieved. As shown in Figure 4.13, the simulations with the discrete vessel model indicate that significant preheating takes place in the prostate capsule vasculature, which forms a possible explanation for the discrepancy. Preheating in the larger pelvic vessels is very moderate, approximately 0.1°C to 0.3°C.

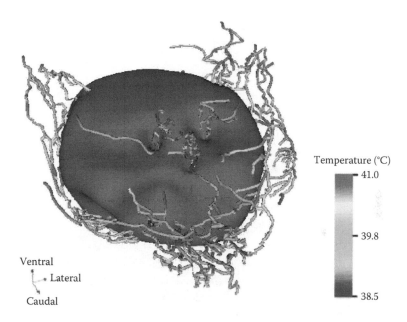

Temperature (°C)

41.0

39.8

38.5

Ventral

Lateral

Caudal

Figure 4.13   The 40.5°C isotemperature surface around the prostate during a regional hyperthermia treatment together with the prostate vessels [38]. Along the vessels, the preheating is depicted in a gray scale.

The conclusion of these works is that perfusion imaging provides important input for thermal modeling and can be used to obtain a lower limit on the prostate and tumor temperature in regional hyperthermia. Also, it is shown that the prostate vasculature plays a crucial role for patient-specific temperature calculations. Currently, the reconstruction of the discrete vessels *in vivo* is not possible. However, a discrete vessel model helps one to understand the mechanics of heat transfer in the prostate during hyperthermia.

4.4.2 Safety Guidelines for Radiofrequency Exposure

Predicting the temperature distribution on an individual patient basis might be cumbersome; still, DIVA offers excellent possibilities to determine the thermal impact of various exposures in a generic way. A nice example is the evaluation of safety guidelines for radiofrequency exposure.

In Van Leeuwen et al. [40], a realistic head model was evaluated for the 3D temperature rise induced by a mobile phone. This was done with the consecutive use of a finite-difference time domain (FDTD) model to predict the absorbed electromagnetic power distribution due to the mobile phone exposure, and DIVA for describing bioheat transfer both by conduction and by blood flow.

Figure 4.14 shows the anatomy of the head, which was reconstructed by manual segmentation from T1-weighted MR images. (T1 weighting is just another contrast weighting of MRI, it is beyond the scope of this text to spend extra time

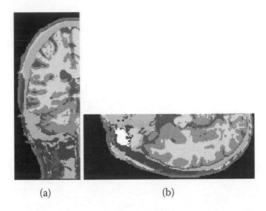

(a)   (b)

Figure 4.14   Segmented anatomy of the head: (a) coronal slice, and (b) transversal slice.

of this topic.) The vasculature was also reconstructed manually from MR angiography; see Figure 4.15a. For a better description of the heat transfer, the vasculature was extended using the program described in Section 4.2.6; the result is shown in Figure 4.15b.

The maximum rise in brain temperature was calculated and was 0.11°C for an antenna with average emitted power of 0.25 watts, the maximum value in common mobile phones; see Figure 4.16. The maximum temperature rise was found to occur at the skin.

The power distribution is characterized by a maximum averaged specific absorption rate (SAR) over an arbitrarily shaped 10 g volume of approximately

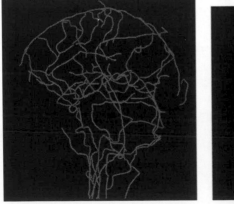

(a) Manual reconstruction of the vessels
tracked in the MRA images.

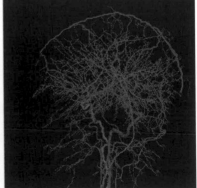

(b) The vasculature of (a) expanded by
artificial generation of new blood
vessels as discussed in Section 4.2.6.

Figure 4.15   The vasculature as used in the thermal simulations. Arteries are light gray, and veins are dark gray. The view is from the side, with the head facing left.

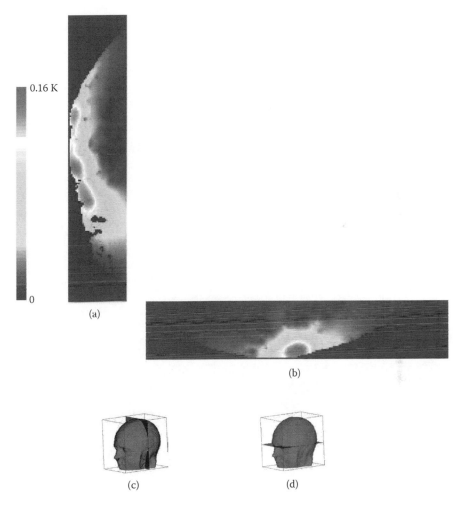

Figure 4.16   Temperature rise in the normal head calculated with discrete vasculature: (a) in a transversal plane, and (b) in a coronal plane. Depictions of the transverse and coronal planes are shown in (c) and (d). Temperatures on the tissue boundary were fixed according to the temperature distribution in the whole-head heat sink simulation.

1.6 W kg^{-1}. Although these power distributions are not in compliance with all proposed safety standards, the temperature rises are far too small to have lasting effects. A logical step would be to leave the SAR limits for the safety guidelines and use the actual temperature rises, as is originally proposed in the standards [41–43].

Flyckt et al. [44] performed a similar study, but focused specifically on the temperature rise in the eye since this is considered a sensitive organ for thermal damage. A detailed model of the orbit, including vasculature (shown in Figure 4.5), was reconstructed from microtome slices [29], and this eye model was inserted in the head model presented in Figure 4.14. The result is shown in Figure 4.17.

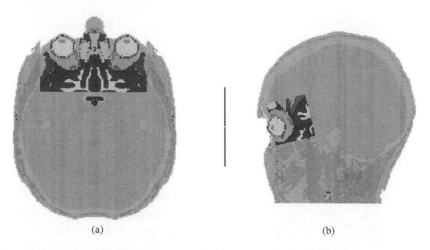

(a)   (b)

Figure 4.17   The original head anatomy from Van Leeuwen et al. [40], as presented in Figure 4.14, in which the detailed eye anatomy from Flyckt et al. [29] is inserted. (a) Transversal slice and (b) sagittal slice also showing the dipole antenna at 5 cm from the cornea. Arbitrary gray-scale coding is used merely to show the insertion of the detailed eye model into the whole-head model.

With this detailed anatomy, the SAR and thermal effects were determined under exposure to a dipole antenna representing a mobile phone operating at 900, 1500, and 1800 MHz with an output power of 1 W. The heterogeneous SAR distribution can be characterized by the peak SARs in the humor: 4.5, 7.7, and 8.4 W kg^{-1} for 900, 1500, and 1800 MHz, respectively, while the average SARs over the whole eyeball are 1.7, 2.5, and 2.2 W kg^{-1}. The maximum temperature rises in the eye due to the exposure are 0.22, 0.27, and 0.25°C for exposure of

Figure 4.18   The entire anatomy of 5.3 × 5.0 × 5.6 cm^3 together with the vessels is shown; also, the temperature distribution for the exposure to 1500 MHz with the vessels gray-scale coded by the blood temperature is shown. Temperatures remain below the body core temperature.

900, 1500, and 1800 MHz, respectively, calculated with DIVA. In Figure 4.18, the temperature distribution for the 1500 MHz exposure is shown; the temperature increase does not rise above the body core temperature, so no thermal damage is expected.

Again, similar to Van Leeuwen et al. [40], even for these artificial and high-exposure conditions, for which the SAR values are not in compliance with safety guidelines, the maximum temperature rises in the eye are too small to give harmful effects. The temperature in the eye also remains below body core temperature.

4.4.3 Evaluation of High-Intensity Focused Ultrasound (HIFU) Scan Strategies

Another application is the evaluation of treatment strategies in which the exact temperature distribution around blood vessels is crucial. An example is the evaluation of HIFU treatment strategies. Thermal underdosage around large blood vessels [4] is a problem for curative ablation procedures. The easy way of solving this is by stopping the blood flow, as can be done during ablation of liver lesions (e.g., Rossi et al. [45]).

With DIVA, it is possible to calculate the temperature distribution during a HIFU treatment of a vascularized tissue. An example is the evaluation of the spiral HIFU trajectory, as proposed by Salomir et al. [46]. The model of the perfused, isolated bovine tongue from Raaymakers et al. [35] was used as a vascularized phantom; see Figure 4.19.

The transient temperature during a counterclockwise spiral scan is shown at 10, 30, 50, 70, and 91 seconds of scanning in Figure 4.20. For comparison, the temperature distribution calculated using the heat sink model [2] is also shown. In the latter, the blood flow is accounted for by a perfusion term, whereas in DIVA the thermal impact of all individual vessels is taken into account. Clearly, the discrete vasculature causes thermal underdosages around the blood vessels. One solution for this is to occlude the feeding blood vessels to the lesion volume by ablation first and then start the ablation of the lesion itself.

Figure 4.19   Reconstructed blood vessels in the bovine tongue from Raaymakers et al. [35], shown in Figure 4.11. The HIFU focus is shown together with the plane of the spiral scan trajectory.

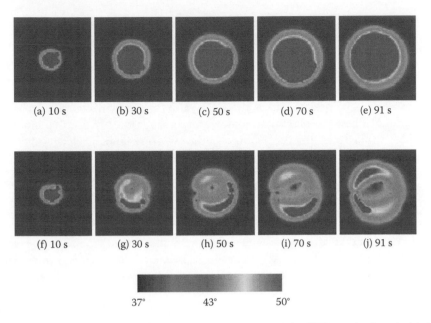

Figure 4.20   The calculated transient temperature distribution during HIFU scanning in an isolated, perfused bovine tongue. The maximum temperature is 80°C; for the sake of clarity, the gray scale is maximized at 50°C. Panels (a) through (e) show the results for the heat sink model (blood flow is accounted for by a perfusion term), and panels (f) through (j) for DIVA (blood flow is accounted for by discrete blood vessels).

4.5 CONCLUSIONS

The inclusion of individual blood vessels in thermal calculations is necessary for a detailed evaluation of the *in vivo* temperature distribution. DIVA offers a flexible way of modeling the temperature distribution in the presence of a detailed, geometrically described vasculature.

Thermal modeling, while taking all discrete blood vessels into account for an individual patient, is hampered by limited data acquisition of the exact geometry of the anatomy, vasculature, and perfusion distribution. This is even more so when the patient, or part of the patient, is heated, as is the case in hyperthermia or ablation therapy. Then also, the temperature dependency of the blood flow and perfusion has to be included in the model.

The main advantage of a discrete vessel thermal model is the possibility to investigate details of the temperature distribution for vascularized tissues in a generic way. Specifically for assessment of thermal safety guidelines and thermal treatment strategies, the generic knowledge of the temperature distribution around blood vessels is essential.

The largest improvement for thermal modeling is expected from improved data acquisition. Better knowledge of, for instance, the (temperature-dependent) perfusion distribution and blood flow, as well as higher-resolution data on the geometry of the vasculature and its connectivity, helps on two sides: the patient-specific modeling can be improved, and at the same time, more realistic data allow the definition of more realistic generic situations, leading to a more reliable calculation of the true temperature response from external heating.

REFERENCES

1. J. J. W. Lagendijk, Hyperthermia treatment planning, *Physics in Medicine and Biology* 45 (2000) R61–R76, topical review.
2. H. H. Pennes, Analysis of tissue and arterial blood temperature in the resting human forearm, *Journal of Applied Physiology* 1 (1948) 93–122.
3. M. M. Chen and K. R. Holmes, Microvascular contributions in tissue heat transfer, *Annals of the New York Academy of Sciences* 335 (1980) 137–150.
4. J. Crezee and J. J. W. Lagendijk, Temperature uniformity during hyperthermia: the impact of large vessels, *Physics in Medicine and Biology* 37(6) (1992) 1321–1337.
5. G. M. J. Van Leeuwen, A. N. T. J. Kotte, B. W. Raaymakers, and J. J. W. Lagendijk, Temperature simulations in tissue with a realistic computer generated vessel network, *Physics in Medicine and Biology* 45(4) (2000) 1035–1049.
6. H. W. Huang, C. L. Chan, and R. B. Roemer, Analytical solutions of Pennes bioheat transfer equation with a blood vessel, *Journal of Biomechanical Engineering: Transactions of the ASME* 116(2) (1994) 208–212.
7. S. Weinbaum, L. M. Jiji, and D. E. Lemons, Theory and experiment for the effect of vascular microstructure on surface tissue heat transfer part I: anatomical foundation and model conceptualization, *Journal of Biomechanical Engineering: Transactions of the ASME* 106 (1984) 321–330.
8. J. W. Baish, P. S. Ayyaswamy, and K. R. Foster, Heat transport mechanisms in vascular tissues: a model comparison, *Journal of Biomechanical Engineering: Transactions of the ASME* 108(4) (1986) 324–331.
9. E. H. Wissler, An analytical solution countercurrent heat transfer between parallel vessels with a linear axial temperature gradient, *Journal of Biomechanical Engineering: Transactions of the ASME* 110 (1988) 254–256.
10. M. Zhu, S. Weinbaum, L. M. Jiji, and D. E. Lemons, On the generalization of the Weinbaum-Jiji bioheat equation to microvessels of unequal size: the relation between the near field and local average tissue temperatures, *Journal of Biomechanical Engineering: Transactions of the ASME* 110 (1988) 74–81.
11. M. Zhu, S. Weinbaum, and L. M. Jiji, Heat exchange between unequal countercurrent vessels asymmetrically embedded in a cylinder with surface convection, *International Journal of Heat and Mass Transfer* 33(10) (1990) 2275–2284.
12. L. Zhu and S. Weinbaum, A model for heat transfer from embedded blood vessels in two-dimensional tissue preparations, *Journal of Biomechanical Engineering: Transactions of the ASME* 117 (1) (1995) 64–73.
13. J. J. W. Lagendijk, M. Schellekens, J. Schipper, and P. M. van der Linden, A three-dimensional description of heating patterns in vascularised tissues during hyperthermia treatment, *Physics in Medicine and Biology* 29(5) (1984) 495–507.
14. Z. P. Chen and R. B. Roemer, The effect of large blood vessels on temperature distributions during simulated hyperthermia, *Journal of Biomechanical Engineering: Transactions of the ASME* 114(4) (1992) 473–481.
15. R. J. Rawnsley, R. B. Roemer, and A. W. Dutton, The simulation of discrete vessel effects in experimental hyperthermia, *Journal of Biomechanical Engineering: Transactions of the ASME* 116 (3) (1994) 256–262.

16. C. L. Chan, Boundary element method analysis for the bioheat transfer equation, *Journal of Biomechanical Engineering: Transactions of the ASME* 114 (1992) 358–365.

17. H. W. Huang, Z. P. Chen, and R. B. Roemer, A counter current vascular network model of heat transfer in tissues, *Journal of Biomechanical Engineering: Transactions of the ASME* 118 (1996) 120–129.

18. J. Mooibroek and J. J. W. Lagendijk, A fast and simple algorithm for the calculation of convective heat transfer by large vessels in three-dimensional inhomogeneous tissues, *IEEE Transactions on Biomedical Engineering* 38(5) (1991) 490–501.

19. M. M. Stanczyk, G. M. J. Van Leeuwen, and A. A. Van Steenhoven, Discrete vessel heat transfer in perfused tissue: model comparison, *Physics in Medicine and Biology* 52(9) (2007) 2379–2391.

20. Y. J. Liu, A. K. Qiao, Q. Nan, and X. Yang, Thermal characteristics of microwave ablation in the vicinity of an arterial bifurcation, *International Journal of Hyperthermia* 22(6) (2006) 491–506.

21. F. Dughiero and S. Curazza, Numerical simulation of thermal disposition with induction heating used for oncological hyperthermic treatment, *Medical and Biological Engineering and Computing* 43 (1) (2005) 40–46.

22. A. N. T. J. Kotte, G. M. J. Van Leeuwen, J. De Bree, J. F. Van der Koijk, J. Crezee, and J. J. W. Lagendijk, A description of discrete vessel segments in thermal modelling of tissues, *Physics in Medicine and Biology* 41(5) (1996) 865–884.

23. A. N. T. J. Kotte, G. M. J. Van Leeuwen, and J. J. W. Lagendijk, Modelling the thermal impact of a discrete vessel tree, *Physics in Medicine and Biology* 44 (1999) 57–74.

24. C. H. Blanchard, G. Gutierrez, J. A. White, and R. B. Roemer, Hybrid finite element–finite difference method for thermal analysis of blood vessels, *International Journal of Hyperthermia* 16(4) (2000) 341–353.

25. W. H. Press, B. P. Flannery, S. A. Teukolsky, and W. T. Vetterling, *Numerical Recipes in C*, Cambridge: Cambridge University Press, 1992.

26. F. P. Incropera and D. P. DeWitt, *Introduction to Heat Transfer*, New York: John Wiley, 1990.

27. W. M. Kays and M. E. Crawford, *Convective Heat and Mass Transfer*, 2nd ed., McGraw-Hill Series in Mechanical Engineering, New York: McGraw-Hill, 1980.

28. J. C. Chato, Fundamentals of bioheat transfer, in M. Gautherie (Ed.), *Thermal Dosimetry and Treatment Planning*, Clinical Thermology, Subseries Thermotherapy, Berlin: Springer-Verlag, 1990, ch. 1, pp. 1–56.

29. V. M. M. Flyckt, B. W. Raaymakers, and J. J. W. Lagendijk, Modelling the impact of blood flow on the temperature distribution in the human eye and orbit: fixed heat transfer coefficients versus the Pennes bioheat model versus discrete blood vessels, *Physics in Medicine and Biology* 51(19) (2006) 5007–5021.

30. G. C. Cormack and B. G. H. Lamberty, Cadaver studies of correlation between vessel size and anatomical territory of cutaneous supply, *British Journal of Plastic Surgery* 39 (1986) 300–306.

31. B. W. Raaymakers, A. N. T. J. Kotte, and J. J. W. Lagendijk, How to apply a discrete vessel model in thermal simulations when only incomplete vessel data is available, *Physics in Medicine and Biology* 45 (2000) 3385–3401.

32. G. M. J. Van Leeuwen, A. N. T. J. Kotte, and J. J. W. Lagendijk, A flexible algorithm for construction of 3-D vessel networks for use in thermal modeling, *IEEE Transactions on Biomedical Engineering* 45 (1998) 596–604.

33. G. M. J. Van Leeuwen, A. N. T. J. Kotte, J. De Bree, J. F. Van der Koijk, J. Crezee, and J. J. W. Lagendijk, Accuracy of geometrical modelling of heat transfer from tissue to blood vessels, *Physics in Medicine and Biology* 42(7) (1997) 1451–1460.

34. G. M. J. Van Leeuwen, A. N. T. J. Kotte, J. Crezee, and J. J. W. Lagendijk, Tests of the geometrical description of blood vessels in a thermal model using counter-current geometries, *Physics in Medicine and Biology* 42 (1997) 1515–1532.

35. B. W. Raaymakers, J. Crezee, and J. J. W. Lagendijk, Modelling individual temperature profiles from an isolated perfused bovine tongue, *Physics in Medicine and Biology* 45 (2000) 765–780.

36. M. R. Horsman, Tissue physiology and the response to heat, *International Journal of Hyperthermia* 22(3) (2006) 197–203.

37. J. Lang, B. Erdmann, and M. Seebass, Impact of nonlinear heat transfer on temperature control in regional hyperthermia, *IEEE Transactions on Biomedical Engineering* 46(9) (1999) 1129–1138.

38. C. A. T. Van den Berg, J. B. Van de Kamer, A. A. C. De Leeuw, C. R. L. P. N. Jeukens, B. W. Raaymakers, M. Van Vulpen, and J. J. W. Lagendijk, Towards patient specific thermal modelling of the prostate, *Physics in Medicine and Biology* 51(4) (2006) 809–825.

39. M. Van Vulpen, A. A. C. De Leeuw, J. B. Van de Kamer, H. Kroeze, T. A. Boon, C. C. Wárlám-Rodenhuis, J. J. W. Lagendijk, and J. J. Battermann, Comparison of intra-luminal versus intra-tumoral temperature measurements in patients with locally advanced prostate cancer treated with the Coaxial TEM system: report of a feasibility study, *International Journal of Hyperthermia* 19(5) (2003) 481–497.

40. G. M. J. Van Leeuwen, J. J. W. Lagendijk, B. J. A. M. Van Leersum, A. P. M. Zwamborn, S. N. Hornsleth, and A. N. T. J. Kotte, Calculation of change in brain temperatures due to exposure to a mobile phone, *Physics in Medicine and Biology* 44 (1999) 2367–2379.

41. IEEE Standards Board, *IEEE Standard for Safety Levels with Respect to Human Exposure to Radio Frequency Electromagnetic Fields, 3 kHz to 300 GHz*, Tech. Rep. IEEE C95.1-1991, New York: Institute of Electrical and Electronics Engineers, 1992.

42. National Radiological Protection Board (NRPB), *Board Statement on Restrictions on Human Exposure to Static and Time Varying Electromagnetic Fields and Radiation*, Documents of the NRPB, Vol. 4(5), London: NRPB, 1993.

43. CENELEC CLC/SC211B, *European Prestandard prENV 50166-2, Human Exposure to Electromagnetic Fields High-Frequency: 10 kHz–300GHz*, Brussels: European Committee for Electrotechnical Standardization, 1995.

44. V. M. M. Flyckt, B. W. Raaymakers, H. Kroeze, and J. J. W. Lagendijk, Calculation of SAR and temperature rise in a high-resolution vascularized model of the human eye and orbit when exposed to a dipole antenna at 900, 1500 and 1800 MHz, *Physics in Medicine and Biology* 52(10) (2007) 2691–2701.

45. S. Rossi, F. Garbagnati, R. Lencioni, H. P. Allgaier, A. Marchiano, F. Fornari, P. Quaretti, G. D. Tolla, C. Ambrosi, V. Mazzaferro, H. E. Blum, and C. Bartolozzi, Percutaneous radio-frequency thermal ablation of nonresectable hepatocellular carcinoma after occlusion of tumor blood supply, *Radiology* 217(1) (2000) 119–126.

46. R. Salomir, F. C. Vimeux, J. A. De Zwart, N. Grenier, and. T. W. Moonen, Hyperthermia by MR-guided focused ultrasound: accurate temperature control based on fast MRI and a physical model of local energy deposition and heat conduction, *Magnetic Resonance in Medicine* 43(3) (2000) 342–347.

5

Numerical Bioheat Transfer in Tumor Detection and Treatment

A. Zhang and L. X. Xu

CONTENTS

5.1 INTRODUCTION

Bioheat and mass transfer is widely used in tumor detection and treatment. For tumor diagnosis, abnormal skin surface temperature distributions have been investigated in relation to breast cancer and its malignancy [1–4]. Also, tumor cells and neovasculature are found to be more sensitive to mildly raised temperatures than are normal tissues [5–7], which provides the possibility for hyperthermia treatment of tumors. Hyperthermia has been used either alone or in adjuvant to radiation or chemotherapy for tumor treatment. Mildly heating the tumor cells up to about 40°C to 45°C for a certain time duration changes cell membrane fluidity and cytoskeletal protein structure and impedes transmembrane transport protein function [7–9]. It decreases the polymerization of RNA

149

and DNA during protein synthesis and affects cellular signaling pathways. These physical and biochemical changes can lead the cells to undergo either necrosis or apoptosis. Further, hyperthermia has been found to induce the immunological response and tumor microenvironment alterations that include changes of blood flow, oxygen supply, pH value, and the like [7,10–18].

Extremely low and high temperatures are also used to ablate undesired tissues [19–22]. These thermal treatments are expected to be minimally invasive, and with limited side effects. Freezing causes both cellular and vascular injury to tumors [23,24]. Direct cellular injury is induced by either the solute effect as cells dehydrate in response to freezing or to intracellular ice formation that tends to disrupt the intracellular organelles and the cell membrane [25]. Freezing also arrests blood flow and causes damage to blood vessels, particularly the microcirculation [26,27], which leads to tissue necrosis due to the absence of blood supply. A freezing-stimulated immunologic response may contribute to further destruction of tumor [28–30]. In radiofrequency (RF) heating, focused ultrasound, or laser ablation, much higher temperatures (normally above 60°C) are used to cause protein coagulation and cell necrosis [11,31,32].

During the treatment, all the physiological changes depend on the thermal history that the target tissue has experienced, and it is essential to monitor the tissue temperature. Although various noninvasive temperature measurement techniques have been attempted using the RF impedance, magnetic resonance imaging (MRI), and so on [33–37], there are still many issues remaining with regard to resolution and sensitivity of the real-time measurement. Accurate prediction of tissue temperatures through bioheat and mass transfer modeling and numerical computations is necessary and has been used to improve the efficacy of the thermally based diagnosis and treatment [38–43].

In this chapter, considering the biophysiological basis for the bioheat and mass transfer modeling, the tumor characteristics are first discussed. It is followed by the numerical study of thermal imaging for breast cancer detection, simulations of various thermal treatment protocols, and thermal damage assessment at the cellular and tissue levels, respectively.

5.2 TUMOR CHARACTERISTICS

Despite the fact that tumor growth is quite different in different organs and even in the same organ, some common characteristics have been found. Tumor growth is characterized by loss of proliferation control. Solid tumors usually originate from benign growth, gradually acquire autonomy, and grow progressively. At a certain stage, they metastasize into new locations [44]. Owing to the progressive growth of the tumor cells and lack of sufficient blood supply, the tumor microenvironment is normally hypoxic and acidic. Tumor cells are found to be more sensitive to mild heat than normal cells. This fact serves as the biological basis for hyperthermia treatment [7,45,46].

Uncontrolled tumor growth is supported by blood vessels, angiogenesis. Blood perfusion is heterogeneous in tumor and sometimes is found to be higher than that in normal tissue. The tumor microvasculature network usually has irregular branching patterns, lacks smooth muscle cells, and has few endothelial cells [47]. By the use of an animal tumor window chamber model [48,49], the angiogenesis development of murine mammary carcinoma 4T1 has been studied [50]. In the first few days, there is no obvious change observed in the tumor-implanted region, and the surrounding host as shown in Figure 5.1. On the seventh day, irregularly formed and tortuous tumor angiogenesis is observed in the tumor periphery. On the tenth day, tumor angiogenesis becomes more abundant and permeates the entire tumor. Histological analysis shows that in the tumor central region, the vessels possess a clear tubular structure embedded in thick connective tissues, and are surrounded by high-density tumor cells. In the peripheral region, they are incomplete, consisting of only a sparse endothelial cell layer. The endothelial cells and erythrocytes can be found in this region, indicating the abundance of newly formed tumor vessels.

The tumor microvessels are normally very leaky, resulting in a much higher interstitial pressure than the normal tissue [51,52]. The vascular wall permeability varies not only from one tumor to another, but also within the same tumor both spatially and temporally [53], dependent on tumor angiogenesis and growth [54–56]. The tumor microvessels are more sensitive to heat than are normal

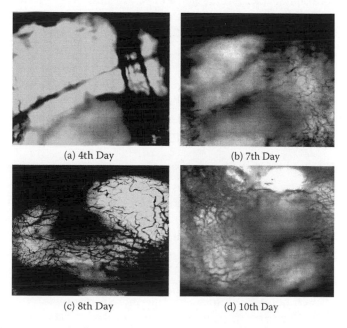

(a) 4th Day   (b) 7th Day

(c) 8th Day   (d) 10th Day

Figure 5.1   The progressive 4T1 tumor angiogenesis. The scale bar is 200 μm.

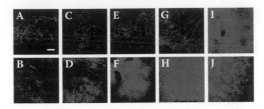

Figure 5.2   Vasculature fluorescent images in the tumor center (upper row) and periphery (lower row) after different treatments [59]. Bar: 200 µm. (A) and (B) after cooling at 1°C for 1 hour; (C) and (D) after cooling at –10°C for 1 hour; (E) and (F) after heating at 42°C for 1 hour; (G) and (H) after cooling at 1°C for a half-hour and subsequent heating at 42°C for a half-hour; (I) and (J) after cooling at –10°C for a half-hour and subsequent heating at 42°C for a half-hour.

vessels, and their permeability of tumor microvessels to nanoparticles can be greatly enhanced by hyperthermia [57,58]. With a water bath unit, the implanted tumor inside the window chamber of the nude mouse model has been exposed to different thermal treatment protocols [50,59,60]. Nanoliposomes are used as the tracer to evaluate the degree of vasculature damage through fluorescence imaging via confocal microscopy.

Shown in Figure 5.2, vessels in the peripheral region of the tumor are found to be more sensitive to either cooling or heating than are those in the center where well-developed vessels appear more resistive to thermal stress. The heating effect is much more significant, and the liposome extravasation is nearly twofold of that after a cooling treatment. After being cooled at –1°C for half an hour and subsequent heating at 42°C for another half-hour, the most severe vascular damage is found both in the center and the periphery. It is significantly different from that of any other treatments. The abrupt increase in fluorescence intensity as the heating starts after prefreezing for 30 minutes at –10°C implies that the vessel wall breaks, possibly due to large thermal stresses experienced during the rapid temperature change. The alternate cooling and heating treatment is much more effective in damaging the tumor vasculature compared to that of cooling or heating alone.

The excessive growth of cells in the tumor is normally accompanied by much higher metabolic rates. Researchers have found that the internal temperature of breast cancer is 2°C to 3°C higher than that of normal breast tissues, mainly due to higher metabolic rate and higher local blood perfusion of the tumor [4,61,62]. To understand their correlation, the metabolic activity of breast cancer cells has been studied with respect to temperature using *in vitro* cell lines (MCF-7 and HS 578 T) and their normal controls (MCF-10a and HST 587 Bst), respectively [62]. The cancer cell lines are found to exhibit greater NAD(P)H autofluorescence than their paired normal cell lines, showing more active metabolism. Relative heat generation in breast cancer cells versus temperature is compared with that reported by Backman [63]. Figure 5.3 shows that the trends are similar, and

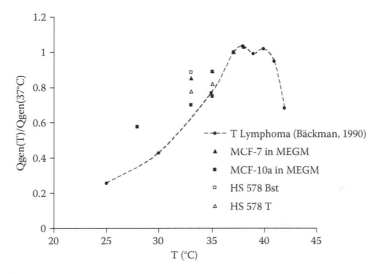

Figure 5.3   Comparison of metabolic energy production with temperature in MCF-7/MCF-10a, HS 578 Bst/HS 578 T breast cancer cell pairs [62] to that reported for T lymphoma in Reference [63].

that maximal heat production occurs in the temperature range from 37°C to 40°C. At temperatures higher than 40°C (i.e., 40°C to 45°C), the metabolism is inhibited, indicating a possible suppression effect on tumor growth through local hyperthermia.

5.3 THERMAL IMAGING OF TUMORS

Mapping the skin surface temperature and relating its distribution to the presence of a tumor inside are called tumor thermography [1,3,64]. Due to its superficial location compared to other solid tumors, thermoimaging of breast cancer has become a possible adjunct technique for diagnosis [2,3], and the technique has been studied by researchers for decades. Breast cancer cells are known to produce excessive nitric oxide, which can cause local vasodilatation and thus an increased blood supply to the tumor region [62,65]. The differences in energy generation between the normal and the cancerous tissues may be expressed as a higher local temperature by up to 2°C to 3°C on the skin surface above the tumor [4]. These higher temperatures can be easily detected by an infrared (IR) camera as an abnormal thermogram.

However, the results have been contradictory and often controversial. In particular, frequent occurrences of false-positive results have prevented thermography from becoming a standard early detection method. Chen et al. [66] have studied the feasibility and limitations of determining interior information from surface temperature measurements. They have performed a two-dimensional (2D) numerical simulation to examine the sensitivity of thermography to changes of the nonuniform perfusion component, which represents a lesion in a normal tissue, with the assumption that the convection heat transfer coefficient on the surface is constant. It is concluded that for a deep lesion, thermography does not provide sufficient sensitivity for detection. Osman and Afifi numerically simulated the three-dimensional (3D) temperature distribution in the breast using the finite element method (FEM) [40]. They correlated the tumor metabolic heat generation rate with the time required for the doubling of the tumor volume as measured by Gautherie [4]. The tumor is modeled as a point source with the heat generation equal to the total metabolic heat production in the tumor. Based on the simulation results, the authors claimed that the abnormal surface temperature variations may not have a direct relation to an underlying tumor. Ng and Sudharsan have reported numerical simulations of thermal processes in human breast using a 3D model based on FEM [67,68]. They presented simulations of a female breast under cold stress by incorporating the vasomotor behavior. Transient solutions are obtained for the rewarming process [68]. They found that blood perfusion, subcutaneous metabolic heat generation, and the surface heat transfer coefficient are the three most significant parameters affecting the skin temperature.

The skin vascular responses to enhanced surface cooling such as forced convection may enhance the difference between the signal and the noise, especially for a more deeply embedded cancer that otherwise might not be detectable. The effect of the forced convection on thermography has been studied using bioheat transfer analysis [61].

The breast is approximated as a hemisphere with a short cylindrical appendage that emulates the boundary region with the chest wall (Figure 5.4a). There have been various bioheat transfer equations among which the Pennes equation is the earliest and the most widely used in modeling [69]. In steady state, the equation takes the form:

$$\lambda_t \nabla^2 T_t + \omega(\rho c)_b (T_a - T_t) + q_{met} = 0 \tag{5.1}$$

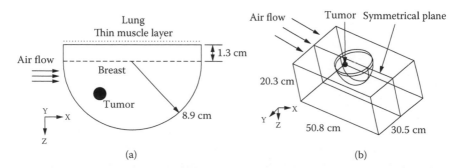

Figure 5.4   (a) Schematic of the symmetrical cross-sectional plane of the breast. (b) Geometry of the tunnel with relevant dimensions. The tumor lies on the symmetrical plane [61].

The tumor is treated as a heat source in the simulation and assumed to have the same thermal conductivity as that of the surrounding tissue. The female breast is primarily composed of fat and glandular tissues. The ratio of fat to glandular tissue varies for different individuals. Gautherie has measured the thermal conductivity of fat and glandular tissues of human breast and reported an average value of $k_t = 0.41$ W/m/K. The range of metabolic heat generation rates of normal tissue and tumor, along with the respective perfusion rates, has been estimated by Gautherie [4]. Accordingly, the metabolic heat generation rates of normal tissue and tumor in steady state are taken to be $q_{m,normal} = 450$ W/m³ and $q_{m,tumor} = 29,000$ W/m³, respectively, and the corresponding perfusion rates are 0.00018 ml/s/ml and 0.009 ml/s/ml. The product of density and specific heat capacity of both human blood and breast tissue is $(\rho c_p)_b = 3.6 * 10^6$ J/m³/K.

The influence of airflow is investigated by calculating the steady-state temperature distribution on the breast skin surface of a patient lying in the prone position with openings in the bed. An air stream is assumed to be flowing underneath the bed, realized by blowing ambient air into the tunnel with uniform velocity normal to the inlet, as shown in Figure 5.4b. Steady-state heat transfer is simulated in both the breast and the air domains. The bioheat transfer is coupled with the forced convection on the skin surface between the two domains. The heat flux at the fluid–surface interface is continuous. The wall of the tunnel is assumed to be insulated. The interface between the breast and the chest shown in Figure 5.4a is also assumed to be adiabatic, since the interior of the breast is not affected by the flow at the surface and is close to the core temperature.

The weak and strong airflows that may be present in any uncontrolled lab environment are represented with velocities of 1 m/s (Re 6500) and 3 m/s (Re 20,000) respectively. Thus, the airflow is strictly laminar. The Grashof number is approximately 100,000, and the ratio Gr/Re^2 is << 1; thus, free convection can be easily neglected. The velocity components in all other directions are zero. The outlet of the tunnel is considered as a distant boundary, that is, it has no effect on the velocities inside the model tunnel, the boundary satisfying the large Peclet number (Pe) and having no inflow condition. The wall of the tunnel and the skin of the breast are impermeable, and the airflow velocity on the surfaces is specified to be zero. Perspiration and subsequent evaporation on the skin were ignored.

The coupled Navier-Stokes equations describing the 3D fluid flow in the tunnel and the energy equations in the air and breast tissue are solved using the finite volume method by FLUENT (Lebanon, New Hampshire). Since the temperature difference between the breast surface and the free stream is small, the air properties are treated as constant. The airflow field is simulated for the given inlet and boundary conditions, and the heat equation is solved for both the breast and air domain. A user-defined function written in C is developed to incorporate the blood perfusion and metabolic heat generation terms in the energy equation.

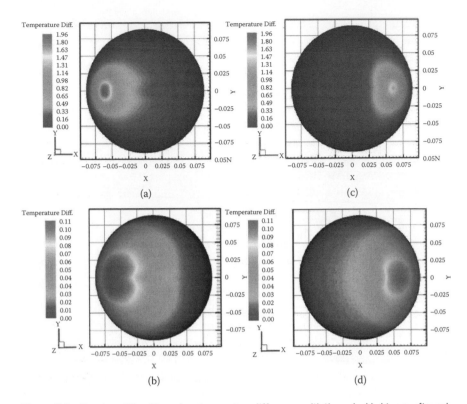

Figure 5.5 Top view of the skin surface temperature differences with the embedded tumor after subtracting the normal: (a) upstream and 2 cm deep tumor; (b) upstream and 5 cm deep tumor; (c) downstream and 2 cm deep tumor; and (d) downstream and 5 cm deep tumor [61].

Two tumor locations in relation to the airflow direction and two tumor depths are considered for a total of four cases as follows: case 1—a tumor on the side facing the oncoming airflow (upstream tumor) at a depth of 2 cm (depth from the skin to the center of the tumor); case 2—an upstream tumor at a depth of 5 cm; case 3—a tumor on the airflow wake side of the breast (downstream tumor) at a depth of 2 cm; and case 4—a downstream tumor at a depth of 5 cm. The temperature differences on the breast surface for the four tumor cases obtained by subtracting the temperature distribution on the normal breast surface are shown in Figure 5.5a,b,c,d.

The highest temperature difference of 1.72°C in Figure 5.5a clearly defines the tumor in case 1. For the much deeper tumor of case 2, the surface temperature difference reduces to 0.10°C, which can still be detected by the current state-of-the-art IR quantum well cameras with a thermal sensitivity of 0.02°C. The downstream tumors of cases 3 and 4 produce maximum temperature differences of 1.34°C and 0.09°C, respectively. The upstream tumors of cases 1 and 2 produce higher temperature rises than those produced by the downstream tumors of cases 3 and 4 because of the enhanced local convection in the upstream. Thus, further enhancement in the heat transfer coefficient can be used to achieve better differentiated thermograms.

The simulation results show that surface cooling plays a very important role in determining the temperature distributions exhibited in thermal images of breast cancer. The nonuniform convective heat transfer caused by a parallel air-flow, used in the past, may give rise to irregularities in the surface temperature distributions that are sufficient to obscure the tumor signatures. By subtracting a normal thermogram, which has similar variations caused by the airflow from that of the tumor, under the same boundary conditions, the obscuration can be removed and the thermogram resolution is significantly improved.

5.4 THERMALLY TARGETED DRUG DELIVERY

The sensitivity of the tumor vasculature to heat has been used in improving the delivery of antitumor drugs [55,57,70]. Local thermal treatment greatly enhances the tumor vascular permeability to nanoparticles while sparing the normal vas-culature [50] to achieve thermally targeted drug delivery. The antitumor efficacy is found to be greatly improved through experimental and numerical studies [71,72]. The thermally enhanced transport of liposomal drugs in tumors is a typical bio-heat and mass transfer problem. The drug distribution throughout the tumor *in vivo*, whether it is adequate for killing tumor cells, is critical. The optimization of the drug delivery process determines the final outcome of the chemotherapeu-tic treatment. Any survival of the tumor cells would result in tumor recurrence or tumor metastasis. Numerical simulations provide effective predictions of the outcome, while *in vivo* experimental observations are difficult.

Several models have been developed to describe the drug delivery process for different chemotherapeutic applications [73–75]. In consideration of the char-acteristics of breast cancer, Lankelma et al. have developed a model to study the drug transport and doxorubicin activity in islets of the breast cancer [76,77]. They assumed that the drug is transported through both transcellular (through the cellular network) and paracellular (through the intercellular interstitium) pathways. In Ward and King [75], the transport and tumor growth under drug administration have been modeled without consideration of the vascular net-work in the tumor and small molecules, drugs, and nutriments transport to the center from the edge of the tumor through diffusion. Magni et al. [78] have used a mathematical model to describe the cancer growth dynamics in response to anticancer agents in xenografts. The model consists of several ordinary differen-tial equations with three parameters describing the untreated growth and two parameters for the drug action [78]. El-Kareh and Secomb have presented models accounting for the cellular pharmacodynamics of drugs [72,79]. They have also numerically studied the intraperitoneal delivery of Cisplatin in tumors and inves-tigated the influence of hyperthermia on the drug penetration distance [80].

Unlike small molecule drug transport, the delivery process of liposomal
drugs is much more complex. The lipo-
some-encapsulated drugs selectively
extravasate into the tumor region [81].
Such an extravasation process has been
found substantially enhanced by hyperthermia [82,83]. The breakage of the liposome releases the antitumor drug that
diffuses into the tumor region. Without
taking the morphology into consideration, EI-Karech and Secomb [84]

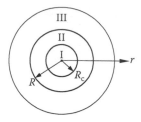

Figure 5.6   Schematic of the tumor. (I) Central region of tumor with the radius R_c; (II) periphery region of tumor ($R_c < r < R$); and (III) normal tissue.

have developed a homogeneous model to compare the different delivery methods
such as bolus injection, continuous infusion, and liposomal delivery of doxorubicin. The heterogeneous distribution of tumor vessels results in heterogeneous
extravasation of liposomal drugs, which could, in turn, significantly influence the
drug concentration gradient [85]. Thus, a homogeneous model is not adequate for
precisely predicting the nanoparticle drug transportation *in vivo*. Besides, the
apoptosis of the tumor cells after the exposure to the antitumor drugs may also
affect the drug penetration [86] and should be taken into consideration.

Accounting for the nonuniform vasculature and its heterogeneous thermal
sensitivity, the tumor is simulated by a sphere with radius R and divided into
two parts: the central region with only mature vessels ($r < R_c$) and the peripheral region with evenly distributed microvasculature ($R_c < r < R$), as shown in
Figure 5.6.

The liposome concentration in plasma decreases exponentially with time
owing to the elimination effect, and is found to follow the biexponential equation
as follows [84]:

$$C_{L,V} = \frac{M}{M_g}\left(A_1 e^{-k_1 t} + A_2 e^{-k_2 t}\right)$$

(5.2)

where $C_{L,V}$ is the liposome concentration in plasma (mg per ml plasma), M is the
dose of drug (mg/m² animal surface area), and M_g is the dose used as the plasma
pharmacokinetic parameters for liposome: A_1, A_2, k_1, and k_2 are fitted [87].

The liposomal drug in the vasculature is transported across the leaky vessel
wall of the tumor vasculature and into the tumor interstitial. Both diffusion and
convection are involved for the transvascular mass exchange [53]. Thus, the total
transport rate of drug J_s (g/s) across a vessel wall can be described by the S-K-K
equation [53].

$$J_s = Q(1-\sigma_f)\bar{C}_s + PA_v(C_b - C_i)$$

(5.3)

The rate of solute transport flux via convection (the first term in Equation 5.3) is proportional to the rate of fluid volumetric flow Q and the average drug concentration in the vessel wall $\overline{C_s}$, where σ_f is the filtration reflection coefficient. The transvascular fluid volume flow Q is proportional to the microvascular hydraulic conductivity L_p and the osmotic pressure across the vessel wall as follows [88]:

$$Q = L_p A_v (p_v - p_i - \sigma(\pi_v - \pi_i)) \tag{5.4}$$

where p_v and p_i are vascular and interstitial hydrodynamic pressures, respectively; π_v and π_i are the colloid-osmotic pressure in the plasma and the interstitium, respectively; and σ is the osmotic reflection coefficient.

The second term in Equation (5.3) describes the rate of diffusive transport, where P is the microvessel wall permeability, A is the surface area of the vessel wall, C_b is the solute concentration in blood, and C_i is the interstitium near the vessel. As the transport across the vessel wall is always a mix of diffusion and convection, an apparent permeability P_{app} is usually used to combine the diffusive and convective fluxes. Thus, the leaked liposomal drugs from the microvessels can be expressed as

$$J_s = P_{L,app} A_v \cdot (C_{L,V} - C_{L,E}) \tag{5.5}$$

where $P_{L,app}$ is the apparent permeability of the tumor vasculature and reported to be 2.0×10^{-8} cm/s at room temperature [89]. A_v is the surface area of the microvessel wall. As the blood vessels in the central region are almost impermeable to the nanoliposomes [50], the source term describing the vascular leakage is neglected in the central region $r < R_c$. When the tumor is heated uniformly at 42°C for 1 hour right after the injection of the liposomal drug, the apparent permeability of the tumor vasculature is greatly enhanced.

Further, if the tumor is pretreated with freezing at –10°C for 30 minutes before being heated at 42°C for another 30 minutes, the vasculature is found to be broken in both the center and peripheral regions, and the liposomes flood throughout the tumor, as shown in Figure 5.2 [59]. Subsequently, the coagulation blocks the vessels, correspondingly trapping the liposomes in the tumor.

The leaked liposomes from the tumor vasculature are transported to the tumor interstitial. Both diffusion (due to concentration gradients) and convection (due to the motion of interstitial fluid) coexist in the interstitial drug transport. The diffusion flux can be determined from Fick's law directly if the diffusivity is known. For macromolecules (molecular weight: $W_t > 1000$), such as nanoliposomes, the sizes of the macromolecules are comparable to the interstitial space and the extracellular matrix (ECM) is a hindrance to their transport. As the size of the particles is increased and becomes comparable to the channels they transport through, significant friction effects should be considered and can be quantified by the viscous tortuosity. Thus, both of the geometric and viscous tortuosities should be considered to obtain an effective diffusivity [90].

By assuming the tissue as a porous media, convection inside the tumor tissue can be described using Darcy's law. The convection flux, J_C, of a solute in the interstitium is then given by

$$J_c = ACR_F u = -ACR_F \frac{k_p}{\eta} \frac{\partial p}{\partial x} \qquad (5.6)$$

where A is the transport area, R_F is the retardation factor (solute convective velocity/solvent convective velocity), u is the convective flow velocity resulting from the pressure gradient in the medium, k_p is the fluid permeability (Darcy's constant), and η is the solvent viscosity. Similarly, as it is hard to separate the diffusion from the convection during liposome transport in the interstitium, an apparent diffusivity D_{app} is usually used. It is defined as $D_{app} = -\frac{J_i}{\partial C/\partial x A}$, where J_i is the total liposome transport flow rate across area A in the interstitium. $\partial C/\partial x$ is the concentration gradient.

Liposomes are assumed to release the drugs following first-order kinetics, with the decay time constant τ_r [84]. Thus, the concentration of the drug encapsulated in liposomes in the tumor peripheral region can be determined by

$$\frac{\partial C_{L,E}}{\partial t} = D_{L,app} \cdot \nabla^2 C_{L,E} + P_{L,app} \cdot A_t (C_{L,V} - C_{L,E}) - \frac{C_{L,E}}{\tau_r} \quad R_c \leq r < R \qquad (5.7)$$

where $C_{L,E}$ is the drug concentration encapsulated in liposomes in the extracellular space (mg/ml tumor volume); $D_{L,app}$ is the apparent diffusivity of liposomal drugs in tumor tissue (cm²/s).

For macromolecules like liposomes, the diffusion is comparable to the convection, and both should be considered in modeling. But for small molecules like the antitumor drug doxorubicin, convection is relatively small [91]. Furthermore, high interstitial pressure in the tumor interior owing to the high vascular hydraulic conductivity results in low convective drug transport. Therefore, convection can be neglected for small-molecule-drug transport in tumor.

If the tumor tissue is considered to be a porous medium, the effective diffusivity of the antitumor drug freed from the liposome D_D can be determined from the drug diffusivity in the interstitial fluid $D_{D,0}$ and the interstitial volume fraction φ_e [92]:

$$D_D = \frac{2\varphi_e}{3 - \varphi_e} D_{D,0} \qquad (5.8)$$

With more and more tumor cells undergoing necrosis or apoptosis during or after the treatment, the interstitial volume fraction φ_e increases [86] and is assumed to change linearly with the tumor cell survival rate S,

$$\varphi_e = 1 - (1 - \varphi_{e0}) \times S \tag{5.9}$$

where φ_{e0} is the initial volume fraction of interstitial space before the treatment. Thus, the transport of free doxorubicin in the interstitial region can be described by

$$\frac{\partial C_{D,E}}{\partial t} = D_D \cdot \nabla^2 C_{D,E} + \frac{C_{L,E}}{\tau_r} - \frac{\partial C_{D,I}}{\partial t}(1 - \varphi_e) \tag{5.10}$$

where $C_{D,E}$ is free Dox concentration in the extracellular space. The third term in Equation (5.10) is the tumor cellular uptake rate of free doxorubicin, where $C_{D,I}$ is the intracellular-bound drug concentration (mg per ml volume of tumor cells), φ_e is the volume fraction of interstitial space, and its value in the tumor center is different from that in the peripheral region.

For the drug uptake process, the transmembrane propagation takes place both passively and actively [93], and many models have been developed [72,79,84,94]. Doxorubicin is believed to transport passively across the cell membrane, and the cellular uptake rate of free drug can be expressed using the passive carrier-mediated transportation equation [84]:

$$\frac{\partial C_{D,I}}{\partial t} = v_{max}\left(\frac{C_{D,E}}{C_{D,E} + k_e} - \frac{C_{D,I}}{C_{D,I} + k_i}\right) \tag{5.11}$$

where v_{max} is the maximal cellular uptake rate, equal to 2.8 µg/ml/min for doxorubicin. The quantities k_i and k_e are constant parameters whose values are 13.7 µg/ml and 0.219 µg/ml, respectively [84].

According to the killing mechanism of the chemodrug, the cell survival rate depends on both drug concentration and the exposure time. It is usually quantified by the area under the concentration–time curve (AUC, the integral of the drug concentration over time), and there exists an exponential relationship. Thus, both the drug concentration (either liposomal or free) and the survival rate of the tumor cells after the treatment can be numerically obtained [95]. The numerical results (shown in Table 5.1) are in good agreement with the experimental findings. One day after the treatment, the drug effect on the peripheral region starts to take effect and the average cell survival rate in this region decreases to 0.74. While very little doxorubicin has reached the central region, the average survival rate in this region is about 0.96. On the third day, the survival rate in the tumor

Table 5.1   Average Tumor Cell Survival Rates at Different Time Intervals after the Thermally Targeted Drug Delivery (42°C for 1 hour)

	1st Day	3rd Day	6th Day
Tumor center	0.96	0.60	0.1
Tumor periphery	0.74	0.21	0.04

Source: Sun et al. [178].

periphery further decreases to an average value of 0.21. Also, in the center, an obvious decrease of the survival rate is observed and the average value is 0.60. Six days later, the survival rates in the peripheral and central tumor become 0.04 and 0.1, respectively. These predictions are confirmed by experimental results reported in Reference [96].

The thermally targeted liposomal drug delivery to a larger tumor of 1 cm in radius is also simulated, and results are shown in Figure 5.7. Both the liposomes and free doxorubicin mainly accumulate in the peripheral region ($0.6\ R < r < R$). The survival rate of the tumor cells in the periphery decreases evidently, and it is almost zero in 6 days after the treatment. The locally imposed hyperthermia greatly enhances the accumulation of both the liposomal and free drug in the peripheral region, inducing more serious tissue damage in comparison to that without heating. However, in the center region, owing to the impermeability of the mature vessels and high interstitial pressure impeding drug diffusion inside the tumor, the drug effect is very limited even with local hyperthermia.

An alternate freezing and heating treatment of tumor has been proposed to completely damage the vasculature in both the tumor peripheral and central regions. Its effect on drug transport has also been evaluated through modeling. If the liposomal drug is injected into blood circulation when heating is turned on right after the prefreezing, then excessive extravasation of liposomes occurs due to the breakage of the vascular bed. Numerical results shown in Figure 5.8 indicate that on the sixth day after the treatment, the tumor of 1 cm in radius is expected to be completely destroyed. This new protocol has the merit of complete damage of tumor vasculature and thus more effective drug delivery via flooding of blood in the tumor. Such a protocol of combining both physical and chemical treatments is expected to have significant clinical applications for more effective drug delivery.

The numerical study has clearly shown that the hyperthermia-aided nanoliposomal drug delivery system can effectively damage small tumors. But for larger tumors (i.e., > 1 cm in radius), hyperthermia can only enhance the accumulation of more antitumor drugs in the peripheral region to increase the corresponding damage. The proposed alternate freezing and heating protocol aids the nanoliposomal drug delivery throughout the tumor and is expected to greatly enhance the treatment effect.

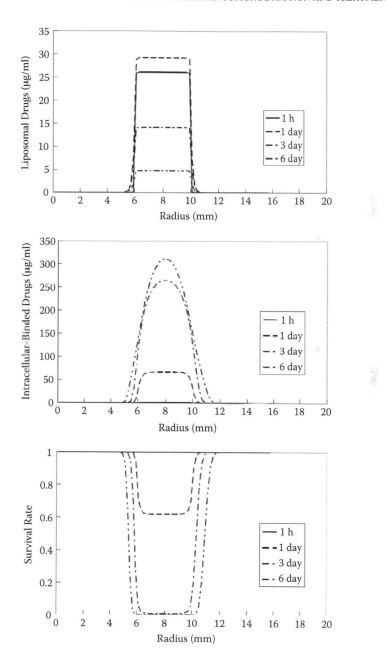

Figure 5.7 Distribution of the liposomal and intracellular-bound doxorubicin, and the cell survival rates at different times after the treatment (1 cm tumor, with local heating at 42°C for 1 hour) [95].

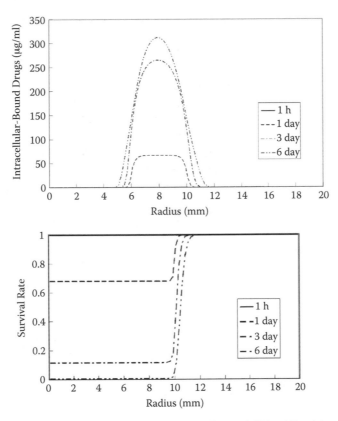

Figure 5.8   Distribution of doxorubicin and the cell survival rates at different time intervals after the treatment (1 cm tumor treated with alternated cooling at −10°C for 30 minutes, followed by heating at 42°C for 30 minutes) [95].

5.5 TUMOR TREATMENT PLANNING AND ASSESSMENT

Numerical simulations are also widely used in planning and the assessment of a thermal treatment for tumor. In thermal therapy, a tumor is either heated or cooled to an extremely low temperature [11,22,97]. Under heating, the tumor temperature can be mildly raised to higher than body temperature (i.e., hyperthermia, or 42°C to 45°C) or much higher to reach direct tissue necrosis (thermal ablation). The low temperature can be as low as −196°C depending on the freezing agent used. Compared with other modalities, thermal therapy has many potential advantages. It is minimally invasive and requires little hospitalization. Further, it is not dose limited and can be carried out repeatedly for recurrent cases. For a successful thermal treatment, the heating or freezing should cover the entire tumor while avoiding overheating or overfreezing of the surrounding normal tissues. Moreover, the thermal dose and treatment frequency should enable complete killing of tumor cells. Thus, precise planning and assessment are essential for the treatment.

A good therapy plan includes treatment location and orientation, treatment protocol setting, thermal dosage, treatment frequency, and so on [33,34,75,98]. Medical images (e.g., via ultrasound [US], computed tomography [CT], or MRI) are commonly used to determine the size, the shape, and the location of a deeply situated tumor in relation to blood vessels and critical structures [99–105]. The imaging technologies are also used for temperature measurement in monitoring the treatment [106–110]. But, the sensitivity and resolution are still remaining issues. For more precise temperature information, invasive measurements by thermocouples and optical thermal sensors have been used. In cryosurgery, imaging of the iceball formation [109,111,112] is another alternative for monitoring. As ultrasound signals are almost completely reflected at the freezing interface, acoustic shadows are present and distal to the ice front, which hinders controlled freezing and risks complications due to the freezing of vital structures in the surrounding tissue. The introduction of advanced imaging guidance such as X-ray CT can provide better visualization of the frozen tissues [113,114]. However, freezing alone does not necessarily execute adequate ablation. For an example, the critical temperature for the ablation of prostate tissue is well below 0°C [114–117]. As the degree of damage to a tumor is determined by the thermal history it experienced, an accurate temperature distribution inside the tissue is therefore necessary for the successful planning of a treatment. Both *in vivo* animal experimental studies, and tissue or tissue equivalents like phantom gel and numerical simulations, have been used for thermal treatment design and effect assessment [118–120], while for *in vivo* treatment of human beings, thermal modeling of the treatment process is the only noninvasive way for temperature prediction and monitoring.

Numerous models have been developed to study the thermal histories of treatments via RF heating, laser ablation, microwave, ultrasound, high-focused ultrasound, and cryosurgery in various tumors [110,121–126]. The Pennes bioheat equation is the most frequently used [127] in modeling given its simplicity and good approximation. In high-temperature treatment, the incident energy such as microwave, radiofrequency, laser, or ultrasound interacts with the biological molecules and deposits thermal energy in tissue. The volumetric heating is treated as the energy source in the bioheat equation. Inhomogeneous heating, irregular tumor geometry and vascular morphology, thermal property, and variations are important factors influencing the temperature distribution. With such complexity, only numerical methods can be used to accurately solve the bioheat transfer problem.

Freezing of biological tissues are quite different from nonliving materials that have a fixed freezing point. Phase change for biological tissues occurs over a temperature range, with an upper limit ranging from 0°C to –1°C and a lower limit from –5°C to –10°C, due to the complex composition and inhomogeneity of tissues [128,129]. Thermal effects of blood perfusion and metabolism also affect the freezing process in tissues. A few analytical solutions to the one-dimensional freezing problems have been applied to cryosurgery [130–135]. These solutions cannot be used to solve multiple freeze-thaw cycles where multiple freezing fronts coexist. Numerical modeling, on the other hand, provides a powerful tool to handle complex geometries and boundary conditions.

For the treatment assessment, the approach can be either phenomenological or mechanistic. Phenomenologically, a critical temperature is normally used for evaluation. For example, in high-temperature ablation, 50°C or 70°C is used for different tumors [136,137]. The tissue damage process is assumed to follow the Arrhenius function from the thermodynamic point of view [138]. Thus, the rate of injury degree can be described using the Henriques model or modified ones [139]. At the cellular level, the concept of thermal dose is widely [140] accepted. And in cryosurgery, critical temperatures, such as −40°C or −60°C, are commonly used for prostate and breast cancer, respectively [116,141]. Based on the mechanism of cryoinjury [142], more and more researchers prefer using the probability of the lethal intracellular ice formation (IIF), P_{IIF}, to evaluate the damaged region after cryoablation [143]. At the tissue level, many researchers have analyzed the mechanical damage of the freezing resulting from the thermal stresses during the fast-freezing process [144–146].

Therefore, for the thermal treatment of tumor with certain protocols, numerically simulated temperature distributions can be performed to predict the degree and volume of the damaged region. On the other hand, for a target tissue of known volume to be treated, an inverse heat transfer problem can be solved to aid the design for the heating protocol. An optimization algorithm has been developed in References [147,148]. The squared difference between the calculated damage region and the desired region is the objective function to be minimized using the algorithm. Results are found sensitive to the bioheat transfer and the damage evaluation models.

5.5.1 High-Temperature Treatment

Numerical models are developed to describe the high-temperature treatment of all kinds of tumors mainly based on the traditional mathematical bioheat transfer analysis techniques [122,149–151]. Properties for tumor tissues are approximated to that of normal tissue and are assumed to be constant [152,153]. The specific rate of absorption (SAR) is usually used to quantify the transformation efficiency of the irradiated energy (either light or electromagnetic waves) to thermal energy. It is determined by the tissue electrical or light properties and is inhomogeneous. The distribution of SAR provides important information for improving the heating element. It can be obtained by either simulation of the transportation of electromagnetic waves [154], photons [155], sound waves [156], or experimental measurements [157]. And high-temperature gas-based ablation therapy, both flow of the vapor (pressure and density gradient) and heat transfer (temperature gradient) inside the tumor, is also studied [158].

In microwave or radiofrequency heating, numerical methods have been used to solve the Maxwell equation of the electromagnetic (EM) field to obtain the SAR in which detailed information on the tissue dielectrical property distribution and their variations with temperature is relatively difficult to obtain. In most investigations, SAR has been measured at various locations by determining the initial rate of temperature rise immediately following a step increase in the applied power [154,159–165]. It can be approximated by taking the temporal gradient of temperature since heat conduction is negligible at the instant that heating power is on. A curve fit is used to obtain an expression for the SAR distribution. Using the SAR expression as the source in the Pennes bioheat transfer equation, the temperature field during the treatment can be obtained numerically through either finite element or finite difference numerical methods [166–169].

With the temperature field information available, the minimum power, Q_{min}, to heat a certain volume of tissue above a given temperature for a period of time can be calculated. For example, the temperature field within the uniformly perfused prostate has been approximately expressed using an empirical equation according to analysis of the numerical results [152]:

$$\Delta T(r,z) - Q_{dep} K e^{-(r-3)/r_d} e^{-z^2/z_d^2}, \quad K = 0.7175\omega^{-0.7237}$$

$$r_d = 9.4637\omega^{-0.2764}, \quad z_d = 21.2766\omega^{-0.1746}$$

(5.12)

where r_d and Z_d are the temperature decay lengths in the radial and axial directions, respectively. The decay length is defined as the distance over which the temperature elevation decreases by a factor of e. Q_{dep} (W) is the RF power level and K (°C/W) is a scale parameter depending on the blood perfusion rate ω. From this equation, the minimum power required (Q_{min}) to heat a 10% volume of the prostatic tissue above 44°C is found to range from 5.5 W to 36.4 W depending on the local blood perfusion rate ($\omega = 0.2$ ml/g/min to 1.5 ml/g/min). The configuration of the applicators is believed to be the main factor that affects the radial decay of the SAR distribution [170]. Changing the distance between the electrodes in a RF hyperthermia system results in different heating patterns in tissue [152].

Blood perfusion is a key factor in simulating the bioheat transfer process. It is closely related to temperature and is also tumor dependent [171]. Mild heating increases blood flow, while higher tissue temperatures cause a quick increase of blood flow followed by vascular damage and blood flow stasis [171–174]. Simulated results show that the minimum power required to achieve a steady state in the prostatic tissue above 44°C is in the range from 5.5 W ($\omega = 0.2$ ml/g/min) to 36.4 W ($\omega = 1.5$ ml/g/min). However, if the perfusion response to the tissue temperature is considered, the required power should be increased sevenfold [152]. Obviously, underestimation of blood perfusion during the treatment would reduce the expected therapeutic efficacy. Many models developed have taken the tumor blood perfusion and its dependence on temperature into consideration [175–178].

For accurate thermal planning, the tissue anatomical structures and vasculature also need to be accounted for [179]. Their influences on temperature distributions during the thermal treatment have been studied by several investigators [180,181]. Unlike hyperthermia, in thermal ablation, much higher temperatures result in immediate tissue necrosis and water evaporation, which correspondently influences the thermal, electrical, and light properties of the tumor tissue. This is quantified by the models developed in References [176,182]. More importantly, flow of the evaporated water vapor influences heat transfer inside the tumor and, correspondently, prediction of the tissue necrosis degree [183].

With the transient temperature distributions under certain given heating protocols, planning can be adjusted according to the correlated tissue damage. Heating to the tissue are usually nonuniform; thus, T_{90}, T_{50}, and T_{10}, the temperatures at which 90%, 50%, and 10% of the tissue volume exceeds, respectively, are frequently expressed clinically for the heating power quantification [182]. As the damage to tissue is not only subjected to the highest temperature it reaches but also the lasting time at the temperature, the concept of thermal dosage is proposed [140]. The thermal dosage is defined as the equivalent heating time at 43°C to reach the same degree of damage and is denoted by CEM43°C (i.e., the cumulative number of equivalent minutes at 43°C) [140,184,185]; while in thermal ablation, the burn injury models using the Arrhenius equation are usually applied for the evaluation of tissue damage [138].

5.5.2 Cryosurgery

Local freezing of tumor tissue is usually imposed by cold metal probes with liquid nitrogen or liquid argon flowing inside. Existing numerical models to characterize the freezing process may be categorized in two approaches. The first approach is based on the front-tracking method, in which latent heat is released at a specific single temperature and the phase change interface is tracked explicitly by adjusting meshes dynamically [186]. Although this method works well for sharp front cases, it is not accurate for describing the freezing process in biological systems where a phase transition region exists, and its application in three-dimensional cases would be cumbersome [187].

The second approach is based on the enthalpy method, where latent heat content is a function of temperature and the energy equation is solved either in the enthalpy form [187,188] or with an effective specific heat to incorporate latent heat [189–191]. The latter approach is currently accepted as the most powerful tool to deal with multidimensional phase change problems. Devireddy et al. [39] incorporated microscale biophysical phenomena into the enthalpy formulation, which can be used to simultaneously predict macroscale heat transfer and microscale biophysical response within tissue. However, this method is computationally expensive even in a simple case, and the results suggest that the macroscale enthalpy formulation alone is sufficiently accurate. In order to relate the enthalpy h_t to tissue temperature, an enthalpy–porosity relationship is used

[192,193]. In biological tissue, the enthalpy is composed of a sensible part, $h_{t,s}$, and a latent heat part, ΔH,

$$h_t = h_{f,s} + \Delta H \tag{5.13}$$

where the sensible enthalpy is expressed by

$$h_{f,s} = h_{f,s}(T_{ref}) + \int_{T_{ref}}^{T} C_p(T)dT \tag{5.14}$$

and the latent heat content is

$$\Delta H = \beta L \tag{5.15}$$

where β is the liquid fraction, which has the value of zero in the solid region and one in the liquid region, and is defined as

$$\beta = \frac{T - T_s}{T_l - T_s} \qquad T_s < T < T_l \tag{5.16}$$

where T_s and T_l are the two temperature bounds of the phase transition region. The solution for temperature is obtained through iterations between the energy balance equation and the liquid fraction equation (Equation 5.16). It is noted that the liquid fraction in biological tissues is nonlinear with respect to temperature. For practical purposes, a linear relation can be used as a first-order approximation given the rather thin phase transition region (1~2 mm) observed experimentally [125].

Other than the phase change of tissue, the geometry and boundary conditions are also important in the simulation of a real cryosurgical situation. Previous studies have made progress toward this goal. Budman et al. [194] were the first to calibrate commercially available cryoprobes in which a liquid nitrogen cryoprobe and a CO_2 Joule-Thomson probe were studied numerically and experimentally. Rewcastle and coworkers [129,195] have successfully calibrated the advanced argon Joule-Thomson cryoprobes in single and multiple configurations, which could be applied to generate three-dimensional temperature distributions around multiple cryoprobes. Also, Jankun et al. [196] have performed a thermal simulation for three-probe freezing in a prostate geometry reconstructed from ultrasound imaging data of a prostate phantom. With the fast development of imaging techniques, a more realistic computational geometry can be constructed based on images of a real prostate clinically, and by calibrating of the temperature distribution on the surfaces of the commercially available cryoprobes and urethral warmer used, more accurate boundary conditions could be obtained [125]. The geometric solid model of a real prostate is generated from sequential MRI images, as shown in Figure 5.9.

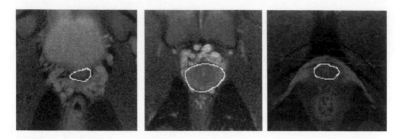

Figure 5.9   Reconstruction of the prostate with a urethra from MRI images. The prostate contours on the MRI slices are drawn manually on the Philips MxView workstation. (Philips Medical Systems, Andover, Massachusetts [193].)

For large tumors, multiple cryoprobes are needed, and the alignment of the probes is one part of the treatment planning. For the prostate, certain guidelines are needed [197], and six parallel cryoprobes have been aligned, as shown in Figure 5.10. Probes 1 and 2 are operated in 100% cryogen flow mode for 2 minutes before probes 3 and 4 are turned on. Two minutes after probes 3 and 4 are activated, probes 5 and 6 are turned on and operated in the 100% mode as well.

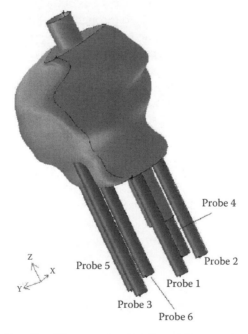

Figure 5.10   The configuration of the cryoprobes. Locations of the cryoprobes can be adjusted in the graphic user interface (GUI) of GAMBIT to fulfill the clinical requirements; the short cylinder in the center is the urethral warmer [193].

Cryoprobes 1 and 2 are deactivated after operating for 8 minutes, and the other probes continue to run for another 2 minutes. The whole procedure takes 10 minutes.

Using the Pennes equation and the functioning of the cryoprobes as the convective boundary, the temperature profiles can be obtained inside the prostate. It has been well established that the final temperature is a dominant factor contributing to freeze injury during prostate cryosurgery [114–117]. But there is some controversy as to the specific value of the critical temperature that ensures complete tissue destruction. Tatsutani et al. [117] claimed that this temperature is dependent on the cooling rate, and around −40°C for cooling rates lower than 5°C/min and around −19°C for cooling rates higher than 25°C/min. Their conclusions are based on cellular experiments on human prostatic adenocarcinoma cells. Larson et al. [114] has concluded that the critical temperature should be around −41.4°C for a double freeze–thaw cycle and −61.7°C for a single freeze–thaw cycle based on *in vivo* investigations of the human prostate.

In addition to the end temperature, the hold time of the critical isotherms has been shown to greatly influence the cellular survival rate of the AT-1 rat prostate tumor [116]. If this finding also holds true for *in vivo* human prostate tumor, then a detailed knowledge of the whole prostate volume or a particular subvolume of prostate versus the hold time would have greater practical significance. Figure 5.11 presents a hold-time tissue volume histogram (HTVH) at the end of the proposed surgery for the whole prostate. It shows that subvolumes of the

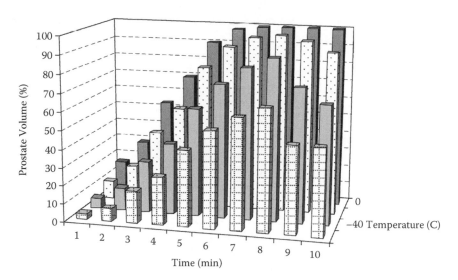

Figure 5.11   Time–temperature volume histogram of the treated prostate. The evolution of the prostate volume enclosed by different critical isotherms is shown [193].

prostate experience considerably different thermal histories and thus may have a distinct therapeutic outcome that needs to be taken into account in treatment planning.

The cooling rate is another factor that determines freezing injury to a variety of living cells via the mechanisms of intracellular ice formation and severe cell dehydration [25]. Since the intracellular ice is much more deleterious, a higher cooling rate (25°C/min) could induce more cellular destruction than a slower freezing rate (5°C/min) in the human prostatic cancer cells [117]. However, it is quite interesting that this dependence of cryoinjury on the cooling rate has not been found in the experiments using AT-1 rat prostate tumor cells [115,116]. It should be noted that only suspended cells have been used in these studies. Freezing of *in vivo* tissues also induces vascular damage, which enhances the cellular destruction in turn. But, cooling rate effects on vascular damage *in vivo* have not been fully studied.

Thawing is considered to be the other important phase of cryosurgery that causes tissue injury [24]. During the thawing process, small ice crystals merge together to form large crystals, which readily disrupt cellular membranes. This phenomenon may enhance tissue destruction induced by the preceding freeze. It is recognized that slow thawing is more destructive than rapid thawing [24]. For this reason, freeze-thaw cycles are usually used in cryosurgery. Though a simulation to include the thawing process is needed, coupling the crystallization model with the present heat transfer model would make the computation extremely expensive and possibly unfeasible. Thus, a tissue injury analysis based on isotherm protocols alone should be treated as a conservative prediction.

A presurgical treatment planning simulation is aimed at modeling the intended cryosurgical process. The actual thermal behavior during cryosurgery may deviate from the simulation results. In a simulation, the prostate and surrounding tissues can be idealized as continuous media with uniform thermophysical properties. Examination of the anatomy suggests that the prostate is in close contact with the surrounding organs, such as the bladder, the obturatorius internus, and the urogenital diaphragm. Also saline is usually injected into the gap between the rectum and prostate during the operation. With a high percentage of water content in their composition, these tissues are expected to have thermal properties close to those of water [198,199]. Thus, it is a reasonable approximation to treat them as a continuous homogeneous medium. During cryosurgery, the temperature evolution also depends on how exactly the cryoprobes are positioned and how closely they are operated compared to the planned positioning and operation. Any difference between the planned cryosurgery and the actual operation would affect the actual temperature distribution. And a postsurgery simulation can then be done to incorporate all these differences to derive the actual treatment delivered.

In future research and development, there are still many issues to be addressed. First, calibration of the cryoprobes operated in partial freezing mode and thawing mode may be useful for some clinical protocols. Second, *in vivo* experimental validation of the simulation results is very important and may be aided by advanced imaging techniques. Third, optimization of the cryosurgical planning may be performed by incorporating an optimization algorithm into the numerical solver. Although some preliminary optimization studies have been performed [200], a computationally efficient numerical solver needs to be found to practically apply optimization for prostate cryosurgery [201]. Finally, integration of the geometric modeling, meshing, numerical solver, visualization, and even optimization in a single software package with a user-friendly interface is required for routine surgical planning of cryosurgery.

5.5.3 Alternate Cold and Heat Treatment

The idea of combined freezing and heating was first proposed by Gage et al. [202]. Its effect on increasing the destruction volume is observed experimentally. Hoffman et al. [203] have performed a histological study of normal tissue treated subsequently by freezing and heating, but the result is inconclusive. A hybrid device has been proposed by Hines-Peralta et al. [204] to enhance the efficiency of a bipolar RF system by placing a freezing unit between the two RF poles. RF heating and cooling based on the Joule-Thomson effect of argon gas are used simultaneously. Hines-Peralta et al.'s [204] preliminary results show that with the addition of freezing in the middle of the RF probe, the treated region is enlarged and the shape is better controlled owing to the change of the electrical impedance caused by the ice formation in tissue. Liu et al. [205] also have designed a cryoprobe system with a vapor-heating feature. They claimed that cooling immediately followed by a rapid heating of the target tissues would improve the treatment effect due to thermal stress.

Further investigation of the biological effect of the alternate cooling and heating treatment on tumors via animal studies has found that this treatment is much more effective than single hyperthermia or cryosurgery in destroying a tumor microvascular network [59], as shown in Figure 5.2. Histopathological analyses further confirm the effect, as rare tumor vessel recurrence and large necrotic tumor tissue areas are found on the seventh day after the treatment [60]. This is further proven by the studies on the cellular effects, as shown in Table 5.2.

Table 5.2 The Death Rate of Human Umbilical Endothelial Cells (HUVEC) and Breast Cancer Cells (MDA-231) under Different Thermal Treatments

Cell Death Rate	Heating for 60 Minutes at 50°C	Cooling for 60 Minutes at −13°C	Cooling for 30 Minutes at −13°C, Then Heating for 30 Minutes at 50°C
HUVECs	17.82%	50.85%	59.4%
MDA-231	31.01%	31.6%	44.51%

Accordingly, a new thermal system has been developed to provide rapid cooling and heating alterations in tissue for more effective cancer treatment [206]. The system facilitates liquid nitrogen (LN$_2$) cryosurgery and RF heating, either alternately or simultaneously. In the cryoheat probe system, the unstable flow and phase change of the liquid nitrogen make the freezing process hard to control [207]. As a consequence, the temperature field is difficult to predict during freezing. Simulation of the flow inside the probe can help investigate the heat transfer performance under different conditions and enable better control of the cryosurgery process [208]. It is found that using a constant wall heat transfer coefficient to simulate the cooling progress under a specific flow is more reasonable. For the two-phase flow of nitrogen inside the cryoprobe, the convective heat transfer coefficient on the probe wall is reported to be about 8 * 10^3 W/m^{2}K. The temperature distribution inside the tumor during the alternated cooling and heating treatment can be obtained through numerical study [178], and results are shown in Figure 5.12. The dynamic temperature changes at different distances from the active probe wall during an alternate cooling and heating treatment. The protocol is 10 minutes freezing followed by 100W-RF heating for 30 minutes with 6.0 m/s nitrogen flow inside the probe to prevent overheating in the surrounding of the probe wall. After the 10 minutes of freezing, an iceball is formed and blood perfusion stops in the area where temperatures are lower than 0°C [209].

As also seen in Figure 5.12, there is a sharp increase of the temperature when the RF heating is turned on. Thermal stresses are expected to occur, and the heterogeneous mechanical properties inside the tumor may lead to more damage than those of a uniform material experiencing the similar thermal history. The alternate cooling and heating treatment also enhances damage to the tumor cells. The thermally induced biological effects can be investigated at both the tissue and cellular levels.

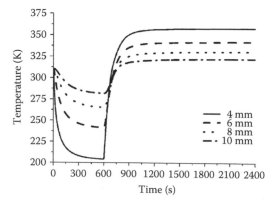

Figure 5.12 Dynamic temperature changes at different distances of 4 mm, 6 mm, 8 mm, and 10 mm, respectively, from the active probe wall during an alternate cooling and heating treatment (10 minutes of cooling followed by 100W-RF heating for 30 minutes with 6.0 m/s nitrogen flow inside the probe) [178].

5.5.4 Treatment Evaluation

The phenomenological method by using a critical temperature (for either high- or low-temperature treatments) is the easiest way to assess the therapeutic outcome and has been frequently used. However, more precisely, the actual damage is related to the thermal history that the tissue has experienced, which includes the heating or freezing rate, hold time, end temperature, and thawing rate. To precisely assess the damage region after a treatment, and then to optimize the treatment protocol, models based on the damage mechanisms have been developed [138,140,176,210]. Among them, there are models for quantifying the tissue damage degree after the treatment and for describing the cellular response to thermal energy input. As vasculature is essential to tumor growth and its damage is also critical for treatment outcome, there are also models studying the vasculature damage after the thermal treatment [208].

5.5.4.1 Tissue Damage Model

The model developed to describe the high-temperature injury is frequently applied for evaluation of tissue damage after the heat treatment. The tissue damage process is assumed to follow the Arrhenius function from the thermodynamics point of view [138]. Thus, the rate of injury degree Ω can be described as

$$\frac{d\Omega}{dt} = \xi e^{-\frac{\Delta E}{R_u T}} \tag{5.17}$$

which can be integrated as

$$\Omega = \xi \int_0^t e^{-\frac{\Delta E}{R_u T}} \, dt \tag{5.18}$$

where ξ is the frequency factor constant, ΔE is the tissue inactivation energy to govern the development of injury, R_u is the universal gas constant, T is the absolute temperature, and t is the thermal treatment time. The corresponding Henriques values of $\xi = 3.1 * 10^{98}$ s^{-1} and $\Delta E = 6.27 * 10^8$ J/kmole are used. As proposed by Moritz and Henriques, the second-degree burn $\Omega = 1$ may be evaluated as tissue necrosis [211].

Figure 5.13 shows the integral damage degree during heating alone for 10, 20, and 30 minutes, respectively, as a function of position inside the tumor. The vertical axis represents the damage degree, Ω, and the horizontal axis represents the short-axis radius from the probe. The cross point of each curve with the line $\Omega = 1$ has been marked as 8.7 mm, 9.4 mm, and 9.75 mm, respectively, in which, where $\Omega > 1$, complete tissue necrosis is assumed.

The quantification by the Henriques model predicts the degree of tissue damage. But it is hard to draw the conclusion that all tumor cells have been killed within the defined region. As any cell survival might result in tumor recurrence

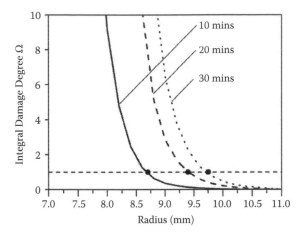

Figure 5.13   Degree of damage during heating alone for 10, 20, and 30 minutes, respectively. 100W-RF heating with 6.0 m/s nitrogen flow inside the probe turned on as the probe temperature exceeds 60°C [178].

or tumor metastasis, accurate predictions of the cell survival rate in the treatment region are necessary and of clinical importance.

5.5.4.2 Cellular Damage Model

The life of a cell is sustained by various metabolic activities. Its death may be an accumulation of inactivation of different reactions and denaturation of chromosomal proteins [5], while the damage of each reaction or protein denaturation also follows the Arrhenius function [212–214]:

$$k_s = \frac{BT}{h} e^{-\Delta G/RT} \tag{5.19}$$

where k_s is the rate of damage; B is the Boltzmann's constant; h is the Planck's constant; and ΔG is the Gibbs free energy of inactivation, which equals $\Delta H - T\Delta s$, where Δs is the entropy of inactivation (J/K/mole) and ΔH is the molecular inactivation energy. Thus, it is approximately that [213,214]

$$k_s = 2.05 \times (10)^{10} \times T \times e^{\Delta s/2} \times e^{-\Delta H/2T} \tag{5.20}$$

The occurrence of a complete inactivation of certain reaction and protein denaturation is related to the damage rate k_s and time t. Δs is an extensive state function that accounts for the effect of irreversibility in a thermodynamic system, and the smallest functional unit in a living system, a cell, can be considered as such a system. As a living cell goes to death, the maximum Δs is expected. In comparison, the differences between different kinds of live cells and the same cells at different temperatures are relatively small and assumed to be negligible. The reported $\Delta s = 374.5$ cal/K/mole for Chinese hamster ovary (CHO) cells [5,215] is used in this study.

The fraction of the reactions or the damaged proteins is as follows [214,216,217]:

$$F(t) = 1 - e^{-k_s t} \qquad (5.21)$$

Assuming the total cellular damage is an accumulation of n discrete and independent protein or reaction inactivations, the survival rate of the cells treated shall follow this equation [212–214]:

$$S_{heat} = 1 - [F(T)]^n = 1 - (1 - e^{-k_s t})^n \qquad (5.22)$$

where S_{heat} is the cell survival rate, and $n = 100$ from Reference [5].

Owing to different cellular sensitivities to heat, the values of inactivation energy ΔH have been reported to be 271,000, 135,600, and 141,000 cal/mole for a mouse tumor [217], mouse ear cells [214], and CHO cells [5,215], respectively. There has been no report of the ΔH value for human breast cancer cell lines, and especially for those exposed to prefreezing before the heat treatment. Experiments have been performed to evaluate the inactivation energy for cells that undergo different thermal treatments. The mortality of MAD-MB-231 cells heated to different temperatures for different time durations with or without prefreezing has been measured by the staining of Trypan blue. By fitting the experimental data to Equation (5.14), the inactivation energy ΔH for the cells to heating and to heating after prefreezing to –13°C for 30 minutes have been found to be 145,149 cal/mole and 143,898 cal/mole, respectively. The prefreezing reduces the inactivation energy about 1251 (0.87%) cal/mole compared to direct heating alone, but its effect is significant given the exponential relationship presented by Equation (5.22). The results suggest that the cells are more sensitive to heat after the pre-cooling process, and possible synergistic effects have been demonstrated of the alternate cooling and heating treatment.

The survival rates of cells undergoing the freezing process can be predicted according to the cryoinjury mechanisms. When the cooling rate is slow, extracellular ice forms and water is transported out of the cell, causing its osmotic dehydration. The consequent increase of intracellular solute concentrations may induce damage of intracellular proteins, the membrane, and enzymatic machinery of the cell. When the cooling rate is fast, there is not enough time for water to be transported out of the cell and intracellular ice forms [25]. Intracellular ice is usually lethal to cells. For tumor cells that are closely packed [42,191], IIF shall be the key factor determining the cell death. Thus, the survival rate of the tumor cells after freezing can be determined from the IIF probability P_{iif},

$$S_{cool} = 1 - P_{iif} \qquad (5.23)$$

where S_{cool} is the cell survival rate during freezing.

The IIF probability, P_{iif}, can be predicted given the thermal history that cells have experienced [143]:

$$P_{iif} = 1 - \exp\left[-\int_{T_{seed}}^{\pi} A\Omega_0 \left(\frac{T}{T_{f_0}}\right)\left(\frac{\eta_0}{\eta}\right)\left(\frac{A}{A_0}\right)\exp\left(-\frac{\kappa_0(T_f/T_{f_0})^4}{\Delta T^2 T^3}\right)\right] \quad (5.24)$$

where T_{seed} is the ice nucleation temperature (K), Ω_0 and κ_0 are the cell-type dependent constants, A is the surface area (m^2), and T_f is the intracellular phase change temperature, which depends on the solute mole fraction and can be determined by analyzing the cellular dehydration process [218–221]. Using both the tissue damage model and the cellular damage model, the survival rates of the tumor cells after different treatments are shown in Figure 5.14. Using the criterion of $\Omega = 1$ in the tissue damage model, the damage region after the 30 minutes of RF heating (assuming all tumor cells completely killed within the region) is predicted to be 9.75 mm, shown by the dash–dot line. However, as predicted by the cellular damage model, the region with incomplete cell killing after the RF treatment is clearly illustrated by the dot line in Figure 5.14 (8.2 mm in radius). In comparison, the tissue damage model highly overestimates the damaged region by nearly 15%. This model is originally proposed for burn injury evaluation after tissue experiencing a high temperature for a short time period. For example, Abraham and Sparrow have built a thermal ablation model that is more accurate by accounting for the phase change (water evaporation),

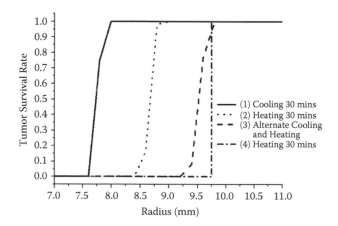

Figure 5.14   The survival rate distributions after different protocols: (1) Solid line, cooling for 30 minutes; (2) dotted line, 30 minutes of the 100W-RF heating and with 6.0 m/s nitrogen flow inside the probe turned on as the probe temperature exceeds 60°C; (3) dashed line, 10 minutes precooling followed by the 100 W-RF heating for 30 minutes and with 6.0 m/s nitrogen flow inside the probe turned on as the probe temperature exceeds 60°C; (4) dash–dot line, the tissue damage degree $\Omega = 1$ after 30 minutes of the 100 W-RF heating and with 6.0 m/s nitrogen flow inside the probe turned on as the probe temperature exceeds 60°C for 30 minutes (from Figure 5.13) [178].

change of blood perfusion, and other thermophysical properties [176]. However, it seems unsuitable for the hyperthermic condition under which tissue temperature is mildly raised relatively over a long period of time.

With the cellular damage model, after cooling alone for 30 minutes, the complete damage region is only about 7.6 mm in radius from the probe, as shown by the solid line. Tumor cell survival rate after the alternate freezing and heating is also determined and shown by the dashed line in Figure 5.14. As seen, the influence of a 10-minute prefreezing before the same heating protocol (100W RF and with 6.0 m/s nitrogen flow protection turned on as the probe temperature exceeds 60°C) for 30 minutes is significant. The complete damage region increases and reaches about 9.2 mm in radius. This is much larger than that caused by either the heating or cooling treatment alone.

5.5.4.3 Vascular Damage

Given the thermal stress through the rapid alternation of cooling and heating and possible vessel wall fractures induced by thermal stresses [205], the corresponding damage to tumor vasculature is also expected to be greater than that of either the cooling or heating treatment alone. This has already been observed in *in vivo* experimental study [59]. The mechanisms of vascular injury under such kind of treatment may be attributed to several important factors, as discussed in the following. In general, tumor vasculature and endothelial cells are more fragile to thermal stress because of their abnormal characteristics and microenvironment [222]. Freezing prior to heating might increase the thermal sensitivity of vascular endothelial cells by altering the intracellular protein nature. Mechanically, in freezing, ice normally first occurs inside vessel lumen, exerting forces on the inner endothelial cells. The endothelial cellular cytoskeleton and intercellular junction could be weakened by dehydration and chilling effects, resulting in cell detachment and wall rupture of the new vessels with a single endothelium layer, especially in the tumor periphery. Study of the endothelial cells' shape change and the subjected stress on the cell membrane surface can provide insight into the damage mechanisms of thermally induced vasculature injury [208]. Endothelial cells are modeled as ellipsoidal balloons whose volume varies according to the osmotic pressure experienced during the freezing process (Figure 5.15).

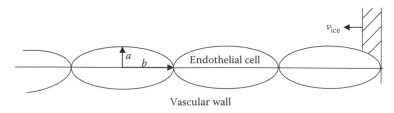

Figure 5.15 Schematic of the endothelium cell during freezing [208]. Parameters *a*, *b*, and v_{ice} are the long and short axes of the endothelial cell, and ice propagation velocity, respectively.

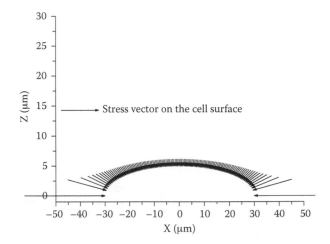

Figure 5.16   Surface stresses over the endothelial cell after freezing to −20°C at a cooling rate of −3°C/min [208]. Arrows of the vectors point in the stress direction while lengths of the vectors illustrate the stress magnitudes.

The cell surface tension has been analyzed to reveal the mechanical effect on the cell shape change during freezing. It is found that after being frozen to −20°C at a cooling rate of −3°C/min, the stresses on the cell surface are much larger than those before freezing, especially at the two ends of the long axes (Figure 5.16): more than 10 times of those before freezing [208]. The greatly enhanced surface stresses by dehydration plus the chilling effects are expected to finally break the cell junction and cause cell detachment from the extracellular matrix.

When cooling and heating are applied alternatively, tumor vessels experience much larger thermal stress over a short period of time than any single cooling or heating treatment, which can induce the vessel wall rupture. In addition, ischemia–reperfusion injuries may also occur, as indicated in frostbite by other studies [23,223]. Besides, it is possible that the alternate cooling and heating treatments of a tumor can trigger a much stronger immunologic response of the body, which further enhances the overall treatment effect of the system [224]. A comprehensive model should be developed to quantify both the direct and indirect cellular damage for more accurate predictions of the treatment outcome in future.

5.6 CONCLUSIONS

The development and applications of numerical bioheat and mass transfer have greatly improved techniques in tumor diagnosis and treatment using thermal physical methods. Together with the advancement of other techniques, especially imaging techniques, more detailed information of the tumor structure, vascular geometry, and even temperature can be obtained for accurate simulations. However, many important parameters such as the thermal, electrical, light, sound, and mechanical properties are dependent not only on individual tumors but also on the phase change in the tumor, and certainly the temperature during the treatment. Measurements of these parameters are difficult, and the estimated values should have a biological basis.

The tumor tissue response to either heating or freezing is essential to the prediction and evaluation of different treatments. The thermal treatment has both direct and indirect influences on the target tumor cells. When subjected to temperature alterations, many complex biomolecular changes take place resulting in either survival or death of the cells, which includes immediate necrosis and apoptosis (programmed death). The apoptosis pathways involve many biological reactions inside the cells that are thermally dependent. Modeling bioheat and mass transfer at the nanoscale level will be of great importance to the understanding of the molecular mechanisms of cellular damage, gene expression (especially for thermally induced apoptosis), the replication process of DNA [225,226], and thus the development of novel treatments of tumors at the molecular level. For the indirect damage to tumor cells through vascular and immunological responses to thermal treatment, new theories are highly desired and should be developed in future.

Treatment of the metastatic malignant tumors is still a difficult problem to solve in medical science. The alternate cooling and heating have shown better therapeutic effects in both preliminary animal studies and theoretical analysis [59,178]. More importantly, the immunity it has activated may provide an effective and systemic treatment for cancers with micrometastasis. Further study on the mechanisms of the enhanced treatment effects by incorporating bioheat and mass transfer analysis will help optimize the treatment protocol.

ACKNOWLEDGMENT

This work has been supported by the National Natural Science Foundation of China (50436030, 50725622, and 50506016).

NOMENCLATURE

A: surface area, cm^2

A_t: the effective surface area of the vasculature per unit tumor volume, cm^{-1}

A_v: surface area of the microvessel wall, cm^2

B: Boltzmann's constant, J/K

C: concentration, mg/ml

c_p: specific heat capacity under constant pressure, J/kg

$\underline{C_S}$:   average drug concentration in the vessel wall, mg/ml

D: diffusivity, cm^2/s

D_D: the effective diffusivity of the antitumor drug freed from the liposome, cm²/s

ΔE: activation energy, J/kmole

F: the fraction of the reactions or the damaged proteins

G: Gibbs free energy, J/kmole

Gr: Grashof number

h: Planck's constant, J s

h_t: enthalpy, J/kg

$h_{t,s}$:   sensible part of enthalpy, J/kg

ΔH: latent heat part of enthalpy, J/kg

J_C: convection flux, mg/s

J_i: liposome transport flow rate in the interstitial, mg/s

J_s: the total transport rate of drug across the vessel wall, mg/s

k_p: fluid permeability constant, cm³/cm

k_s: rate of damage, s⁻¹

L: latent heat of water solidification, kJ/kg

L_p: microvascular hydraulic conductivity, ml/s/N

M: drug dose, mg/m²

M_g: plasma pharmacokinetic parameters, mg/m²

n: the number of activations

P_{iif}:   intracellular ice formation (IIF) probability

$P_{L,app}$:   apparent permeability of the tumor vasculature, cm/s

P: microvessel permeability, cm/s

p: pressure, Pa

Pe: Peclet number

Q: the rate of fluid volumetric flow, ml/s

Q_{dep}:   deposition power

q: heat generation rate, W/kg

R, r: radius, m

r_d: temperature decay lengths in the radial direction, m

Re: Reynolds number

R_u: universal gas constant, J/K/mol

S: cell survival rate

S_{heat}:   cell survival rate after heating

S_{cool}:   cell survival rate after freezing

s: entropy, J/kg

T: temperature, K

t: time, s

T_f: intracellular phase change temperature, K

T_{seed}:   ice nucleation temperature, K

u: convective flow velocity, m/s

x: X coordinate, m

Z_d: temperature decay lengths in the axial direction, m

GREEK SYMBOLS

β:	liquid fraction
η:	solvent viscosity, Pa s
λ:	thermal conductivity, W/m/K
v_{max}:	maximal cellular uptake rate, μg/ml/s
ξ:	frequency factor constant, s^{-1}
π:	colloid-osmotic pressure, N/cm^2
ρ:	density, g/cm^3
σ:	osmotic reflection coefficient
σ_f:	filtration reflection coefficient
τ_r:	decay time constant, s
φ_e:	interstitial volume fraction
Ω:	rate of injury degree
ω:	blood perfusion rate, ml/s/ml

SUBSCRIPTS

a:	artery blood
app:	apparent
b:	blood
C:	center
D:	antitumor drug freed from the liposome as doxorubicin
E:	in the extracellular space
I:	intracellular bound
i:	interstitial
L:	liposome
l:	liquid
met:	metabolism
ref:	reference value
s:	solid
t:	tissue
v:	vascular
0:	initial value

REFERENCES

1. R. Lawson, Implications of surface temperatures in the diagnosis of breast cancer, *Can Med Assoc J*, vol. 75, pp. 309–311, 1956.
2. D. L. Harris, W. P. Greening, and P. M. Aichroth, Infra-red in the diagnosis of a lump in the breast, *Br J Cancer*, vol. 20, pp. 710–721, 1966.
3. Y. R. Parisky, K. A. Skinner, and R. Cothren, Computerized thermal breast imaging revisited: An adjunct tool to mammography, *Proc 20th IEEE EMBS*, vol. 20, p. 2, 1998.

4. M. Gautherie, Thermopathology of breast cancer: Measurement and analysis of in vivo temperature and blood flow, *Ann N Y Acad Sci*, vol. 335, pp. 383–415, 1980.

5. W. C. Dewey, L. E. Hopwood, S. A. Sapareto, and L. E. Gerweck, Cellular responses to combinations of hyperthermia and radiation, *Radiology*, vol. 123, pp. 463–474, 1977.

6. T. E. Dudar and R. K. Jain, Differential response of normal and tumor microcirculation to hyperthermia, *Cancer Res*, vol. 44, pp. 605–612, 1984.

7. B. Hildebrandt, P. Wust, O. Ahlers, A. Dieing, G. Sreenivasa, T. Kerner, R. Felix, and H. Riess, The cellular and molecular basis of hyperthermia, *Crit Rev Oncol Hematol*, vol. 43, pp. 33–56, 2002.

8. E. Dikomey and J. Franzke, Effect of heat on induction and repair of DNA strand breaks in x-irradiated CHO cells, *Int J Radiat Biol*, vol. 61, pp. 221–233, 1992.

9. J. L. Roti, H. H. Kampinga, R. S. Malyapa, W. D. Wright, R. P. Vanderwaal, and M. Xu, Nuclear matrix as a target for hyperthermic killing of cancer cells, *Cell Stress Chap*, vol. 3, pp. 245–255, 1998.

10. S. K. Calderwood, J. R. Theriault, and J. Gong, How is the immune response affected by hyperthermia and heat shock proteins? *Int J Hyperthermia*, vol. 21, pp. 713–716, 2005.

11. C. J. Diederich, Thermal ablation and high-temperature thermal therapy: Overview of technology and clinical implementation, *Int J Hyperthermia*, vol. 21, pp. 745–753, 2005.

12. L. F. Fajardo, A. B. Schreiber, N. I. Kelly, and G. M. Hahn, Thermal sensitivity of endothelial cells, *Radiat Res*, vol. 103, pp. 276–285, 1985.

13. D. S. Rappaport and C. W. Song, Blood flow and intravascular volume of mammary adenocarcinoma 13726a and normal tissues of rat during and following hyperthermia, *Int J Radiat Oncol Biol Phys*, vol. 9, pp. 539–547, 1983.

14. S. A. Shah and J. A. Dickson, Effect of hyperthermia on the immunocompetence of vx2 tumor-bearing rabbits, *Cancer Res*, vol. 38, pp. 3523–3531, 1978.

15. S. A. Shah and J. A. Dickson, Effect of hyperthermia on the immune response of normal rabbits, *Cancer Res*, vol. 38, pp. 3518–3522, 1978.

16. C. W. Song, Effect of local hyperthermia on blood flow and microenvironment: A review, *Cancer Res*, vol. 44, pp. 4721s–4730s, 1984.

17. C. W. Song, M. S. Kang, J. G. Rhee, and S. H. Levitt, The effect of hyperthermia on vascular function, pH, and cell survival, *Radiology*, vol. 137, pp. 795–803, 1980.

18. F. Stewart and A. Begg, Blood flow changes in transplanted mouse tumours and skin after mild hyperthermia, *Br J Radiol*, vol. 56, pp. 477–482, 1983.

19. J. Baust, A. A. Gage, H. Ma, and C. M. Zhang, Minimally invasive cryosurgery: technological advances, *Cryobiology*, vol. 34, pp. 373–384, 1997.

20. A. A. Gage, Cryosurgery in the treatment of cancer, *Surg Gynecol Obstet*, vol. 174, pp. 73–92, 1992.

21. P. R. Stauffer, Evolving technology for thermal therapy of cancer, *Int J Hyperthermia*, vol. 21, pp. 731–744, 2005.

22. P. R. Stauffer and S. N. Goldberg, Introduction: Thermal ablation therapy, *Int J Hyperthermia*, vol. 20, pp. 671–677, 2004.

23. N. E. Hoffmann and J. C. Bischof, The cryobiology of cryosurgical injury, *Urology*, vol. 60, pp. 40–49, 2002.

24. A. A. Gage and J. Baust, Mechanisms of tissue injury in cryosurgery, *Cryobiology*, vol. 37, pp. 171–186, 1998.

25. P. Mazur, Freezing of living cells: Mechanisms and implications, *Am J Physiol*, vol. 247, pp. C125–C142, 1984.

26. M. H. Bourne, M. W. Piepkorn, F. Clayton, and L. G. Leonard, Analysis of microvascular changes in frostbite injury, *J Surg Res*, vol. 40, pp. 26–35, 1986.
27. H. M. Carpenter, L. A. Hurley, E. Hardenbergh, and R. B. Williams, Vascular injury due to cold, *Arch Path & Lab Med*, vol. 92, pp. 153–161, 1971.
28. R. J. Ablin, *An appreciation and realization of the concept of cryoimmunology*. St. Louis, MO: Quality Medical Publishing, 1995.
29. J. P. Johnson, Immunologic aspects of cryosurgery: Potential modulation of immune recognition and effector cell maturation, *Clin Dermatol*, vol. 8, pp. 39–47, 1990.
30. M. S. Sabel, M. A. Nehs, G. Su, K. P. Lowler, J. L. Ferrara, and A. E. Chang, Immunologic response to cryoablation of breast cancer, *Breast Cancer Res Treat*, vol. 90, pp. 97–104, 2005.
31. A. N. Mirza, B. D. Fornage, N. Sneige, H. M. Kuerer, L. A. Newman, F. C. Ames, and S. E. Singletary, Radiofrequency ablation of solid tumors, *Cancer J*, vol. 7, pp. 95–102, 2001.
32. C. M. Pacella, G. Bizzarri, G. Francica, A. Bianchini, S. De Nuntis, S. Pacella, A. Crescenzi, S. Taccogna, G. Forlini, Z. Rossi, J. Osborn, and R. Stasi, Percutaneous laser ablation in the treatment of hepatocellular carcinoma with small tumors: Analysis of factors affecting the achievement of tumor necrosis, *J Vasc Interv Radiol*, vol. 16, pp. 1447–1457, 2005.
33. N. J. McDannold and F. A. Jolesz, Magnetic resonance image-guided thermal ablations, *Top Magn Reson Imaging*, vol. 11, pp. 191–202, 2000.
34. R. J. Stafford and J. D. Hazle, Magnetic resonance temperature imaging for focused ultrasound surgery: A review, *Top Magn Reson Imaging*, vol. 17, pp. 153–163, 2006.
35. P. M. Meaney, M. W. Fanning, K. D. Paulsen, D. Lit, S. A. Pendergrass, Q. Fang, and K. L. Moodie, Microwave thermal imaging: Initial in vivo experience with a single heating zone, *Int J Hyperthermia*, vol. 19, pp. 617–641, 2003.
36. P. M. Meaney, K. D. Paulsen, M. W. Fanning, D. Li, and Q. Fang, Image accuracy improvements in microwave tomographic thermometry: Phantom experience, *Int J Hyperthermia*, vol. 19, pp. 534–550, 2003.
37. T. Varghese and M. J. Daniels, Real-time calibration of temperature estimates during radiofrequency ablation, *Ultrason Imaging*, vol. 26, pp. 185–200, 2004.
38. E. R. Cosman, B. S. Nashold, and J. Ovelman-Levitt, Theoretical aspects of radiofrequency lesions in the dorsal root entry zone, *Neurosurgery*, vol. 15, pp. 945–950, 1984.
39. R. V. Devireddy, D. J. Smith, and J. C. Bischof, Effect of microscale mass transport and phase change on numerical prediction of freezing in biological tissues, *J Heat Trans-T ASME*, vol. 124, pp. 365–374, 2002.
40. M. M. Osman and E. M. Afify, Thermal modeling of the malignant woman's breast, *J Biomech Eng*, vol. 110, pp. 269–276, 1988.
41. B. Rubinsky, Microscale heat transfer in biological systems at low temperatures, *Exp Heat Transfer*, vol. 10, pp. 1–29, 1997.
42. B. Rubinsky and D. E. Pegg, A mathematical model for the freezing process in biological tissue, *Proc R Soc Lond B Biol Sci*, vol. 234, pp. 343–358, 1988.
43. J. C. Bischof and B. Rubinsky, Microscale heat and mass transfer of vascular and intracellular freezing in the liver, *ASME J Heat Transfer*, vol. 115, pp. 1029–1035, 1993.
44. W. H. Clark, Tumor progression and the nature of cancer, *Br J Cancer*, vol. 64, pp. 631–644, 1991.
45. K. Engin, D. B. Leeper, A. J. Thistlethwaite, L. Tupchong, and J. D. McFarlane, Tumor extracellular pH as a prognostic factor in thermoradiotherapy, *Int J Radiat Oncol Biol Phys*, vol. 29, pp. 125–132, 1994.

46. J. A. Koutcher, D. Barnett, A. B. Kornblith, D. Cowburn, T. J. Brady, and L. E. Gerweck, Relationship of changes in pH and energy status to hypoxic cell fraction and hyperthermia sensitivity, *Int J Radiat Oncol Biol Phys*, vol. 18, pp. 1429–1435, 1990.

47. F. Hammersen, B. Endrich, and K. Messmer, The fine structure of tumor blood vessels. I. Participation of non-endothelial cells in tumor angiogenesis, *Int J Microcirc Clin Exp*, vol. 4, pp. 31–43, 1985.

48. I. P. Torres Filho, B. Hartley-Asp, and P. Borgstrom, Quantitative angiogenesis in a syngeneic tumor spheroid model, *Microvasc Res*, vol. 49, pp. 212–226, 1995.

49. Q. Huang, S. Shan, R. D. Braun, J. Lanzen, G. Anyrhambatla, G. Kong, M. Borelli, P. Corry, M. W. Dewhirst, and C. Y. Li, Noninvasive visualization of tumors in rodent dorsal skin window chambers, *Nat Biotechnol*, vol. 17, pp. 1033–1035, 1999.

50. P. Liu, A. Zhang, Y. Xu, and L. X. Xu, Study of non-uniform nanoparticle liposome extravasation in tumour, *Int J Hyperthermia*, vol. 21, pp. 259–270, 2005.

51. Y. Boucher, L. T. Baxter, and R. K. Jain, Interstitial pressure gradients in tissue-isolated and subcutaneous tumors: Implications for therapy, *Cancer Res*, vol. 50, pp. 4478–4484, 1990.

52. J. R. Less, M. C. Posner, Y. Boucher, D. Borochovitz, N. Wolmark, and R. K. Jain, Interstitial hypertension in human breast and colorectal tumors, *Cancer Res*, vol. 52, pp. 6371–6374, 1992.

53. R. K. Jain, Transport of molecules across tumor vasculature, *Cancer Metastasis Rev*, vol. 6, pp. 559–593, 1987.

54. H. F. Dvorak, L. F. Brown, M. Detmar, and A. M. Dvorak, Vascular permeability factor/vascular endothelial growth factor, microvascular hyperpermeability, and angiogenesis, *Am J Pathol*, vol. 146, pp. 1029–1039, 1995.

55. R. K. Jain, Delivery of molecular and cellular medicine to solid tumors, *Adv Drug Deliv Rev*, vol. 46, pp. 149–168, 2001.

56. G. Kong, R. D. Braun, and M. W. Dewhirst, Characterization of the effect of hyperthermia on nanoparticle extravasation from tumor vasculature, *Cancer Res*, vol. 61, pp. 3027–3032, 2001.

57. G. Kong and M. W. Dewhirst, Hyperthermia and liposomes, *Int J Hyperthermia*, vol. 15, pp. 345–370, 1999.

58. A. T. Lefor, S. Makohon, and N. B. Ackerman, The effects of hyperthermia on vascular permeability in experimental liver metastasis, *J Surg Oncol*, vol. 28, pp. 297–300, 1985.

59. Y. Shen, P. Liu, A. Zhang, and L. X. Xu, Tumor microvasculature response to alternated cold and heat treatment, *Proc 27th IEEE EMBS*, 2005.

60. Y. Shen, P. Liu, A. Zhang, and L. X. Xu, Study on tumor microvasculature damage induced by alternate cooling and heating, *Ann Biomed Eng*, vol. 36, pp. 1409–1419, 2007.

61. L. Hu, A. Gupta, J. P. Gore, and L. X. Xu, Effect of forced convection on the skin thermal expression of breast cancer, *ASME J Biomech Eng*, vol. 126, pp. 204–211, 2004.

62. D .M. Sabados, L. X. Xu, and J. P. Gore, Study of cell metabolic activities using fluorescence microscopy, *Proceedings of the 2003 ASME Summer Heat Transfer Conference*, Las Vegas, NV, 2003.

63. P. Backman, Effects of experimental factors on the metabolic rate of t-lymphoma cells as measured by microcalorimetry, *Thermochim Acta*, vol. 172, pp. 123–130, 1990.

64. J. P. Gore and L. X. Xu, *Thermal imaging for biological and medical diagnostics*, vol. 17, T. Vo-Dinh (Ed.), Boca Raton, FL: CRC Press, 2003.

65. L. L. Thomsen, D. W. Miles, L. Happerfield, L. G. Bobrow, R. G. Knowles, and S. Moncada, Nitric oxide synthase activity in human breast cancer, *Br J Cancer*, vol. 72, pp. 41–44, 1995.

66. M. M. Chen, C. O. Pederson, and J. C. Chato, On the feasibility of obtaining three-dimensional information from thermographic measurement, *ASME J Biomech Eng*, vol. 99, pp. 58–64, 1977.

67. E. Y. Ng and N. M. Sudharsan, Numerical uncertainty and perfusion induced instability in bioheat equation: Its importance in thermographic interpretation, *J Med Eng Technol*, vol. 25, pp. 222–229, 2001.

68. E. Y. Ng and N. M. Sudharsan, Effect of blood flow, tumour and cold stress in a female breast: A novel time-accurate computer simulation, *Proc Inst Mech Eng [H]*, vol. 215, pp. 393–404, 2001.

69. H. H. Pennes, Analysis of tissue and arterial blood temperatures in the resting human forearm, 1948, *J Appl Physiol*, vol. 85, pp. 5–34, 1998.

70. M. Habeck, Cancer drug delivery is hot stuff, *Drug Discov Today*, vol. 6, pp. 754–756, 2001.

71. F. C. Szoka, Liposomal drug delivery: Current status and future propects, *Membrane Fusion*, pp. 845–890, 1991.

72. A. W. El-Kareh and T. W. Secomb, A mathematical model for cisplatin cellular pharmacodynamics, *Neoplasia*, vol. 5, pp. 161–169, 2003.

73. B. Ribba, K. Marron, Z. Agur, T. Alarcon, and P. K. Maini, A mathematical model of doxorubicin treatment efficacy for non-Hodgkin's lymphoma: Investigation of the current protocol through theoretical modelling results, *Bull Math Biol*, vol. 67, pp. 79–99, 2005.

74. A. R. Tzafriri, E. I. Lerner, M. Flashner-Barak, M. Hinchcliffe, E. Ratner, and H. Parnas, Mathematical modeling and optimization of drug delivery from intratumorally injected microspheres, *Clin Cancer Res*, vol. 11, pp. 826–834, 2005.

75. J. P. Ward and J. R. King, Mathematical modelling of drug transport in tumour multicell spheroids and monolayer cultures, *Math Biosci*, vol. 181, pp. 177–207, 2003.

76. J. Lankelma, R. F. Luque, H. Dekker, W. Schinkel, and H. M. Pinedo, A mathematical model of drug transport in human breast cancer, *Microvasc Res*, vol. 59, pp. 149–161, 2000.

77. J. Lankelma, R. Fernandez Luque, H. Dekker, and H. M. Pinedo, Simulation model of doxorubicin activity in islets of human breast cancer cells, *Biochim Biophys Acta*, vol. 1622, pp. 169–178, 2003.

78. P. Magni, M. Simeoni, I. Poggesi, M. Rocchetti, and G. De Nicolao, A mathematical model to study the effects of drugs administration on tumor growth dynamics, *Math Biosci*, vol. 200, pp. 127–151, 2006.

79. A. W. El-Kareh and T. W. Secomb, Two-mechanism peak concentration model for cellular pharmacodynamics of doxorubicin, *Neoplasia*, vol. 7, pp. 705–713, 2005.

80. T. W. Secomb and A. W. El-Kareh, A theorectical model for intraperitoneal delivery of cisplatin and the effect of hyperthermia on drug penetration distance, *Neoplasia*, vol. 6, pp. 117–127, 2004.

81. Y. Yoshioka, Y. Tsutsumi, H. Kamada, T. Kihira, S. Tsunoda, Y. Yamamoto, T. Okamoto, H. Shibata, Y. Mukai, T. Taniai, T. Shimizu, M. Kawamura, Y. Abe, S. Nakagawa, and T. Mayumi, Selective enhancer of tumor vascular permeability for optimization of cancer chemotherapy, *Biol Pharm Bull*, vol. 27, pp. 437–439, 2004.

82. M. H. Gaber, N. Z. Wu, K. Hong, S. K. Huang, M. W. Dewhirst, and D. Papahadjopoulos, Thermosensitive liposomes: Extravasation and release of contents in tumor microvascular networks, *Int J Radiat Oncol Biol Phys*, vol. 36, pp. 1177–1187, 1996.

83. G. Kong, R. D. Braun, and M. W. Dewhirst, Hyperthermia enables tumor-specific nanoparticle delivery: Effect of particle size, *Cancer Res*, vol. 60, pp. 4440–4445, 2000.

84. A. W. El-Kareh and T. W. Secomb, A mathematical model for comparison of bolus injection, continuous infusion, and liposomal delivery of doxorubicin to tumor cells, *Neoplasia*, vol. 2, pp. 325–338, 2000.

85. J. Lankelma, H. Dekker, F. R. Luque, S. Luykx, K. Hoekman, P. van der Valk, P. J. van Diest, and H. M. Pinedo, Doxorubicin gradients in human breast cancer, *Clin Cancer Res*, vol. 5, pp. 1703–1707, 1999.

86. S. H. Jang, M. G. Wientjes, D. Lu, and J. L. Au, Drug delivery and transport to solid tumors, *Pharm Res*, vol. 20, pp. 1337–1350, 2003.

87. A. Gabizon, R. Catane, B. Uziely, B. Kaufman, T. Safra, R. Cohen, F. Martin, A. Huang, and Y. Barenholz, Prolonged circulation time and enhanced accumulation in malignant exudates of doxorubicin encapsulated in polyethylene-glycol coated liposomes, *Cancer Res*, vol. 54, pp. 987–992, 1994.

88. O. Kedem and A. Katchalsky, Thermodynamic analysis of the permeability of biological membranes to non-electrolytes, *Biochim Biophys Acta*, vol. 27, pp. 229–246, 1958.

89. F. Yuan, M. Leunig, S. K. Huang, D. A. Berk, D. Papahadjopoulos, and R. K. Jain, Microvascular permeability and interstitial penetration of sterically stabilized (stealth) liposomes in a human tumor xenograft, *Cancer Res*, vol. 54, pp. 3352–3356, 1994.

90. A. Pluen, Y. Boucher, S. Ramanujan, T. D. McKee, T. Gohongi, E. di Tomaso, E. B. Brown, Y. Izumi, R. B. Campbell, D. A. Berk, and R. K. Jain, Role of tumor-host interactions in interstitial diffusion of macromolecules: Cranial vs. subcutaneous tumors, *Proc Natl Acad Sci USA*, vol. 98, pp. 4628–4633, 2001.

91. A. W. El-Kareh and T. W. Secomb, Theoretical models for drug delivery to solid tumors, *Crit Rev Biomed Eng*, vol. 25, pp. 503–571, 1997.

92. A. W. El-Kareh, S. L. Braunstein, and T. W. Secomb, Effect of cell arrangement and interstitial volume fraction on the diffusivity of monoclonal antibodies in tissue, *Biophys J*, vol. 64, pp. 1638–1646, 1993.

93. B. Szachowicz-Petelska, Z. Figaszewski, and W. Lewandowski, Mechanisms of transport across cell membranes of complexes contained in antitumour drugs, *Int J Pharm*, vol. 222, pp. 169–182, 2001.

94. K. T. Luu and J. A. Uchizono, P-glycoprotein induction and tumor cell-kill dynamics in response to differential doxorubicin dosing strategies: A theoretical pharmacodynamic model, *Pharm Res*, vol. 22, pp. 710–715, 2005.

95. A. Zhang, X. P. Mi, and L. X. Xu, Study of the thermally targeted nano-particle drug delivery for rumor tumor therapy, *Proc Micro/Nanoscale Heat Transfer International Conference, ASME*, Taiwan, 2008

96. P. Liu and L. X. Xu, Enhanced efficacy of anti-tumor liposomal doxorubicin by hyperthermia, *Proc 28th IEEE EMB*, vol. 1, pp. 4354–4357, 2006.

97. R. W. Habash, R. Bansal, D. Krewski, and H. T. Alhafid, Thermal therapy, part 1: An introduction to thermal therapy, *Crit Rev Biomed Eng*, vol. 34, pp. 459–489, 2006.

98. C. Villard, L. Soler, and A. Gangi, Radiofrequency ablation of hepatic tumors: Simulation, planning, and contribution of virtual reality and haptics, *Comput Methods Biomech Biomed Engin*, vol. 8, pp. 215–227, 2005.

99. D. Tanaka, K. Cleary, D. Stewart, B. Wood, M. Mocanu, et al., Volumetric treatment planning and image guidance for radiofrequency ablation of hepatic tumors, in *Medical Imaging 2003: Visualization, Image-Guided Procedures and Display*, R. L. Galloway (Ed.), vol. 5029, pp. 528–534, 2003.

100. K. Tim, P. Tobias, W. Andreas; R. Felix, Z. Stephan, and P. Heinz-Otto, Workflow oriented software support for image guided radiofrequency ablation of focal liver malignancies, in *Medical Imaging 2007: Visualization and Image-Guided Procedures*, K. R. Cleary and M. I. Miga (Eds.), *Proceedings of the SPIE*, vol. 6059, pp. 650919, Bellingham, WA: SPIE, 2007.

101. W. Andreas, K. Tim, R. Felix, P. Tobias, Z. Stephan, and P. Heinz-Otto, Workflow oriented software support for image guided radiofrequency ablation of focal liver malignancies, in *Medical Imaging 2007: Visualization and Image-Guided Procedures*, K. R. Cleary and M. I. Miga (Eds.), *Proceedings of the SPIE*, vol. 6059, pp. 650919, Bellingham, WA: SPIE, 2007.

102. D. R. Daum, N. Smith, N. McDannold, and K. Hynynen, MR-guided non-invasive thermal coagulation of in vivo liver tissue using an ultrasonic phased array, in *Proceedings of the SPIE: Int Soc Opt Eng*, vol. 3594, pp. 185–193, Bellingham, WA: SPIE, 1999.

103. H. Rhim, Review of Asian experience of thermal ablation techniques and clinical practice, *Int J Hyperthermia*, vol. 20, pp. 699–712, 2004.

104. S. N. Goldberg, G. S. Gazelle, and P. R. Mueller, Thermal ablation therapy for focal malignancy: A unified approach to underlying principles, techniques, and diagnostic imaging guidance, *AJR Am J Roentgenol*, vol. 174, pp. 323–331, 2000.

105. S. N. Goldberg and D. E. Dupuy, Image-guided radiofrequency tumor ablation: Challenges and opportunities—part I, *J Vasc Interv Radiol*, vol. 12, pp. 1021–1032, 2001.

106. T. L. Boaz, J. S. Lewin, Y. C. Chung, J. L. Duerk, M. E. Clampitt, and J. R. Haaga, MR monitoring of MR-guided radiofrequency thermal ablation of normal liver in an animal model, *J Magn Reson Imaging*, vol. 8, pp. 64–69, 1998.

107. E. Atalar and C. Menard, MR-guided interventions for prostate cancer, *Magn Reson Imaging Clin N Am*, vol. 13, pp. 491–504, 2005.

108. S. Permpongkosol, R. E. Link, L. R. Kavoussi, and S. B. Solomon, Temperature measurements of the low-attenuation radiographic ice ball during ct-guided renal cryoablation, *Cardiovasc Intervent Radiol*, vol. 31, 2007.

109. S. G. Silverman, K. Tuncali, D. F. Adams, E. vanSonnenberg, K. H. Zou, D. F. Kacher, P. R. Morrison, and F. A. Jolesz, MR imaging-guided percutaneous cryotherapy of liver tumors: Initial experience, *Radiology*, vol. 217, pp. 657–664, 2000.

110. M. S. Breen, L. Chen, and D. L. Wilson, Image-guided laser thermal ablation therapy: A comparison of modeled tissue damage using interventional MR temperature images with tissue response, *Prog Biomed Optics Imag: Med Imaging*, pp. 516–523, 2004.

111. A. Gupta, M. E. Allaf, L. R. Kavoussi, T. W. Jarrett, D. Y. Chan, L. M. Su, and S. B. Solomon, Computerized tomography guided percutaneous renal cryoablation with the patient under conscious sedation: Initial clinical experience, *J Urol*, vol. 175, pp. 447–452; discussion 452–523, 2006.

112. W. B. Shingleton, P. Farabaugh, M. Hughson, and P. E. Sewell Jr., Percutaneous cryoablation of porcine kidneys with magnetic resonance imaging monitoring, *J Urol*, vol. 166, pp. 289–291, 2001.

113. J. C. Saliken, B. J. Donnelly, and J. C. Rewcastle, The evolution and state of modern technology for prostate cryosurgery, *Urology*, vol. 60, pp. 26–33, 2002.

114. T. R. Larson, D. W. Robertson, A. Corica, and D. G. Bostwick, In vivo interstitial temperature mapping of the human prostate during cryosurgery with correlation to histopathologic outcomes, *Urology*, vol. 55, pp. 547–552, 2000.

115. J. C. Bischof, D. Smith, P. V. Pazhayannur, C. Manivel, J. Hulbert, and K. P. Roberts, Cryosurgery of dunning at-1 rat prostate tumor: Thermal, biophysical, and viability response at the cellular and tissue level, *Cryobiology*, vol. 34, pp. 42–69, 1997.

116. D. J. Smith, W. M. Fahssi, D. J. Swanlund, and J. C. Bischof, A parametric study of freezing injury in at-1 rat prostate tumor cells, *Cryobiology*, vol. 39, pp. 13–28, 1999.

117. K. Tatsutani, B. Rubinsky, G. Onik, and R. Dahiya, Effect of thermal variables on frozen human primary prostatic adenocarcinoma cells, *Urology*, vol. 48, pp. 441–447, 1996.

118. A. L. Denys, T. De Baere, V. Kuoch, B. Dupas, P. Chevallier, D. C. Madoff, P. Schnyder, and F. Doenz, Radio-frequency tissue ablation of the liver: In vivo and ex vivo experiments with four different systems, *Eur Radiol*, vol. 13, pp. 2346–2352, 2003.

119. J. M. Lee, J. K. Han, J. M. Chang, S. Y. Chung, S. H. Kim, J. Y. Lee, and B. I. Choi, Radiofrequency ablation in pig lungs: In vivo comparison of internally cooled, perfusion and multitined expandable electrodes, *Br J Radiol*, vol. 79, pp. 562–571, 2006.

120. C. Kim, A. P. O'Rourke, D. M. Mahvi, and J. G. Webster, Finite-element analysis of ex vivo and in vivo hepatic cryoablation, *IEEE Trans Biomed Eng*, vol. 54, pp. 1177–1185, 2007.

121. V. D. Ambrosio, F. Dughiero, and M. Forzan, Numerical models of RF-thermal ablation treatments, *Intl J App Electr Mech*, vol. 25, pp. 429–433, 2007.

122. P. C. Johnson and G. M. Saidel, Thermal model for fast simulation during magnetic resonance imaging guidance of radio frequency tumor ablation, *Ann Biomed Eng*, vol. 30, pp. 1152–1161, 2002.

123. B. Sohrab, G. Farzan, J. Amin, and B. Ashkan, Numerical simulation of ultrasound thermotherapy of brain with a scanned focus transducer, *J Acoust Soc Am*, vol. 117, p. 2412, 2005.

124. B. Sohrab, G. Farzan, B. Ashkan, and J. Amin, Ultrasound thermotherapy of breast: Theoretical design of transducer and numerical simulation of procedure, *Jap J Appl Phys Part 1: Reg. Papers Brief Commun. Rev. Papers*, vol. 45, pp. 1856–1863, 2006.

125. J. Zhang, G. A. Sandison, J. Y. Murthy, and L. X. Xu, Numerical simulation for heat transfer in prostate cancer cryosurgery, *J Biomech Eng*, vol. 127, pp. 279–294, 2005.

126. Q. Nan, X. Yang, L. Li, Y. Liu, and D. Hao, Numerical simulation on microwave ablation with a water-cooled antenna, *Proceedings of the Bioinf Biomed Eng. ICBBE: The 1st International Conference*, 2007.

127. H. H. Pennes, Analysis of tissue and arterial blood temperatures in the resting human forearm, *J Appl Physiol*, vol. 85, pp. 5–34, 1948.

128. G. Comini and S. Del Giudice, Thermal aspects of cryosurgery, *J Heat Transfer*, vol. 98, pp. 543–549, 1976.

129. J. C. Rewcastle, G. A. Sandison, K. Muldrew, J. C. Saliken, and B. J. Donnelly, A model for the time dependent three-dimensional thermal distribution within iceballs surrounding multiple cryoprobes, *Med Phys*, vol. 28, pp. 1125–1137, 2001.

130. T. E. Cooper and G. J. Trezek, Analytical prediction of the temperature field emanating from a cryogenic surgical cannula, *Cryobiology*, vol. 7, pp. 79–83, 1970.

131. T. E. Cooper and G. J. Trezek, Rate of lesion growth around spherical and cylindrical cryoprobes, *Cryobiology*, vol. 7, pp. 183–190, 1970.

132. T. E. Cooper and G. J. Trezek, On the freezing of tissue, *J Heat Transfer*, vol. 94, pp. 251–253, 1972.

133. B. Rubinsky and A. Shitzer, Analysis of a Stefan-like problem in a biological tissue around a cryosurgical probe, *J Heat Trans-T ASME*, pp. 514–519, 1976.

134. H. Budman, A. Shitzer, and J. Dayan, Analysis of the inverse problem of freezing and thawing of a binary solution during cryosurgical processes, *ASME J Biomech Eng*, vol. 117, pp. 193–202, 1995.

135. Y. Rabin and A. Shitzer, Exact solution to the one-dimensional inverse-Stefan problem in nonideal biological tissues, *ASME J Heat Transfer*, vol. 117, pp. 425–431, 1995.

136. B. Fischer, P. Jais, D. Shah, S. Chouairi, M. Haissaguerre, S. Garrigues, F. Poquet, L. Gencel, J. Clementy, and F. I. Marcus, Radiofrequency catheter ablation of common atrial flutter in 200 patients, *J Cardiovasc Electrophysiol*, vol. 7, pp. 1225–1233, 1996.

137. B. Nilsson, X. Chen, S. Pehrson, and J. H. Svendsen, The effectiveness of a high output/short duration radiofrequency current application technique in segmental pulmonary vein isolation for atrial fibrillation, *Europace*, vol. 8, pp. 962–965, 2006.

138. Y. I. Cho, Bioengineering heat transfer, *Adv Heat Transfer*, pp. 222–242, 313–357, 1992.

139. F. C. Henriques, Studies of thermal injury. V. The predictability and the significance of thermally induced rate processes leading to irreversible epidermal injury, *Arch Pathol*, vol. 43, pp. 489–502, 1947.

140. S. A. Sapareto and W. C. Dewey, Thermal dose determination in cancer therapy, *Int J Radiat Oncol Biol Phys*, vol. 10, pp. 787–800, 1984.

141. P. W. Whitworth and J. C. Rewcastle, Cryoablation and cryolocalization in the management of breast disease, *J Surg Oncol*, vol. 90, pp. 1–9, 2005.

142. D. Gao and J. K. Critser, Mechanisms of cryoinjury in living cells, *ILAR J*, vol. 41, pp. 187–196, 2000.

143. M. Toner, E. G. Cravalho, and M. Karel, Thermodynamics and kinetics of intracellular ice formation during freezing of biological cells, *J Appl Physiol*, vol. 67, pp. 1582–1593, 1990.

144. Y. Rabin and P. S. Steif, Analysis of thermal stresses around a cryosurgical probe, *Cryobiology*, vol. 33, pp. 276–290, 1996.

145. Y. Rabin and P. S. Steif, Thermal stress modeling in cryosurgery, *Int J Sol Str*, vol. 37, pp. 2363–2375, 2000.

146. P. S. Steif, M. C. Palastro, and Y. Rabin, Analysis of the effect of partial vitrification on stress development in cryopreserved blood vessels, *Med Eng Phys*, vol. 29, pp. 661–670, 2007.

147. F. S. Gayzik, E. P. Scott, and T. Loulou, Experimental validation of an inverse heat transfer algorithm for optimizing hyperthermia treatments, *ASME J Biomech Eng*, vol. 128, pp. 505–515, 2006.

148. T. Loulou and E. P. Scott, Thermal dose optimization in hyperthermia treatments by using the conjugate gradient method, *Numer Heat Transf: Pt A*, vol. 42, pp. 661–683, 2002.

149. F. Ghalichi, S. Behnia, A. Jafari, and A. Bonabi, Numerical simulation of ultrasound thermotherapy of brain with a scanned focus transducer, *J Acoust Soc Am*, vol. 117, pp. 2412, 2005.

150. S. A. Baldwin, A. Pelman, and J. L. Bert, A heat transfer model of thermal balloon endometrial ablation, *Ann Biomed Eng*, vol. 29, pp. 1009–1018, 2001.

151. L. Wang, S. L. Jacques, and L. Zheng, Monte Carlo modeling of light transport in multi-layered tissues, *Comp Meth Pro Biomed*, vol. 47, pp. 131–146, 1995.

152. L. Zhu and L. X. Xu, Evaluation of the effectiveness of transurethral radio frequency hyperthermia in the canine prostate: Temperature distribution analysis, *ASME J Biomech Eng*, vol. 121, pp. 584–590, 1999.

153. S. Tungjitkusolmun, S. T. Staelin, D. Haemmerich, J. Z. Tsai, J. G. Webster, F. T. Lee Jr., D. M. Mahvi, and V. R. Vorperian, Three-dimensional finite-element analyses for radio-frequency hepatic tumor ablation, *IEEE Trans Biomed Eng*, vol. 49, pp. 3–9, 2002.

154. G. B. Gentili, F. Gori, and M. Leoncini, Electromagnetic and thermal models of a water-cooled dipole radiating in a biological tissue, *IEEE Trans Biomed Eng*, vol. 38, pp. 98–103, 1991.

155. M. N. Iizuka, I. A. Vitkin, M. C. Kolios, and M. D. Sherar, The effects of dynamic optical properties during interstitial laser photocoagulation, *Phys Med Biol*, vol. 45, pp. 1335–1357, 2000.

156. R. M. Arthur, W. L. Straube, J. D. Starman, and E. G. Moros, Noninvasive temperature estimation based on the energy of backscattered ultrasound, *Med Phys*, vol. 30, pp. 1021–1029, 2003.

157. T. Z. Wong, J. W. Strohbehn, K. M. Jones, J. A. Mechling, and B. S. Trembly, SAR patterns from an interstitial microwave antenna-array hyperthermia system, *IEEE Trans Microwave Theory Tech*, vol. 34, pp. 560–567, 1986.

158. E. Sparrow and J. Abraham, Simulation of gas-based, endometrial-ablation therapy, *Ann Biomed Eng*, vol. 36, pp. 171–183, 2008.

159. M. A. Astrahan, M. D. Sapozink, D. Cohen, G. Luxton, T. D. Kampp, S. Boyd, and Z. Petrovich, Microwave applicator for transurethral hyperthermia of benign prostatic hyperplasia, *Int J Hyperthermia*, vol. 5, pp. 283–296, 1989.

160. M. Astrahan, K. Imanaka, G. Jozsef, F. Ameye, L. Baert, M. D. Sapozink, S. Boyd, and Z. Petrovich, Heating characteristics of a helical microwave applicator for transurethral hyperthermia of benign prostatic hyperplasia, *Int J Hyperthermia*, vol. 7, pp. 141–155, 1991.

161. G. T. Martin, M. G. Haddad, E. G. Cravalho, and H. F. Bowman, Thermal model for the local microwave hyperthermia treatment of benign prostatic hyperplasia, *IEEE Trans Biomed Eng*, vol. 39, pp. 836–844, 1992.

162. L. X. Xu, E. Rudie, and K. R. Holmes, Transurethral thermal therapy (t3) for the treatment of benign prostatic hyperplasia (BPH) in the canine: Analysis using Pennes bioheat transfer, *Adv Biol Heat Mass Trans Biotechnol*, vol. HTD-268, pp. 31–35, 1993.

163. D. Y. Yuan, J. W. Valvano, E. N. Rudie, and L. X. Xu, 2-D finite difference modeling of microwave heating in the prostate, *Adv Biol Heat Mass Trans Biotech*, HTD-322, pp. 107–115, 1995.

164. L. Zhu, L. X. Xu, D. Y. Yuan, and E. N. Rudie, Electromagnetic (EM) quantification of microwave antenna for the transurethral prostatic thermotherapy, *Adv Biol Heat Mass Trans Biotech ASME*, HTD-337/BED-34, pp. 17–20, 1996.

165. L. Zhu, L. X. Xu, and N. Chencinski, Quantification of the 3-D electromagnetic power absorption rate in tissue during transurethral prostatic microwave thermotherapy using heat transfer model, *IEEE Trans Biomed Eng*, vol. 45, pp. 1163–1172, 1998.

166. K. Beop-Min, S. L. Jacques, S. Rastegar, S. Thomsen, and M. Motamedi, Nonlinear finite-element analysis of the role of dynamic changes in blood perfusion and optical properties in laser coagulation of tissue, *Sel Topics Quantum Electr, IEEE*, vol. 2, pp. 922–933, 1996.

167. A. Hirata, T. Asano, and O. Fujiwara, FDTD analysis of human body-core temperature elevation due to RF far-field energy prescribed in the icnirp guidelines, *Phys Med Biol*, vol. 52, pp. 5013–5023, 2007.

168. K. D. Paulsen, J. W. Strohbehn, and D. R. Lynch, Theoretical temperature distributions produced by an annular phased array-type system in CT-based patient models, *Radiat Res*, vol. 100, pp. 536–552, 1984.

169. T. Z. Wong, J. A. Mechling, E. L. Jones, and J. W. Strohbehn, Transient finite element analysis of thermal methods used to estimate sar and blood flow in homogeneously and nonhomogeneously perfused tumour models, *Int J Hyperthermia*, vol. 4, pp. 571–592, 1988.

170. J. W. Strohbehn, Temperature distributions from interstitial RF electrode hyperthermia systems: Theoretical predictions, *Int J Radiat Oncol Biol Phys*, vol. 9, pp. 1655–1667, 1983.

171. M. R. Horsman, Tissue physiology and the response to heat, *Int J Hyperthermia*, vol. 22, pp. 197–203, 2006.
172. J. Patterson and R. Strang, The role of blood flow in hyperthermia, *Int J Radiat Oncol Biol Phys*, vol. 5, pp. 235–241, 1979.
173. C. W. Song, M. S. Patten, L. M. Chelstrom, J. G. Rhee, and S. H. Levitt, Effect of multiple heatings on the blood flow in rif-1 tumours, skin and muscle of C3H mice, *Int J Hyperthermia*, vol. 3, pp. 535–545, 1987.
174. M. R. Horsman, B. Gyldenhof, and J. Overgaard, *Hyperthermia-induced changes in the vascularity and hisropathology of a murine tumor and its surrounding normal tissue*, vol. 2, Rome: Tor Vergata, 1996.
175. M. K. Jain and P. D. Wolf, A three-dimensional finite element model of radiofrequency ablation with blood flow and its experimental validation, *Ann Biomed Eng*, vol. 28, pp. 1075–1084, 2000.
176. J. P. Abraham and E. M. Sparrow, A thermal-ablation bioheat model including liquid-to-vapor phase change, pressure- and necrosis-dependent perfusion, and moisture-dependent properties, *Int J Heat Mass Trans*, vol. 50, pp. 2537–2544, 2007.
177. Z. Liu, M. Ahmed, A. Sabir, S. Humphries, and S. N. Goldberg, Computer modeling of the effect of perfusion on heating patterns in radiofrequency tumor ablation, *Int J Hyperthermia*, vol. 23, pp. 49–58, 2007.
178. J. Q. Sun, A. Zhang, and L. X. Xu, Evaluation of alternate cooling and heating for tumor treatment (accepted), *Int J Heat Mass Trans*, 2007.
179. J. J. Lagendijk, Hyperthermia treatment planning, *Phys Med Biol*, vol. 45, pp. R61–76, 2000.
180. J. J. Lagendijk, J. Crezce, and J. Mooibrock, *Principles of Treatment Planning*, vol. 1, New York: Springer, 1995.
181. J. W. Hand, R. W. Lau, J. J. Lagendijk, J. Ling, M. Burl, and I. R. Young, Electromagnetic and thermal modeling of SAR and temperature fields in tissue due to an RF decoupling coil, *Magn Reson Med*, vol. 42, pp. 183–192, 1999.
182. W. C. Dewey, Arrhenius relationships from the molecule and cell to the clinic, *Int J Hyperthermia*, vol. 10, pp. 457–483, 1994.
183. J. P. Abraham, E. M. Sparrow, and S. Ramadhyani, Numerical simulation of a BPH thermal therapy: A case study involving TUMT, *ASME J Biomech Eng*, vol. 129, pp. 548–557, 2007.
184. E. Jones, D. Thrall, M. W. Dewhirst, and Z. Vujaskovic, Prospective thermal dosimetry: The key to hyperthermia's future, *Int J Hyperthermia*, vol. 22, pp. 247–253, 2006.
185. M. W. Dewhirst, B. L. Viglianti, M. Lora-Michiels, M. Hanson, and P. J. Hoopes, Basic principles of thermal dosimetry and thermal thresholds for tissue damage from hyperthermia, *Int J Hyperthermia*, vol. 19, pp. 267–294, 2003.
186. A. Weill, A. Shitzer, and P. Bar-Yoseph, Finite element analysis of the temperature field around two adjacent cryo-probes, *ASME J Biomech Eng*, vol. 115, pp. 374–379, 1993.
187. V. Alexiades and A. D. Solomon, *Mathematical modeling of melting and freezing processes*, Washington, D.C.: Hemisphere, 1993.
188. V. R. Voller and C. R. Swaminathan, Generalized source-based method for solidification phase change, *Numer Heat Trans: Pt B*, vol. 19, pp. 175–189, 1991.
189. L. J. Hayes, K. R. Diller, H. J. Chang, and H. S. Lee, Prediction of local cooling rates and cell survival during the freezing of a cylindrical specimen, *Cryobiology*, vol. 25, pp. 67–82, 1988.
190. Y. Rabin and A. Shitzer, Numerical solution of the multidimensional freezing problem during cryosurgery, *J Biomech Eng*, vol. 120, pp. 32–37, 1998.

191. A. Zhang, L. X. Xu, G. A. Sandison, and J. Zhang, A microscale model for prediction of breast cancer cell damage during cryosurgery, *Cryobiology*, vol. 47, pp. 143–154, 2003.

192. Fluent, Fluent 6.0 user's guide, Lebanon, N.H.: Fluent, 2002.

193. J. Y. Zhang, G. A. Sandison, J. Y. Murthy, and L. X. Xu, Numerical simulation for heat transfer in prostate cancer cryosurgery, *ASME J Biomech Eng*, vol. 127, pp. 279–294, 2005.

194. H. Budman, A. Shitzer, and S. Del Giudice, Investigation of temperature fields around embedded cryoprobes, *ASME J Biomech Eng*, vol. 108, pp. 42–48, 1986.

195. J. C. Rewcastle, G. A. Sandison, L. J. Hahn, J. C. Saliken, J. G. McKinnon, and B. J. Donnelly, A model for the time-dependent thermal distribution within an iceball surrounding a cryoprobe, *Phys Med Biol*, vol. 43, pp. 3519–3534, 1998.

196. M. Jankun, T. J. Kelly, A. Zaim, K. Young, R. W. Keck, S. Selman, and J. Jankun, Computer model for cryosurgery of the prostate, in *Rev Rhum Engl*, 4th ed., pp. 193–199, 1999.

197. W. S. Wong, Temperature monitored tcap™: Time procedure outline, in *Endocare Inc. documentation*, Irvine, CA: Endocare, 2001.

198. J. C. Chato, *Selected thermophysical properties of biological materials*, New York: Plenum Press, 1985.

199. F. A. Duck, *Physical properties of tissue: A comprehensive reference book*, London: Academic Press, 1990.

200. R. Baissalov, G. A. Sandison, B. J. Donnelly, J. C. Saliken, J. G. McKinnon, K. Muldrew, and J. C. Rewcastle, A semi-empirical treatment planning model for optimization of multiprobe cryosurgery, *Phys Med Biol*, vol. 45, pp. 1085–1098, 2000.

201. G. A. Sandison, Future directions for cryosurgery computer treatment planning, *Urology*, vol. 60, pp. 50–55, 2002.

202. A. M. Gage, M. Montes, and A. A. Gage, Destruction of hepatic and splenic tissue by freezing and heating, *Cryobiology*, vol. 19, pp. 172–179, 1982.

203. N. E. Hoffmann, B. H. Chao, and J. C. Bischof, Cryo/hyper or both? Investigating combination cryo/hyperthermia in the dorsal skin flap chamber, *Proc ASME Adv Heat Mass Trans Biotech*, pp. 157–159, 2000.

204. A. Hines-Peralta, C. Y. Hollander, S. Solazzo, C. Horkan, Z. J. Liu, and S. N. Goldberg, Hybrid radiofrequency and cryoablation device: Preliminary results in an animal model, *J Vasc Interv Radiol*, vol. 15, pp. 1111–1120, 2004.

205. J. Liu, Y. Zhou, T. Yu, L. Gui, Z. Deng, and Y. Lv, Minimal invasive probe system capable of performing both cryosurgery and hyperthermia treatment on target tumor in deep tissues, *Min Invas Ther*, vol. 13, pp. 47–57, 2004.

206. J. Sun, X. Luo, A. Zhang, and L. X. Xu, A new thermal system for tumor treatment, *Conf Proc IEEE Eng Med Biol Soc*, vol. 1, pp. 474–477, 2005.

207. B. Rubinsky, Cryosurgery, *Annu Rev Biomed Eng*, vol. 2, pp. 157–187, 2000.

208. A. Zhang, L. X. Xu, G. A. Sandison, and S. Cheng, Morphological study of endothelial cells during freezing, *Phys Med Biol*, vol. 51, pp. 6047–6060, 2006.

209. G. Schuder, G. Pistorius, M. Fehringer, G. Feifel, M. D. Menger, and B. Vollmar, Complete shutdown of microvascular perfusion upon hepatic cryothermia is critically dependent on local tissue temperature, *Br J Cancer*, vol. 82, pp. 794–799, 2000.

210. J. R. Lepock, Cellular effects of hyperthermia: Relevance to the minimum dose for thermal damage, *Int J Hyperthermia*, vol. 19, pp. 252–266, 2003.

211. A. R. Moritz and F. C. Henriques, Studies of thermal injury. II. The relative importance of time and surface temperature in the causation of cutaneous burns, *Am J Pathol*, vol. 23, pp. 695–720, 1947.

212. E. L. Carstensen, M. W. Miller, and C. A. Linke, Biological effects of ultrasound, *J Bio Phys*, vol. 2, pp. 173–192, 1974.

213. F. H. Johnson, J. Eyring, and M. Palissar, *The kinetic basis of molecular biology*, New York: Wiley, 1954.

214. C. C. Church, Thermal dose and the probability of adverse effects from HIFU, *Proceedings of the AIP Conference*, 2007, vol. 911, pp. 131–137, 2007.

215. A. Westra and W. C. Dewey, Variation in sensitivity to heat shock during the cell-cycle of Chinese hamster cells *in vitro*, *Int J Radiat Biol*, vol. 19, pp. 467–477, 1971.

216. H. A. Johnson and M. Pavelec, Thermal noise in cells. A cause of spontaneous loss of cell function, *Am J Pathol*, vol. 69, pp. 119–130, 1972.

217. J. Overgaard, Pathology of heat damage, in *International Symposium on Cancer Therapy Hyperthermia Radiation*, Washington, D.C., pp. 28–30, 1975.

218. R. E. Pitt, Cryobiological implications of different methods of calculating the chemical potential of water in partially frozen suspending media, *Cryoletters*, vol. 11, pp. 227–240, 1990.

219. D. Y. Gao, J. J. McGrath, J. Tao, C. T. Benson, E. S. Critser, and J. K. Critser, Membrane transport properties of mammalian oocytes: A micropipette perfusion technique, *J Reprod Fertil*, vol. 102, pp. 385–392, 1994.

220. R. L. Levin, E. G. Cravalho, and C. E. Huggins, A membrane model describing the effect of temperature on the water conductivity of erythrocyte membranes at subzero temperatures, *Cryobiology*, vol. 13, pp. 415–429, 1976.

221. M. S. Berrada and J. C. Bischof, Evaluation of freezing effects on human microvascular-endothelial cells (HMEC), *Cryoletters*, vol. 22, pp. 353–366, 2001.

222. L. F. Fajardo and S. D. Prionas, Endothelial cells and hyperthermia, *Int J Hyperthermia*, vol. 10, pp. 347–353, 1994.

223. N. Zook, J. Hussmann, R. Brown, R. Russell, J. Kucan, A. Roth, and H. Suchy, Microcirculatory studies of frostbite injury, *Ann Plast Surg*, vol. 40, pp. 246–253; discussion 254–255, 1998.

224. J. X. Dong, P. Liu, A. Zhang, and L. X. Xu, Immunologic response induced by alternated cooling and heating of breast cancer, *Proceedings of the 29th IEEE EMBS*, 2007.

225. H. Li, J. Huang, J. Lv, H. An, X. Zhang, Z. Zhang, C. Fan, and J. Hu, Nanoparticle PCR: Nanogold-assisted PCR with enhanced specificity, *Angew Chem Int Ed Engl*, vol. 44, pp. 5100–5103, 2005.

226. M. Li, Y. C. Lin, C. C. Wu, and H. S. Liu, Enhancing the efficiency of a pcr using gold nanoparticles, *Nucl Acids Res*, vol. 33, p. e184, 2005.

6

Thermal Interactions between Blood and Tissue

Development of a Theoretical Approach in Predicting Body Temperature during Blood Cooling and Rewarming

L. Zhu, T. Schappeler, C. Cordero-Tumangday, and A. J. Rosengart

CONTENTS

6.1 INTRODUCTION

The human thermoregulatory system is capable of maintaining body core temperature near 37°C over a wide range of environmental conditions and during exercise. This is largely facilitated by the vascular system of the body. In addition to its primary functions of mass transport of body metabolisms and regulation of systemic blood pressure, the human vascular system plays an important role in systemic thermoregulation. The blood, which can be viewed as the heat transport medium contained within the vasculature, exerts a dual influence on the thermal energy balance within the body. It acts as either a source or sink of thermal energy by redistributing and balancing the temperature differences between the arterial blood and the individual organ tissue. This dynamic tissue heat transfer is fundamental as some body organs give rise to abundant amounts

197

of heat during exercise (i.e., muscles), whereas others are vitally dependent on nutrient supply (i.e., the brain). Further, heat redistribution is life saving during changing environmental conditions, as in warming of the body during exposure to cold, or it can be disadvantageous, for example, by counteracting therapeutic temperature elevations for cancer treatment with increased blood perfusion of the target region.

To maintain a normal, or euthermic, body temperature, the vasculature facilitates the redistribution and transfer of heat throughout the body, preserving a steady core temperature for all vital organs and making the human body relatively insensitive to environmental temperature changes. Theoretical calculations have shown that if the body would solely depend on heat conduction and not utilize blood-induced heat redistribution, the body would reach its steady-state temperature at about 80°C [1], which, of course, is incompatible with survival. Heat redistribution and the consistency of an euthermic core temperature of 37°C rely on two main mechanisms. First, adjustments in cardiac output or blood recirculation time induced by central and local thermoregulatory responses allow rapid changes in heat "turnover"; and, second, regulation of blood perfusion of the body surface and the skin and subcutaneous tissue permits either enhanced evaporative loss or preservation of heat as demanded by internal and external environmental temperature changes. If those compensatory mechanisms are exhausted, such as in heat stroke, the body experiences a critical decline in its ability to regulate temperature and will decompensate if not treated. The clinical implications of thermoregulatory failure are easily underscored by the fact that, despite aggressive medical care, about 25% of heat stroke victims experience organ failure and 7% to 14% experience permanent neurological deficits [2].

Active control of body temperature is increasingly employed therapeutically in several clinical scenarios, most commonly to protect the brain from the consequences of either primary injury (e.g., head trauma, or stroke) or secondary injury (e.g., after cardiac arrest with brain hypoperfusion). Mild to moderate hypothermia, during which brain temperature is reduced to 30°C to 35°C, has been studied, among others, as an adjunct treatment for protection from cerebral ischemia during cardiac bypass injury [3], carotid endarterectomy [4], and resection of aneurysms [5], and it is also commonly employed in massive stroke and traumatic brain injury patients [6,7]. Even minute reductions in brain temperature as small as 1°C and, importantly, the avoidance of any hyperthermia can substantially reduce ischemic cell damage [8,9] and improve outcome [10]. Some of the beneficial effects of brain hypothermia include decrease in cerebral edema formation and intracranial pressures, reduction of tissue oxygen demands [11], and amelioration of numerous deleterious cellular biochemical mechanisms, including calcium shift, excitotoxicity, lipid peroxidation, and other free-radical reactions [12].

Introduction of brain hypothermia is induced via whole-body (systemic) cooling as currently no efficient and safe cooling device for targeted, selective brain cooling exists, although novel approaches and investigations are ongoing [13].

To achieve systemic cooling, many methods have been advocated and demonstrated various clinical success. Current strategies are either relatively safe but only modestly effective (e.g., antipyretics [14,15], blankets and garments [16], and peripheral infusion of a coolant [17]), or more effective in temperature reductions but invasive and elaborate (e.g., endovascular cooling or extracorporeal heat exchange) [18,19]. Unfortunately, practical limitations in procedural accessibility and nursing care availability often prohibit the use of more complicated and invasive fever reduction methods, highlighting the ongoing need for safer, simpler, and more cost-effective cooling strategies. In clinical practice, the most effective way to control the temperature in critically ill patients is to directly cool the blood of major veins using intravascular cooling catheters [20]. Cooling rates as high as 5°C/hour can be achieved depending on the cooling capacity of the device and the patient's cooling response [19,21–23].

The effect of blood flow on heat transfer in living tissue has been addressed previously [1], and, because of the complex geometry of the vascular system, two theoretical approaches are currently used to assess the blood flow effects in biological systems. The first comprises vascular models that consider blood vessels as rigid tubes incorporated in organ tissues. However, despite the availability of detailed information on vascular geometry, only several large blood vessels are generally considered in these models, neglecting the remaining vascular tree in order to reduce the complexity. Nevertheless, with the advance of computational techniques in recent years, models simulating more complex vascular networks have progressed rapidly and already demonstrated great potentials in accurate and point-to-point blood and tissue temperature mapping [24]. Second, an approach that is different from modeling the vasculature and evaluates flow effects in biological systems is employed in continuum models, which average the influence of flow and temperature on controlled blood and tissue volume. Thus, in the considered tissue region, the blood flow effect is treated by either adding an additional term or changing some of the thermophysical parameters in the traditional heat conduction equation. The continuum models are simple to use as long as one or two representative parameters related to blood flow are available. The limitation of the continuum models is their inability to map point-to-point temperature variation along blood vessels. The two most widely employed equations are the Pennes bioheat equation [25] and the Weinbaum-Jiji equation [26]. In the Pennes bioheat equation, the blood flow effect is modeled as a source term, and its strength is proportional to the local blood perfusion rate and the temperature difference between blood and tissue. In the Weinbaum-Jiji equation, blood flow is modeled as an enhancement in tissue thermal conductivity. In the Pennes equation, blood temperature is considered to be the same as the body core temperature; in the Weinbaum-Jiji equation, on the other hand, the effect of the blood temperature serves as the boundary condition of the tissue domain. As outlined above, due to the complexity of the vasculature, continuum models appear more favorable in simulating the temperature field of the human body.

In either continuum model (Pennes or Weinbaum-Jiji), blood temperature is an input to the governing equation of the tissue temperature. However, in situations in which the blood temperature is actively lowered or increased, both continuum models seem inadequate to account for the tissue–blood thermal interactions and to accurately predict the expected body temperature changes.

In this study, we develop a new simple theoretical model to study blood–tissue thermal interaction in the human body. In addition to using the Pennes bioheat equation to simulate the body temperature distribution during blood cooling and rewarming, a heat transfer equation for the blood temperature is developed to account for the energy balance among blood and tissue as well as for the influence of external cooling or rewarming. The new approach provides the solution procedures of solving for both the body and blood temperatures during active blood temperature modifications. A sample calculation of the transient body temperature is conducted using the theoretical approach to show its feasibility and accuracy. So far, the response pattern of an individual patient to manipulations of systemic temperature changes, such as vascular cooling or rewarming using intravenous fluids or endovascular catheters, has not been well delineated in the literature. A new model simulating the cooling and rewarming process and temperature interactions between the blood and tissue (organs), as presented in this study, will help predict precisely these changes and is, therefore, of both clinical and scientific importance.

6.2 MATHEMATICAL FORMULATION

The mathematical formulation consists of simulation of both body temperature distribution and energy balance of the blood compartment of the body. The Pennes bioheat equation is used to simulate the body temperature, while a lumped system analysis is implemented to predict temperature change of the blood during clinical applications.

6.2.1 Body Temperature Distribution

The human body can be modeled as a simple cylinder or a combination of components representing torso, head, and limbs. As shown in Figure 6.1, the body is exposed to evaporation, convection, and radiation heat transfer on the skin surface. The Pennes bioheat equation simplifies the vasculature by modeling blood flow as a source term. The metabolism is considered as a heat source within the body tissue, and the blood perfusion in tissue is modeled as a heat source or sink depending on whether the arterial temperature is higher or lower than the local tissue temperature. The governing equation for the temperature field in the body tissue can be written as

$$\rho_t c_t \frac{\partial T_t}{\partial t} = k_t \nabla^2 T_t + q_m + \rho_b c_b \omega (T_a - T_t) \qquad (6.1)$$

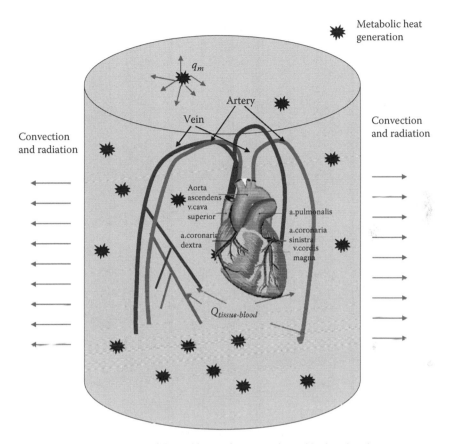

Figure 6.1   Schematic diagram of thermal interaction among tissue, blood, and environment.

where subscripts t and b refer to tissue and blood, respectively; T_t is the body tissue temperature; ρ is density; c is specific heat; k_t is the thermal conductivity of tissue; q_m is the volumetric heat generation rate (W/m³) due to metabolism; and ω is the local blood perfusion rate. The above governing equation can be solved once the boundary conditions and initial condition are prescribed. The boundary at the skin surface is modeled as a convection boundary subject to an environment temperature of T_{air} and a convection coefficient of h. The coefficient h can be considered as the overall heat transfer coefficient related to the combined thermal resistance due to convection, radiation, evaporation at the skin surface, as well as conductive resistance of clothes. The Pennes bioheat equation has been used extensively in the past to model tissue temperature fields for various clinical applications and has been considered as an accurate description of tissue temperature field.

We would like to emphasize that in Equation (6.1), the arterial temperature T_a is used as an input. To predict the body temperature response to change of the arterial blood temperature during clinical applications, one needs to understand the thermal interaction between blood and tissue. Unfortunately, the Pennes equation alone would not predict the thermal interaction. In this study, we will analyze steady-state temperature distribution and predict the overall energy balance of the tissue–blood heat exchange.

Blood interaction with tissue is described by the Pennes perfusion source term (the third term on the right side of Equation 6.1). Depending on the local tissue temperature, it may act as either a heat source or a heat sink to tissue during circulation. Let's consider that in a steady-state situation, the body can maintain a stable body temperature and blood temperature. Maintaining a stable blood temperature implies that the total heat transfer between the blood and tissue should be equal to zero during steady state. It is suggested that heat loss from blood to tissue in the periphery tissue region would be compensated by heat gain in the central region. Based on the Pennes bioheat equation, the rate of the total heat loss from the blood to the tissue during steady state is

$$Q_{blood-tissue,0} = \iiint\limits_{body\ volume} \rho_b c_b \omega (T_{a0} - T_{t0}) dV_{body} = \rho_b c_b \bar{\omega}(T_{a0} - \bar{T}_{t0}) V_{body} = 0 \qquad (6.2)$$

where subscript 0 represents steady state, and V_{body} is the body volume. Equation (6.2) implies that both density and specific heat are constant. In Equation (6.2), $\bar{\omega}$ is the volumetric-average blood perfusion rate, defined as

$$\bar{\omega} = \frac{1}{V_{body}} \iiint\limits_{body\ volume} \omega\, dV_{body} \qquad (6.3)$$

where $\bar{T}_t$ is the weighted-average tissue temperature defined by Equation (6.2), and is given by

$$\rho c \bar{\omega}(T_{a0} - \bar{T}_t) V_{body} = \iiint\limits_{body\ volume} \rho c \omega (T_{a0} - T_{t0}) dV_{body} \qquad (6.4)$$

Note that $\bar{T}_t$ may not be equal to the volumetric-average tissue temperature (T_{avg}) if ω is not uniform in the body tissue. Equation (6.2) implies that during steady state, the arterial blood temperature, T_a, should be the same as the weighted-average tissue temperature, $\bar{T}_t$. The average body temperature $\bar{T}_t$ can be predicted by solving Equation (6.1) and calculated by Equation (6.4). Steady-state thermal regulation in the human body ensures the energy balance to maintain stable body and blood temperatures. In this study, we assume that the body maintains its steady-state temperature via adjusting the overall heat transfer coefficient on the skin surface. The overall heat transfer coefficient can be changed by altering the evaporation rate at the skin surface and/or adding or removing clothes to change the conductive resistance. However, during a transient heat transfer process such as blood cooling or rewarming, the total rate of heat loss from blood to tissue is no longer equal to zero, and is given by

$$Q_{blood-tissue}(t) = \iiint\limits_{body\ volume} \rho_b c_b \omega (T_a(t) - T_t(r,t)) dV_{body} = \rho_b c_b \overline{\omega} [T_a(t) - \overline{T_t}(t)] V_{body} \neq 0$$

(6.5)

6.2.2 Energy Balance in Blood

During clinical applications, external heating or cooling of the blood can be implemented to manipulate the body temperature. In this study, the blood in the human body is represented as a lumped system. We assume that a typical value of the blood volume of the body, V_b, is approximately five liters. External heating or cooling approaches can be implemented via an intravascular catheter or intravenous fluid infusion. A mathematical expression of the energy absorbed or removed per unit time is determined by the temperature change of the blood, and is written as

$$\rho_b c_b V_b [T_a(t + \Delta t) - T_a(t)]/\Delta t \approx \rho_b c_b V_b \frac{dT_a}{dt}$$

(6.6)

where $T_a(t)$ is the blood temperature at time t, and $T_a(t + \Delta t)$ is at time $t + \Delta t$; ρ_b is the blood density; and c_b is the specific heat of blood. In the mathematical model, we propose that the energy change in blood is due to energy added or removed by external heating or cooling (Q_{ext}) and heat loss to the body tissue in the systemic circulation ($Q_{blood-tissue}$). Therefore, the governing equation for the blood temperature can be written as

$$\rho_b c_b V_b \frac{dT_a}{dt} = Q_{ext}(T_a, t) - Q_{blood-tissue}(t)$$

$$= Q_{ext}(T_a, t) - \rho_b c_b \overline{\omega} V_{body} (T_a - \overline{T_t})$$

(6.7)

where Q_{ext} can be a function of time and the blood temperature due to thermal interaction between blood and the external cooling approach, T_a is the arterial temperature, $\overline{T_t}$ is the weighted-average tissue temperature, and $\overline{\omega}$ can be a function of time. Equation (6.7) cannot be solved alone since $\overline{T_t}$ is determined by solving the Pennes bioheat equation. One needs to solve Equations (6.1) and (6.7) simultaneously.

6.2.3 Numerical Method of Solving Equation (6.7)

The governing equation for the blood temperature is a first-order ordinary differential equation and can be solved numerically since most of the parameters in Equation (6.7) are a function of time, and solving for the blood temperature depends on solving for the whole-body temperature field. The time derivative on the left side of Equation (6.7) is discretized using finite differences. Although either an implicit or explicit method can be used, an undesirable feature of the explicit method is that it imposes a restriction on the time step to avoid oscillation in the solution, which is physically impossible. In this study, Equation (6.7) is discretized using either the implicit scheme or the explicit scheme as follows:

$$implicit: \quad \rho_b c_b V_b \frac{T_a^{P+1} - T_a^P}{\Delta t} = Q_{ext}\left(T_a^{P+1}\right) - \rho_b c_b \overline{\omega} V_{body}\left(T_a^{P+1} - \overline{T}_t^P\right) \quad (6.8a)$$

$$explicit: \quad \rho_b c_b V_b \frac{T_a^{P+1} - T_a^P}{\Delta t} = Q_{ext}\left(T_a^P\right) - \rho_b c_b \overline{\omega} V_{body}\left(T_a^P - \overline{T}_t^P\right) \quad (6.8b)$$

where Δt is the time interval for solving the transient blood temperature, and the superscript P denotes the time dependence of temperature. Superscript P and $P+1$ represent temperatures associated with the previous and new times, respectively. Note that Equation (6.8a,b) is the discretized equation for T_a only; the new $(P+1)$ blood temperature can be easily determined by rearranging the equation. The finite element method (FEMLAB®; COMSOL, Burlington, Massachusetts) is used to solve Equation (6.1) for the body temperature distribution. The following procedures are the general approach implemented in solving Equations (6.1) and (6.8).

1. Select a proper temperature of blood, T_{b0}, as the input to the Pennes bioheat equation. Use the FEMLAB software to solve for the steady-state temperature field of the body before the clinical treatment. Adjust the overall heat transfer coefficient h so that the blood temperature T_{a0} is equal to the weighted-average body temperature $\overline{T}_{t0}$. The steady-state temperature fields (T_{a0} and $\overline{T}_{t0}$) serve as the initial temperature fields of the transient heat transfer process.
2. Substitute the initial temperature values (time step P) into Equation (6.8), and determine the blood temperature at the next time step ($P+1$).
3. The newly determined blood temperature at time step $P+1$ is then used as the input to the Pennes bioheat equation to solve for $\overline{T}_t^P$.
4. Steps 2 and 3 are repeated to solve for the transient temperature fields of the subsequent time steps.

6.3 RESULTS

In this sample calculation, we apply the model to a blood-cooling application, during which coolant is pumped into the inner tube of a catheter inserted into the femoral vein and advanced to the veno-vera. Once the coolant reaches the catheter, it flows back from the outer layer of the catheter and out of the cooling device.

This cooling device has been used in clinical trials in recent years as an effective approach to decrease the temperature of the body for stroke or head injury patients. Based on previous research of this device, the cooling capacity of the device is around 100 W.

All the finite element calculations, including the finite element mesh generation for Equation (6.1), were performed on FEMLAB 3.1, operated on a Pentium IV processor of 2.79 GHz speed and using 1 GB of memory under a Windows XP SP2 Professional Operating System. The numerical model was obtained by applying the Galerkin formulation to Equation (6.1). The total number of tetrahedral elements of the finite element mesh was around 135,000. The time-dependent problem was solved using an adaptive time-stepping scheme wherein the convergence criterion was kept at 10^{-6}. Mesh independency was checked by increasing the number of elements in the calculation domain by 100% over the current mesh. The finer mesh showed less than 1% difference in the temperature field. The central processing unit (CPU) time for calculating each time step ($\Delta t = 60$ seconds) is approximately 8 minutes.

A typical male human body has a body weight of 81 kg and a volume of 0.074 m^3. We consider two body geometries and test whether the developed model yields similar results. As shown in Figure 6.2, the body can be modeled as a simple cylinder that is 0.23 m in diameter and 1.8 m tall. A more realistic body geometry is illustrated by Figure 6.3, where the body consists of limbs, a torso (internal organs and muscle), a neck, and a head. The limbs and neck are modeled as cylinders consisting of muscle. Note that the body geometry can be modeled more realistically if one includes a skin layer and a fat layer in each compartment. However, since our objective is to illustrate the principle and feasibility of the developed model, those details are neglected in the sample calculation. The simple

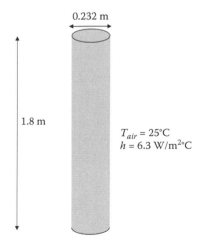

Figure 6.2   Schematic diagram of a simplified human body geometry.

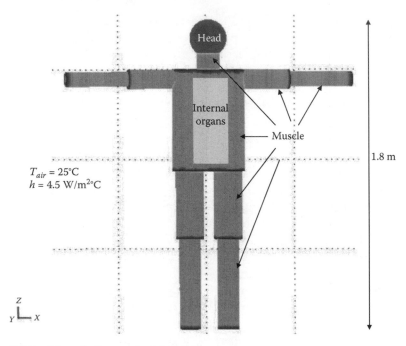

Figure 6.3   Schematic diagram of a detailed human body geometry.

geometry results in a body surface area of 1.31 m², while the detailed geometry has a body surface area of 1.8 m². Based on a previous study by Mosteller [27], the body surface area is usually calculated by the following formula:

$$surface\ area\ (m^2) = \sqrt{[height\ (cm)\ x\ weight\ (kg)]/3600} \qquad (6.9)$$

For a realistic human body with a body weight of 81 kg and a volume of 0.074 m³, the calculated body surface area based on Equation (6.9) is around 2.01 m². The detailed body geometry used in this study agrees relatively well with the realistic body surface area. Table 6.1 gives the geometrical parameters used in both models.

The average heart stroke volume is 0.7 liters, and the average heart beats 75 beats/minute. Based on the total body mass of 81 kg, one can determine the average blood perfusion rate as 6.773 ml/minute, 100 g tissue, or 0.0011 (s⁻¹). For the simple one-compartment model, the average metabolic heat generation rate is estimated based on food consumption of 2000 kCal/day and is equal to 1250 W/m³. The metabolic heat generation rate and local blood perfusion rate used in the detailed body model are listed in Table 6.2. The physical and physiological properties are obtained from the literature [1,28–31].

The body is exposed to an evaporation, convection, and radiation environment with an overall heat transfer coefficient h and a room temperature of 25°C.

Table 6.1   Geometrical Parameters Used in Both Models

	Simple Geometry	Detailed Geometry
Height (m)	1.8	1.8
Diameter (m)	0.232	—
Head volume (m³)	n/a	0.004206
Neck volume (m³)		0.001287
Internal organ volume (m³)		0.019503
Torso (without internal organs) volume (m³)		0.019406
Upper-arm volume (each; m³)		0.002085
Lower-arm volume (each; m³)		0.001622
Upper-leg volume (each; m³)		0.00695
Lower-leg volume (each; m³)		0.005212
Total volume V_{body} (m³)	0.0761424	0.076142
Total blood volume V_b (m³)	0.005	0.005
Total body mass m_{body} (kg)	80.7	80.7

The clothes covering the human subject can be modeled as a conductive resistance and incorporated into the overall heat transfer coefficient. Adding the conductive resistance of clothes will result in an overall heat transfer coefficient smaller than that due to evaporation, convection, and radiation alone. In this sample calculation, ω, h, and T_{air} are assumed unchanged during the cooling process. The initial blood temperature is assigned as 37°C.

The temperature field in the simplified human body can be considered as three-dimensional, and the steady-state temperature can be determined by the FEMLAB software using the finite element method. During steady state, the convection heat transfer coefficient h is adjusted to 6.3 W/m²°C and 4.5 W/m²°C for the simple and detailed models, respectively, so that the average body temperature $\overline{T}_{t0}$ is equal to the blood temperature T_{a0}. The different values of h used in

Table 6.2   Physical and Physiological Parameters Used in Both Models

		Simple Geometry	Detailed Geometry
Thermal conductivity k_t		0.5 W/m°C	0.5 W/m°C
Density ρ_b or ρ_t		1060 kg/m³	1060 kg/m³
Specific heat c_b, or c_t		3800 J/kg°C	3800 J/kg°C
	Muscle	0.001129 (1/s)	0.0005 (1/s)
Blood perfusion rate ω	Head	n/a	0.0083333 (1/s)
	Internal organ	n/a	0.001266 (1/s)
Metabolic heat	Muscle	1249.7 W/m³	553.5 W/m³
generation rate q_m	Head	n/a	9225 W/m³
	Internal organ	n/a	1401.5 W/m³

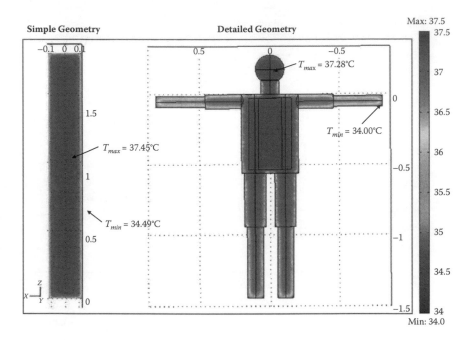

Figure 6.4   Initial steady-state temperature contours of the human body.

those two models can be explained by the different body surface areas. From the energy balance of the entire body during steady state, the metabolic heat generation inside the body has to be dissipated from the body surface via convection and radiation. Figure 6.4 shows the steady-state temperature distribution in the body ranging from 37.28°C inside the brain to 34°C at the fingertip. The left image of the temperature contours is the temperature distribution using the simple geometry. The maximum body temperature is usually slightly higher than the arterial temperature due to metabolic heat generation. The maximum temperature usually occurs inside the brain tissue due to its large metabolic heat generation rate. Note that the detailed body geometry gives more realistic temperature contours in the body, while the simple model still correctly predicts the maximum and minimum temperatures in tissue. The simulated steady-state temperature distribution is used as the initial temperature field for the simulation of the transient heat transfer process.

Since the cooling capacity of the cooling device is around 100 W, one models the external cooling to the blood Q_{ext} as −100 W. One can rewrite the discretized equation for T_a using the explicit scheme (Equation 6.8b) as follows:

$$T_a^{P+1} = T_a^P\left(1 - \frac{\rho_b c_b \overline{\omega} V_{body}\Delta t}{\rho_b c_b V_b}\right) + \frac{Q_{ext}\Delta t}{\rho_b c_b V_b} + \overline{T}_t^P\left(\frac{\rho_b c_b \overline{\omega} V_{body}\Delta t}{\rho_b c_b V_b}\right) \qquad (6.10)$$

Note that the coefficient of the first term on the right side of Equation (6.10) has to be positive to avoid oscillation of the solution. This restriction requires

$$1 - \frac{\rho_b c_b \overline{\omega} V_{body} \Delta t}{\rho_b c_b V_b} \geq 0 \quad or \quad \Delta t \leq \frac{\rho_b c_b V_b}{\rho_b c_b \overline{\omega} V_{body}} = \frac{V_b}{\overline{\omega} V_{body}} \tag{6.11}$$

In this sample calculation, Δt is selected as 60 seconds, which satisfies the constraint in Equation (6.11).

The discretized equation for T_a using the implicit scheme (Equation 6.8a) can be rewritten as

$$T_a^{P+1} = \frac{\frac{Q_{ext} \Delta t}{\rho_b c_b V_b} + T_a^P + \overline{T}_t^P \left(\frac{\rho_b c_b \overline{\omega} V_{body} \Delta t}{\rho_b c_b V_b} \right)}{\left(1 + \frac{\rho_b c_b \overline{\omega} V_{body} \Delta t}{\rho_b c_b V_b} \right)} \tag{6.12}$$

The time step is also selected as 60 seconds when the implicit scheme is applied.

Figure 6.5 shows the simulated temperature of the arterial blood during the first 20 minutes of the cooling. The observed difference between the implicit and explicit schemes used in both body geometries suggests that the time step selected should be shorter than 60 seconds to minimize the error associated with the approximation of the time derivative in the numerical method. Up to a 0.74°C decrease in the arterial temperature is achieved during the first 20 minutes using the implicit scheme. One notices the large initial temperature drop, and the temperature decay rate is slowed down and stable at approximately 0.019°C/min in the detailed model. Based on the cooling rate, it is expected that the arterial blood temperature will decrease to 35.5°C after 1 hour and 34.4°C after 2 hours.

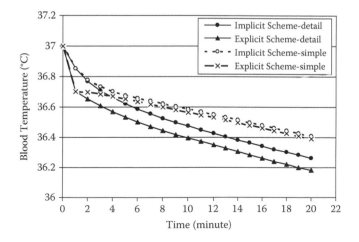

Figure 6.5 Simulated blood temperature changes during the cooling using both implicit and explicit schemes.

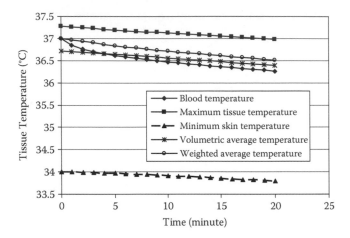

Figure 6.6   Temperature decays during the cooling process using the detailed geometry and implicit scheme.

In clinical applications of whole-body cooling for stroke or head injury patients, physicians are more interested in the temperature of the body tissue. The maximum tissue temperature, the minimum tissue temperature at the skin surface, the volumetric-average body temperature (T_{avg}), and the weighted-average body temperature ($\overline{T}_t$) are plotted in Figure 6.6. The difference between the volumetric-average body temperature and the weighted-average body temperature is due to their different definitions. All tissue temperatures decrease almost linearly with time, and after 20 minutes, the cooling results in approximately a 0.3~0.5°C tissue temperature drop. The cooling rate of the skin temperature is smaller (0.2°C per 20 min). As shown in Figure 6.7a, the initial cooling rate of the blood temperature in the detailed model is very high (~0.14°C/min), and then it decreases gradually until it is stabilized after approximately 20 minutes. On the other hand, cooling the entire body (the volumetric-average body temperature) starts slowly and gradually catches up. It may be due to the inertia of the body mass in responding to the cooling of the blood. Figure 6.7a also illustrates that after the initial cooling rate variation, the stabilized cooling rates of all temperatures approach each other, and they are approximately 0.019°C/min or 1.15°C/hour. The cooling rates of all temperatures in the simple model are plotted in Figure 6.7b, where the cooling rates converge to approximately 0.018°C/min only after 10 minutes. The simulated results demonstrate the feasibility of inducing mild body hypothermia (34°C) within 3 hours using the cooling approach.

The first term on the right side of Equation (6.8) represents the maximum cooling rate of the body if the blood does not gain heat from the tissue. It is estimated that the cooling rate induced by this term is approximately equal to 18.9°C/hour, which is much higher than the achieved cooling rate of the body. It is understandable since the cooling capacity induced by the external cooling device

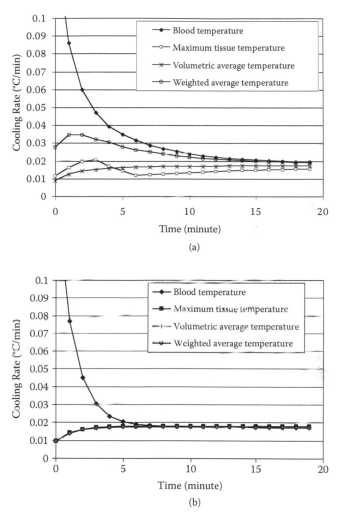

Figure 6.7   Induced cooling rates of the blood temperature, the maximum temperature, the volumetric average temperature, and the weighted average temperature. (a) The detailed model, and (b) the simple model.

cools not only the blood but also the entire body. Therefore, the second term on the right side of Equation (6.8) representing blood–tissue thermal exchange is mainly dependent on the temperature difference between the blood and the tissue. During cooling, blood is colder than the average body temperature; therefore, the second term on the right side of Equation (6.8) is positive to account for the heat transfer from the tissue to the blood. As shown in Figure 6.8, the initial temperature difference is zero since the thermal balance is established during the initial steady state (Equation 6.2). Later, the temperature difference changes and becomes stabilized as 2.5°C. After the difference between T_a and $\overline{T}_t$ is stabilized

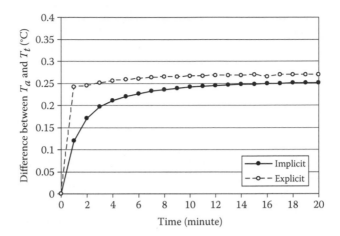

Figure 6.8   Difference between the blood temperature and the weighted average body temperature during cooling.

(approximately 20 minutes after the initiation of the cooling), the right side of Equation (6.8) becomes a constant. This is the direct result of the assumptions that the external cooling capacity (Q_{ext}) and the metabolic heat generation rate, as well as all the physiological parameters such as the blood perfusion rate and convective coefficient, are kept the same during the cooling simulation. Once the temperature difference is unchanged, it is expected that the cooling rate of the blood temperature governed by Equation (6.8) would become a constant. One can then extrapolate the current simulated data to a later time duration.

The effect of simplification of the body geometry on the transient body temperature is illustrated in Figure 6.9. The implicit scheme is used with a time step

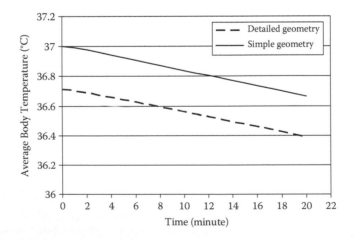

Figure 6.9   The effect of model geometry on the volumetric average body temperature.

of 60 seconds. Although the two geometries are quite different, the yielded cooling rates of the volumetric-average body temperature (T_{avg}) during the cooling are very similar. Both geometries have the same body weight and volume. The results imply that it is the overall energy balance that determines the temperature reduction of the body tissue. The limitation of the simple model lies in its inability to model any realistic thermal regulation during the cooling process.

6.4 DISCUSSION

The provided sample calculations identified the feasibility of applying the model for predicting the temperature distribution in the human body during blood cooling and rewarming. The accuracy of the model can be confirmed by comparing the predicted results from the model to those using a simple lumped analysis of the whole body induced by an external, intravascular cooling device with a capacity of 100 W. The simple lumped analysis of the whole body assumes that the heat transfer through the skin surface is unchanged during the cooling. It is estimated by the following equation:

$$\text{Cooling rate} = -100 / (\rho_t V_{body}\, c_t) = -100 / (1060 \times 0.076142 \times 3800)$$

$$= 3.2605 \times 10^{-4}\,°C/\text{second} = 0.0196°C/\text{min} \qquad (6.13)$$

The estimated cooling rate is very close to the rate predicted by the developed model employing both the implicit and explicit schemes. Decreasing the time step would further improve the accuracy of the model; however, this would increase the CPU time.

The theoretical models developed by Wissler and other investigators [28–31] similarly introduced the whole body as a combination of multiple compartments. However, a major difference between the current model and previous approaches is the assessment of the thermal interactions between the blood and tissue in each compartment. The majority of previously published studies introduced a pair of countercurrent artery and vein with their respective branching (flow system), and then modeled the temperature variations along this flow system to derive the heat transfer between the blood vessels and tissue within each flow segment. Such an approach is computationally intensive although the models are capable of delineating the temperature decay along the artery and the rewarming by the countercurrent vein. The modeling approach introduced here greatly simplifies the theoretical simulations and, therefore, requires less computational resources and time. In our approach, the vasculature is modeled as a lumped system only varying as a function of time, and the blood–tissue thermal interactions are evaluated by integrating the Pennes perfusion source term over the entire body tissue. The combining of these modeling steps is based on two assumptions. First, several previous theoretical and experimental studies have suggested that the thermal equilibration length of the major supply artery is much longer than its physical length.

This implies that the temperature of the arterial blood in each tissue compartment should be very close to the body core temperature. Second, we chose to neglect the rewarming effect by the venous blood due to countercurrent heat exchange. Although recent studies by the authors and others [32,33] suggested a 30% rewarming rate originating from the venous system and, therefore, multiplying the Pennes perfusion source term by a correction coefficient of 0.7, there are currently no rigorously designed animal studies validating such theoretical reasoning. This lack of animal data originates mainly from the difficulty in measuring with precision the local *in vivo* blood perfusion rate. Should, however, this correction coefficient value become experimentally verified and available, the Pennes perfusion term can easily be modified to also account for the effect of countercurrent venous rewarming in tissue. We feel the current model using the Pennes perfusion term and lumped system of the blood is simple to use in comparison with these previous whole-body models while providing meaningful and accurate theoretical estimates.

The presented sample calculations suggest that the newly developed model is accurate in describing the overall thermal balance between blood and body tissue during clinical applications involving blood cooling. Recent animal and clinical studies have demonstrated the effectiveness of improving patient outcomes by inducing systemic cooling and/or actively avoiding hyperthermic (febrile) episodes after brain injury [34–39]. Controlling fever has become a more important clinical treatment task since it is known that more than 60% of brain injury victims developed fever within 72 hours, and the 3-month mortality rate in these patients is significantly higher than it is for those who remained normothermic (1% versus 15.8%, respectively) [40]. Theoretical models have suggested that the actual cooling rate of the volumetric-average body temperature depends largely on the cooling capacity of the cooling device. Among the currently utilized strategies for direct blood cooling are intravascular cooling by placing a cooling catheter within a larger vein of the trunk, or, less invasively, using a peripheral infusion scheme, the intravenous bolus infusion of saline with ice slurry [17,41,42]. Based on these calculations and our clinical experience, a cooling capacity of 100 W delivered intravenously to a patient of average build will result in a ~1.2°C/hour cooling rate. For example, intravascular cooling can achieve within less than 3 hours systemic and, hence, brain, hypothermia of 34°C, or, similarly, reduce critical fever of 40°C to euthermia temperatures. For instance, an intravenous cooling catheter manufactured by Innercool® (Innercool Therapies, San Diego, California) can deliver a cooling capacity of up to 200 W (personal communication, August 2007) and, therefore, is capable of rapid systemic cooling within even shorter time frames. As expected, the cooling capacity of intravenous infusion regimens is much smaller than that of the intravenous cooling catheter. This is mainly due to limitations in the infusion rate and luminal diameter of quickly accessible, peripheral veins. Recent theoretical evaluations by our group suggest that the cooling capacity is approximately 40 W per hour when using 50% ice slurry at an infusion rate of 450 ml/hour [17]. Therefore, the cooling rate induced by intravenous infusion of 50% ice slurry-saline solutions can be extrapolated to be ~0.5°C/hour, less than half than that expected from a 100 W cooling catheter.

Another application of the developed and here-proposed model involves the representation of the rewarming process (i.e., after induced systemic hypothermia or for the treatment for full-body hypothermia). Rapid rewarming in those applications may result in dangerous rebound intracranial pressure elevation and cerebral perfusion pressure reduction, ultimately worsening outcome in brain injury. Several clinical investigations [41,43] have emphasized the importance of gradual rewarming to minimize this clinical problem. In support, a recent animal study performed by Diao and Zhu [44] suggested that a fast rewarming rate in normal hypothermic rats results in local blood perfusion–metabolism mismatch, which may explain the clinical worsening observed in some patients exposed to accelerated rewarming. Conveniently, the current model can be applied to assist in the design of an optimal rewarming strategy for clinical practice. In a research paper published by our group [45], a theoretical model was developed to simulate targeted brain hypothermia induced by an interstitial cooling device in the human neck. This previous model can be combined with the present whole-body model for designing the temperature elevation strategy in the interstitial cooling device to achieve desired rewarming rates in both the body core and the brain.

In the presented sample calculations, we also found that the simple geometry model yields very similar cooling rates to those predicted by a more detailed model. However, the simple model is not capable of simulating blood redistribution in the body since local blood perfusion and/or metabolic heat generation rates are a function of the tissue location. Both models can be used in applications when the physiological parameters, such as local blood perfusion rate, metabolic heat generation rate, or environmental thermal conditions, vary during blood cooling or rewarming. Previous studies have shown that the temperature dependence of local blood perfusion is based on a coupled relationship between blood perfusion and metabolism. For example, it has been calculated that hypothermia decreases the cerebral metabolic rate by an average value of 7% for the first 1°C reduction in temperature, while it is reduced by 50% of normal baseline when the temperature reduction approaches 10°C [46]. Our newly developed models are capable of including the dynamic responses of physiological parameter changes during the simulation.

Although the model was developed for applications involving blood cooling or rewarming, the detailed geometry can also be used to accurately predict body temperature changes during exercise. It is well known that strenuous exercise increases cardiac output, redistributes blood flow from internal organs to muscle, increases metabolism in exercising muscle, and enhances heat transfer to the skin. Our detailed model can be easily modified to also include a skin layer and a fat layer in the compartments of the limbs. Further, redistribution of blood flow from the internal organs to the musculature can be modeled as changes of the local blood perfusion rate in the respective compartments, and the enhanced skin heat transfer can be adjusted for by inducing evaporation at the skin surface and/or taking off clothes to increase the overall heat transfer coefficient (h). Therefore, one can use the detailed model to accurately delineate important clinical scenarios such as heat stroke, and to predict body temperature elevations during heavy exercise and/or heat exposures.

In summary, in this study a theoretical model is developed to simulate the transient body temperature distribution and blood temperature during blood cooling and rewarming. It is a relatively simple but accurate whole-body model for an improved prediction of body temperature changes during various clinical scenarios and applications. The model's predictive strength has been illustrated by sample calculations with intravascular cooling. With minor modifications, the detailed body geometry can be adjusted easily to also describe body temperature changes during physiological and medical conditions such as exercise or rewarming of a hypothermic patient.

NOMENCLATURE

c: specific heat capacity (J/kgK)
k: thermal conductivity (W/mK)
h: overall heat transfer coefficient (W/m²K)
m: mass (kg)
q_m: local metabolic heat generation rate (W/m³)
Q: heat transfer per unit time (W)
t: time (seconds)
T: temperature (°C or K)
T_a: temperature of the blood (°C or K)
T_{avg}:   volumetric-average body temperature (°C or K)
T_{max}:   maximum temperature in body tissue (°C or K)
T_t: tissue temperature (°C or K)
$\bar{T}_t$: weighted-average body temperature (°C or K)
V: volume (m³)

GREEK SYMBOLS

ρ: density (kg/m³)
ω: local volumetric blood perfusion rate per unit volume of tissue (1/second)
$\bar{\omega}$:   volumetric average blood perfusion rate per unit volume of tissue (1/second)

SUBSCRIPT

a: artery
b: blood
m: metabolism
avg: average
t: tissue
0:    steady-state or initial condition

REFERENCES

1. L. Zhu, Bioheat Transfer, in *Standard Handbook of Biomedical Engineering and Design* (M. Kutz, editor-in-chief), Chapter 2, pp. 2.3–2.29, New York: McGraw-Hill, 2002.
2. A. Bouchama, Heatstroke: A New Look at an Ancient Disease, *Intensive Care Medicine*, vol. 21, pp. 623–625, 1995.
3. N. A. Nussmeier, A Review of Risk Factors for Adverse Neurological Outcome after Cardiac Surgery, *Journal of Extra-Corporeal Technology*, vol. 34, pp. 4–10, 2002.
4. S. W. Jamieson, D. P. Kapelanski, N. Sakakibara, G. R. Manecke, P. A. Thistlethwaite, K. M. Kerr, R. N. Channick, P. F. Fedullo, and W. R. Auger, Pulmonary Endarterectomy: Experience and Lessons Learned in 1,500 Cases, *Annals of Thoracic Surgery*, vol. 76(5), pp. 1457–1462, 2003.
5. K. R. Wagner and M. Zuccarello, Local Brain Hypothermia for Neuroprotection in Stroke Treatment and Aneurysm Repair, *Neurological Research*, vol. 27, pp. 238–245, 2005.
6. D. W. Marion, Y. Leonov, M. Ginsberg, L. M. Katz, P. M. Kochanek, A. Lechleuthner, E. M. Nemoto, W. Obrist, P. Safar, F. Sterz, S. A. Tisherman, R. J. White, F. Xiao, and H. Zar, Resuscitative Hypothermia, *Critical Care Medicine*, vol. 24(2), pp. S81–S89, 1996.
7. D. W. Marion, L. F. Penrod, S. F. Kelsey, W. D. Obrist, P. M. Kochanek, A. M. Palmer, S. R. Wisniewski, and S. T. DeKosky, Treatment of Traumatic Brain Injury with Moderate Hypothermia, *New England Journal of Medicine*, vol. 336, pp. 540–546, 1997.
8. R. S. Clark, P. M. Kochanek, D. W. Marion, J. K. Schiding, M. White, A. M. Palmer, and S. T. DeKosky, Mild Posttraumatic Hypothermia Reduces Mortality after Severe Controlled Cortical Impact in Rats, *Journal of Cerebral Blood Flow and Metabolism*, vol. 16(2), pp. 253–261, 1996.
9. C. T. Wass, W. L. Lanier, R. E. Hofer, B. W. Scheithauer, and A. G. Andrews, Temperature Changes of ≥1°C Alter Functional Neurological Outcome and Histopathology in a Canine Model of Complete Cerebral Ischemia, *Anesthesia*, vol. 83, pp. 325–335, 1995.
10. J. Reith, H. S. Jorgensen, P. M. Pedersen, H. Nakayama, H. O. Raaschou, L. L. Jeppesen, and T. S. Olsen, Body Temperature in Acute Stroke: Relation to Stroke Severity, Infarct Size, Mortality, and Outcome, *Lancet*, vol. 347, pp. 422–425, 1996.
11. U. M. Illievich, M. H. Zornow, K. T. Choi, M. S. Scheller, and M. A. Strnat, Effects of Hypothermic Metabolic Suppression on Hippocampal Glutamate Concentrations after Transient Global Cerebral Ischemia, *Anesthesia and Analgesia*, vol. 78(5), pp. 905–911, 1994.
12. K. H. Polderman, Keeping a Cool Head: How to Induce and Maintain Hypothermia, *Critical Care Medicine*, vol. 32(12), pp. 2558–2560, 2004.

13. L. Zhu and A. J. Rosengart, Cooling Penetration into Normal and Injured Brain via Intraparenchymal Brain Cooling Probe: Theoretical Analyses, *Heat Transfer Engineering*, vol. 29(3), pp. 284–294, 2008.

14. D. W. Dippel, E. J. van Breda, H. M. van Gemert, H. B. van der Worp, R. J. Meijer, L. J. Kappelle, and P. J. Koudstaal, Effect of Paracetamol (Acetaminophen) on Body Temperature in Acute Ischemic Stroke: A Double-Blind, Randomized Phase II Clinical Trial, *Stroke*, vol. 32, pp. 1607–1612, 2001.

15. G. Sulter, J. W. Elting, N. Mauritus, G. J. Luyckx, and J. De Keyser, Acetylsalicylic Acid and Acetaminophen to Combat Elevated Body Temperature in Acute Ischemic Stroke, *Cerebrovascular Disease*, vol. 17, pp. 118–122, 2003.

16. S. Mayer, C. Commichau, N. Scarmeas, M. Presciutti, J. Bates, and D. Copeland, Clinical Trial of an Air-Circulating Cooling Blanket for Fever Control in Critically Ill Neurological Patients, *Neurology*, vol. 56, pp. 292–298, 2001.

17. A. J. Rosengart, L. Zhu, T. Schappeler, and F. D. Goldenberg, Fever Control in Hospitalized Stroke Patients Using Simple Intravenous Fluid Regimens: A Theoretical Evaluation, *Journal of Clinical Neuroscience*, in press, 2007.

18. A. Piepgras, H. Roth, L. Schurer, R. Tillmans, M. Quintel, P. Herrmann, and P. Schmiedek, Rapid Active Internal Core Cooling for Induction of Moderate Hypothermia in Head Injury by Use of an Extracorporeal Heat Exchanger, *Neurosurgery*, vol. 42, pp. 311–317, 1998.

19. A. G. Doufas, O. Akca, A. Barry, D. A. Petrusca, M. I. Suleman, N. Morioka, J. J. Guarnaschelli, and D. I. Sessler, Initial Experience with a Novel Heat-Exchanging Catheter in Neurosurgical Patients, *Anesthesia and Analgesia*, vol. 95, pp. 1752–1756, 2002.

20. M. W. Dae, D. W. Gao, P. C. Ursell, C. A. Stillson, and D. I. Sessler, Safety and Efficacy of Endovascular Cooling and Re-warming for Induction and Reversal of Hypothermia in Human-Sized Pigs, *Stroke*, vol. 34, pp. 734–738, 2003.

21. D. Georgiadis, S. Schwarz, R. Kollmar, and S. Schwab, Endovascular Cooling for Moderate Hypothermia in Patients with Acute Stroke: First Results of a Novel Approach, *Stroke*, vol. 32, pp. 2550–2553, 2001.

22. B. Inderbitzen, S. Yon, J. Lasheras, J. Dobak, J. Perl, and G. K. Steinberg, Safety and Performance of a Novel Intravascular Catheter for Inducing and Reversal of Hypothermia in a Porcine Model, *Neurosurgery*, vol. 50, pp. 364–370, 2002.

23. W. J. Mack, J. Huang, C. Winfree, G. Kim, M. Oppermann, J. Dobak, B. Iderbitzen, S. Yon, S. Popilskis, J. Lasheras, R. R. Sciacca, D. J. Pinsky, and E. S. Connolly, Ultrarapid, Convection-Enhanced Intravascular Hypothermia: A Feasibility Study in Nonhuman Primate Stroke, *Stroke*, vol. 34, pp. 1994–1999, 2003.

24. B. W. Raaymakers, J. Crezee, and J. J. W. Lagendijk, Modeling Individual Temperature Profiles from an Isolated Perfused Bovine Tongue, *Physics in Medicine and Biology*, vol. 45, pp. 765–780, 2000.

25. H. H. Pennes, Analysis of Tissue and Arterial Blood Temperatures in the Resting Human Forearm, *Journal of Applied Physiology*, vol. 1, pp. 93–122, 1948.

26. S. Weinbaum and L. M. Jiji, A New Simplified Bioheat Equation for the Effect of Blood Flow on Local Average Tissue Temperature, *ASME Journal of Biomechanical Engineering*, vol. 107, pp. 131–139, 1985.

27. R. D. Mosteller, Simplified Calculation of Body Surface Area, *New England Journal of Medicine*, vol. 317, p. 1098, 1994.

28. G. Fu, A Transient 3-D Mathematical Thermal Model for the Clothed Human, Ph.D. Dissertation, Kansas State University, 1995.

29. M. D. Salloum, A New Transient Bioheat Model of the Human Body and Its Integration to Clothing Models, M.S. Thesis, American University of Beirut, 2005.

30. C. E. Smith, A Transient Three-Dimensional Model of the Thermal System, Ph.D. Dissertation, Kansas State University, 1991.

31. E. H. Wissler, *Mathematical Simulation of Human Thermal Behavior Using Whole Body Models, Heat and Mass Transfer in Medicine and Biology*, Chapter 13, pp. 325–373, New York: Plenum, 1985.

32. S. Weinbaum, L. X. Xu, L. Zhu, and A. Ekpene, A New Fundamental Bioheat Equation for Muscle Tissue, Part I: Blood Perfusion Term, *ASME Journal of Biomechanical Engineering*, vol. 121, pp. 1–12, 1997.

33. L. Zhu, L. X. Xu, Q. He, and S. Weinbaum, A New Fundamental Bioheat Equation for Muscle Tissue, Part II: Temperature of SAV Vessels, *ASME Journal of Biomechanical Engineering*, vol. 124, pp. 121–132, 2002.

34. G. Azzimondi, L. Bassein, F. Nonino, L. Fiorani, L. Vignatelli, and R. G. D'Alessandro, Fever in Acute Stroke Worsens Prognosis: A Prospective Study, *Stroke*, vol. 26(11), pp. 2040–2043, 1995.

35. G. Boysen and H. Christensen, Stroke Severity Determines Body Temperature in Acute Stroke, *Stroke*, vol. 32(2), pp. 413–417, 2001.

36. J. Castillo, F. Martinez, R. Leira, J. M. Prieto, M. Lema, and M. Noya, Mortality and Morbidity of Acute Cerebral Infarction Related to Temperature and Basal Analytic Parameters, *Cerebrovascular Diseases*, vol. 4, pp. 56–71, 1994.

37. B. Hindfelt, The Prognostic Significance of Subfebrility and Fever in Ischaemic Cerebral Infarction, *Acta Neurologica Scandinavica*, vol. 53(1), pp. 72–79, 1976.

38. L. P. Kammersgaard, H. S. Jorgensen, J. A. Rungby, J. Reith, H. Nakayama, U. J. Weber, J. Houth, and T. S. Olsen, Admission Body Temperature Predicts Long-Term Mortality after Acute Stroke: The Copenhagen Stroke Study, *Stroke*, vol. 33(7), pp. 1759–1762, 2002.

39. A. Terent and B. Anderson, Prognosis for Patients with Cerebrovascular Stroke and Transient Ischemic Attacks, *Journal of Medical Science*, vol. 86(1), pp. 63–74, 1981.

40. J. Castillo, A. Davalos, J. Marrugat, and M. Noya, Timing for Fever-Related Brain Damage in Acute Ischemic Stroke, *Stroke*, vol. 29(12), pp. 2455–2460, 1998.

41. S. A. Bernard, B. M. Jones, and M. Buist, Experience with Prolonged Induced Hypothermia in Severe Head Injury, *Critical Care*, vol. 3, pp. 167–172, 1999.

42. I. Virkkunen, A. Yli-Hankala, and T. Silfvast, Induction of Therapeutic Hypothermia after Cardiac Arrest in Pre-hospital Patients Using Ice-Cold Ringer's Solution: A Pilot Study, *Resuscitation*, vol. 62(3), pp. 299–302, 2004.

43. T. Shiozaki, H. Sugimoto, M. Taneda, H. Yoshida, A. Iwai, T. Yoshioka, and T. Sugimoto, Effect of Mild Hypothermia on Uncontrollable Intracranial Hypertension after Severe Head Injury, *Journal of Neurosurgery*, vol. 79, pp. 363–368, 1993.

44. C. Diao and L. Zhu, Temperature Distribution and Blood Perfusion Response in Rat Brain during Selective Brain Cooling, *Medical Physics*, vol. 33(7), pp. 2565–2573, 2006.

45. Y. Wang and L. Zhu, Selective Brain Hypothermia Induced by an Interstitial Cooling Device in Human Neck: Theoretical Analyses, *European Journal of Applied Physiology*, vol. 101(1), pp. 31–40, 2007.

46. E. Bering, Effect of Body Temperature Change on Cerebral Oxygen Consumption of the Intact Monkey, *American Journal of Physiology*, vol. 200, pp. 417–419, 1961.

7

Experimental and Numerical Investigation on Simulating Nanocryosurgery of Target Tissues Embedded with Large Blood Vessels

Z.-S. Deng, J. Liu, J.-F. Yan, Z.-Q. Sun, and Y.-X. Zhou

CONTENTS

7.1 INTRODUCTION

Cryosurgery, sometimes referred to as cryotherapy or cryoablation, is the use of extreme cold produced by cryogenic agents to destroy abnormal or diseased tissues. The application of cryosurgery includes treatment of many kinds of cancer and some noncancerous diseased tissues. Good results have been achieved in

221

cryosurgical treatment of many types of tumors [1]. However, local recurrences at the site of ablation close to large blood vessels have been reported at rates from 5% to 44% or even higher [2,3]. In order to perform a successful cryosurgery for tumors embedded with large blood vessels, it is very necessary to understand the effects of large vessels on the temperature responses of biological tissues subjected to freezing [4-6].

From the viewpoint of heat transfer, a large blood vessel (also termed a thermally significant vessel) denotes a vessel larger than 0.5 mm in diameter [7]. Anatomically, tumors are often situated close to or embedded with large blood vessels, since a tumor's quick growth ultimately depends on nutrients supplied by its blood vessel network. As is well known, the blood flowing through large blood vessels acts as a heat source or heat sink and plays an important role in affecting temperature profiles of cooled or heated tissues [4,8]. During cryosurgery, the blood flow inside a large vessel represents a source that heats the nearby frozen tissues and, thereby, limits freezing lesions during cryosurgery. Under this condition, a part of the vital tumor cells may remain in the cryolesion and lead to recurrence of tumors after cryosurgical treatment. More specifically, tumor cell survival in the vicinity of large blood vessels is often correlated with tumor recurrence after treatment [9]. Consequently, it is difficult to implement an effective cryosurgery when a tumor is contiguous to a large blood vessel. In fact, the heating effects of large blood vessels on the surrounding tumor tissues during cryosurgery can be eliminated by vascular exclusion [3,10] in which vascular inflow occlusion is performed by clamping the entrance of the large vessels. However, the vascular occlusion requires a major surgical procedure, which will exclude one of the main merits of minimally invasive percutaneous cryosurgery.

In our previous works [4-6], both theoretical and experimental investigations had been conducted to study the thermal effect of large blood vessels during cryosurgery. The results suggested that the flowing blood in the large vessels can produce steep temperature gradients and inadequate cooling to the frozen tissues and, therefore, may seriously contribute to failed killing of tumor during cryosurgery. Up to now, simple but effective approaches to totally destroy tumor cells in the vicinity of large blood vessels are still not available. Based on our previous investigation on nanoparticle-enabled cryosurgery (termed "nanocryosurgery") [11,12], nanoparticles with high thermal conductivity were also introduced in this study with the purpose of performing a successful cryosurgical treatment for a tumor embedded with large blood vessels. Different from nanoparticle-enabled hyperthermia and drug delivery [13–16], the basic principle of nanocryosurgery is to introduce adjuvant nanoparticle suspensions with high thermal conductivity into tumors, which can significantly improve the freezing efficiency of a conventional cryosurgical procedure [17]. For the same reason, the highly thermal-conductive nanoparticle suspensions (introduced into tumor tissues in the vicinity of large blood vessels before cryosurgery) would evidently enhance the freezing effect of cryoprobes to the target tumor tissues surrounding large vessels. To illustrate the effectiveness of nanocryosurgery in the treatment of tumors embedded with large blood vessels, the thermal effects of large vessels during both nanocryosurgery and conventional cryosurgery were investigated in this study through simulated experiments in phantom gels and *in vitro* porcine liver tissues and numerical simulation on *in vivo* tissues.

7.2   EXPERIMENTAL PROCEDURES

During cryosurgical experiments, the HR-II Medical Infrared Imaging System (made by the Institute of Optics and Electronics of North China, Beijing, China) was used to record the transient temperature distribution at the tissue surface, and thermocouples were employed to monitor the temperature responses inside the tissue. The infrared thermographic equipment has a discrimination of 0.1°C at ambient temperatures and 0.5°C down to −40°C, which represents the lower limit of the equipment, and the angular resolution is better than two miniradians. The infrared camera was placed in front of the object to be tested, and the vertical distance between the camera and the object was about 80 cm.

For all experiments, the large blood vessels were simulated by a 1 mm OD/0.8 mm ID Teflon tube. Water (25°C) was used to simulate the blood for conceptual illustration in this study. But for practical purposes, the simulating fluid could take a glycerol solution containing glycerol at a specific concentration and distilled water as the blood analog fluid, which has been routinely adopted in previous blood simulations. This feature will be considered in the near future for additional tests. The circuit pressure used to drive the water flow was supplied by means of a 1 meter head of water. In this way, the mass flow rate of water was first measured to be 3.1 g/min. Then, the flow velocity of water can be calculated from the flow rate, and the value was about 10 cm/s. A liquid nitrogen–based cryosurgical system [18] (developed by the Technical Institute of Physics and Chemistry, Chinese Academy of Sciences, Beijing, China) was applied to supply the cooling power, in which two cryoprobes with 5 mm diameters were selected to perform the simulated cryosurgery. One minute before freezing, a data logger connected with thermocouples was turned on to collect temperatures inside the tissue. The total freezing time was 20 minutes. Then, the freezing was stopped, and thawing commenced at room temperature.

7.2.1   Phantom Study

The medium used to simulate the biological tissue was a semitransparent gel phantom composed of 10% gelatin and 90% water by weight. The tissue phantom was formed and contained in a cubical organic glass box (the inner dimensions were 16 × 16 × 5 cm). The Teflon tube passed through 1 mm diameter holes drilled on the two opposite panels of the organic glass box. For all phantom experiments, the distance between the two cryoprobes was 25 mm, and the insertion depth of the cryoprobes was 35 mm into the gel phantom.

The vertical distances between the blood vessel and the two cryoprobes were equal to each other.

Four different cases were investigated for the simulated experiments with tissue phantom: the case of a single large blood vessel without introduced nanoparticles, the case of a single large blood vessel with adjuvantly introduced nanoparticles, the case of parallel countercurrent vessel pairs without loaded nanoparticles, and the case of parallel countercurrent vessel pairs with adjuvantly loaded nanoparticles. Photographs of the experimental setups are shown in Figures 7.1 and 7.2, respectively. For the case of a single large blood vessel, the distance between the blood vessel and the back surface of the phantom was 10 mm. The locations of thermocouples 1, 2, and 3 are shown in Figures 7.1A,D, and the insertion depths of thermocouples 1, 2, and 3 were all 35 mm. For the case of parallel countercurrent vessel pairs, the distance between the artery and the back

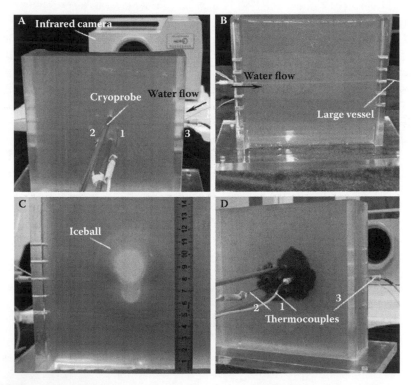

Figure 7.1   Photographs of the experimental setup for tissue phantom with a single large blood vessel in which (A, B, and C) are for the case without loaded nanoparticles: (A) shows the front view of the phantom from where the cryoprobes were inserted, (B) shows the back view of the phantom from where the infrared camera was placed at a distance of 80 cm, (C) shows the iceball formed after 20 minutes of freezing, and (D) is for the case with adjuvantly loaded nanoparticles.

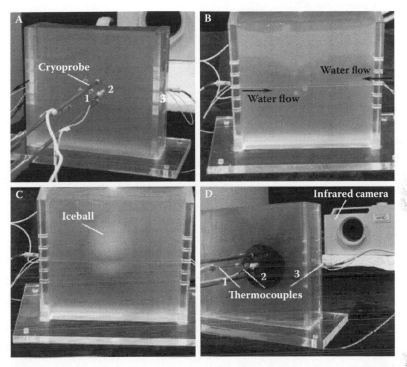

Figure 7.2   The experimental setup for the tissue phantom with parallel countercurrent vessel pairs in which (A, B, and C) are for the case without loaded nanoparticles: (A) shows the front view of the phantom from where the cryoprobes were inserted, (B) shows the back view of the phantom from where the infrared camera was placed at a distance of 80 cm, (C) shows the iceball formed after 20 minutes of freezing, and (D) is for the case with adjuvantly loaded nanoparticles.

surface of the phantom was 20 mm, and the distance between the vein and the back surface of the phantom was 10 mm.

The process of making the countercurrent vessels was as follows: first, the Teflon tube was guided through the 1 mm diameter holes drilled on the two opposite panels; then, it was swerved at the outside of the box; and after that, it was guided through another two holes with the same diameters, which were 10 mm apart from the former holes. The locations of thermocouples 1, 2, and 3 are shown in Figures 7.2A,D, in which the location of thermocouple 3 was close to the inlet of the vein. The insertion depths of thermocouples 1, 2, and 3 were also 35 mm. For the cases with adjuvantly loaded nanoparticles, a 20 milliliter (ml) aqueous suspension of Fe_3O_4 nanoparticles (20% wt.) was first introduced into the target area before the phantom gel solidified in the cubical organic glass box. Then, the solid mixtures of phantom gel and Fe_3O_4 nanoparticles were formed at the corresponding region (i.e., the black area, as shown in Figure 7.1D and Figure 7.2D).

7.2.2 *In Vitro* Tissue Study

Although the tissue phantom has thermal properties similar to those of biological tissue, it does differ from real biological tissue, and it is hard to simulate the other properties of biological tissue such as anisotropy. In order to more realistically simulate the thermal effect of a large blood vessel during an actual cryosurgery, similar simulated experiments inside *in vitro* tissue were also performed. *In vitro* porcine liver tissues were selected as experimental material. For all *in vitro* tissue experiments, the distance between two cryoprobes was 35 mm, and the insertion depths of the cryoprobes were about 25 mm in liver tissues. The depths of large blood vessels were taken as 5 mm from the surface of liver tissues. The vertical distances between the blood vessel and the two cryoprobes were equal to each other.

Four cases were investigated for the simulated experiments with *in vitro* porcine liver tissue, which corresponded to those for the phantom experiments. Photographs of the experimental setups are shown in Figures 7.3 and 7.4. For

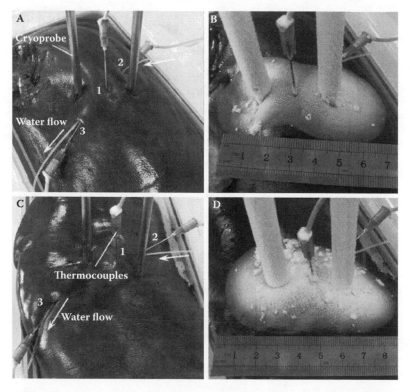

Figure 7.3 The experimental setup for *in vitro* porcine liver tissue with a single large blood vessel in which (A) and (B) are for the case without introduced nanoparticles, and (B) shows the iceball formed after 20 minutes of freezing. Parts (C) and (D) are for the case with adjuvantly introduced nanoparticles, and (D) shows the iceball formed after 20 minutes of freezing.

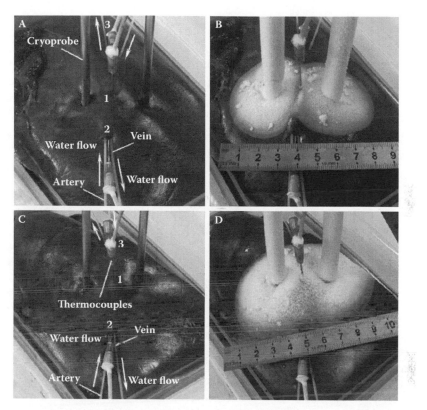

Figure 7.4 The experimental setup for *in vitro* porcine liver tissue with parallel countercurrent vessel pairs in which (A) and (B) are for the case without introduced nanoparticles, and (B) shows the iceball formed after 20 minutes of freezing. Parts (C) and (D) are for the case with adjuvantly introduced nanoparticles, and (D) shows the iceball formed after 20 minutes of freezing.

the case of a single large blood vessel, the locations of thermocouples 1, 2, and 3 were shown in Figures 7.3A,C, in which thermocouples 2 and 3 were inserted along the direction of the blood vessel with the same depth as the vessel. The insertion depth of thermocouple 1 was 5 mm, and the vertical distance between thermocouple 1 and the blood vessel was about 2 mm. For the case of parallel countercurrent vessel pairs, the locations of thermocouples 1, 2, and 3 are shown in Figures 7.4A,C, in which thermocouples 2 and 3 are inserted along the direction of the blood vessels at the same depth as the blood vessels. The insertion depth of thermocouple 1 was also 5 mm, and the vertical distance between thermocouple 1 and the vein vessel (shown in Figures 7.4A,C) was about 2 mm. For the cases with adjuvantly loaded nanoparticles, a 10 ml aqueous suspension of Fe_3O_4 nanoparticles (20% wt.) was injected into the target area of the liver tissue before freezing.

7.3 NUMERICAL MODELING

In this study, two typical vascular models were applied to simulate the phase-change heat transfer of biological tissues embedded with large blood vessels (shown in Figure 7.5), which include (a) a model with a single artery passing through the tumor (SATT); and (b) a model with countercurrent vessel pairs passing through the tumor (CVTT). The boundary conditions are a constant temperature of 37°C on all surfaces of the parallelepiped. Similar models and boundary conditions were first developed to study the effects of large blood vessels on temperature distributions during hyperthermia treatment by Chen and Roemer [19]. In these models, the entire computation domain consists of unfrozen tissue, frozen tissue, and large blood vessel domains.

Generally, it is extremely difficult to quantify the thermal behavior of a biological body by means of distinguishing its temperature in local tissue from that of the vascular network. A simple, albeit intuitive, way is to introduce the temperature T to characterize the overall thermal state in a specific position of the tissue. As is well known, in the capillary bed the condition of very slow flow with a superposed oscillating component favors almost complete thermal equilibrium between the bloodstream and the surrounding tissue. Since the precapillary and capillary beds are the major sites for exchange of heat in tissue, it is reasonable to assume the equality in postcapillary blood and tissue temperatures. It was based on this justification that the Pennes bioheat equation was established.

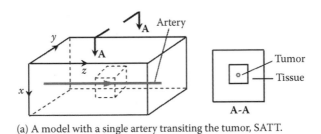

(a) A model with a single artery transiting the tumor, SATT.

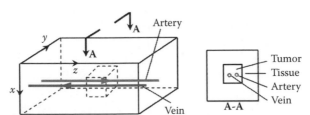

(b) A model with countercurrent vessel pairs transiting the tumor, CVTT.

Figure 7.5   Illustration of two typical vascular models (not to scale).

By introducing a series of effective quantities including effective heat capacity $\tilde{C}$, effective thermal conductivity $\tilde{k}$, effective blood perfusion $\tilde{\omega}_b$, and effective metabolic heat generation $\tilde{Q}_m$, the uniform heat transfer equation for the unfrozen and frozen biological tissues based on the Pennes equation can be written as follows [20]:

$$\tilde{C}\frac{\partial T}{\partial t} = \nabla \cdot \tilde{k}\nabla T - \tilde{\omega}_b C_b T + \tilde{Q}_m + \tilde{\omega}_b C_b T_a, \quad X \in \Omega \tag{7.1}$$

where C_b is the heat capacity of blood; X contains the Cartesian coordinates x, y, and z; Ω denotes the computational domain; T is the tissue temperature; and T_a is the arterial temperature; and [20]

$$\tilde{C}(T) = \begin{cases} C_f, & T < T_{ml} \\ \dfrac{Q_l}{(T_{mu} - T_{ml})} + \dfrac{C_f + C_u}{2}, & T_{ml} \leq T \leq T_{mu} \\ C_u, & T > T_{mu} \end{cases} \tag{7.2}$$

$$\tilde{k}(T) = \begin{cases} k_f, & T < T_{ml} \\ (k_f + k_u)/2, & T_{ml} \leq T \leq T_{mu} \\ k_u, & T > T_{mu} \end{cases} \tag{7.3}$$

$$\tilde{\omega}_b(T) = \begin{cases} 0, & T < T_{ml} \\ 0, & T_{ml} \leq T \leq T_{mu} \\ \omega_b, & T > T_{mu} \end{cases} \tag{7.4}$$

$$\tilde{Q}_m(T) = \begin{cases} 0, & T < T_{ml} \\ 0, & T_{ml} \leq T \leq T_{mu} \\ Q_m, & T > T_{mu} \end{cases} \tag{7.5}$$

where C_u and C_f are the volumetric heat capacities of unfrozen and frozen tissues, respectively; Q_l is the latent heat of tissue released when the water content of tissue was frozen; k_u and k_f are the thermal conductivities of unfrozen and frozen tissues, respectively; ω_b is the blood perfusion of unfrozen tissue; and Q_m is the metabolic heat generation of unfrozen tissue [20]. In this study, the biological tissues are treated as nonideal materials, freezing over a temperature range

(T_{ml}, T_{mu}), where T_{ml} and T_{mu} are, respectively, the lower and upper phase-transition temperatures of tissue, aiming to reflect a relatively real clinical situation.

The basic governing equation for the mean blood temperature in the large vessels, T_b, at any position z along a transiting vessel is [5]

$$C_b \frac{\partial T_b}{\partial t} = \frac{hP}{S}(T_{wb} - T_b) - C_b v \frac{\partial T_b}{\partial z} \qquad (7.6)$$

where $h = Nu \cdot k_b / D$ is the convective heat transfer coefficient between the blood and the vessel wall, D is the diameter of the vessel, P is the perimeter of the vessel, S is the cross-sectional area of the vessel, v is the mean blood velocity in the vessel, Nu is the Nusselt number, and T_{wb} is the wall temperature of the vessel. In Equation (7.6), conduction in the fluid in the z direction (flow direction) is neglected for large flow rates. The sign of the blood velocity v is assigned as positive (i.e., the velocities for blood in arteries and veins are positive and negative, respectively).

The boundary conditions at the surfaces of cryoprobes are prescribed according to the probe tip and probe shank, respectively:

$$T = T_w \quad \text{at probe tip} \qquad (7.7)$$

$$k \frac{\partial T}{\partial n} = 0 \quad \text{at probe shank} \qquad (7.8)$$

The reason for using adiabatic boundary conditions at the probe shank is that the thermal leakage of the probe shank can be approximated to be zero due to the application of vacuum heat insulation in the shank.

In the SATT and CVTT models, since the artery supplies arterial blood to the tumor, the boundary condition for T_b in the artery can be taken as

$$T_b = 37°C, \quad z = 0 \qquad (7.9)$$

In the CVTT model, since the vein collects venous blood in other parts of the body flowing through the tumor, the boundary condition for T_b in the vein can be approximated as

$$T_b \approx 37°C, \quad z = z_0 \qquad (7.10)$$

where z_0 is the dimension in the z direction.

The boundary conditions at the interface of the two regions respectively related to Equations (7.1) and (7.6) can be included in the finite difference formulations for the four neighboring tissue nodes outside the vessel; readers are referred to Deng and Liu [5] for more details. The description will not be repeated here for the sake of brevity.

The initial temperature field in tissue is simplified as

$$T_0(x, y, z) = 37°C \tag{7.11}$$

A finite difference algorithm developed in our previous study [5] is applied to solve the above problem with phase-change heat transfer in biological tissues embedded with large blood vessels. The description and derivation of the algorithm are omitted here for the sake of brevity. Again, readers are referred to Deng and Liu [5] for more details. In calculations, the grid resolution is $\Delta x = \Delta y = \Delta z = 0.002$ m and $\Delta t = 0.1$ s, and the tissue domain is prescribed in a rectangular geometry with $10 \times 10 \times 20$ cm in the x, y, and z directions, respectively. Since the cross-sectional mean blood temperature is used, only one line of finite difference nodes is needed to represent the blood inside a large vessel. If in a large vessel, there is one node at which the temperature is below the mean value at which tissue freezes during the calculations, the blood flow velocity is then set as zero (i.e., this vessel has been frozen).

7.4 RESULTS AND DISCUSSION

7.4.1 Experimental Results

In the experimental study, temperatures inside tissues were measured with copper–constantan thermocouples (which were fabricated using copper wire and constantan wire, both with a diameter of 0.1 mm) connected to an Agilent 34970A Data Logger (Agilent, Santa Clara, California). For convenient use in biological tissue, the thermocouple was packaged by a medical injection pin with an outer diameter of 0.6 mm and an inner diameter of 0.4 mm, in which the injection pin and the thermocouple head were welded together. The temperature distributions at the tissue surface were recorded by an infrared imaging system. Before the experiments, the thermocouples were calibrated and an accuracy of ±0.1°C was obtained, and the infrared thermographic system was correlated using a mixture of ice and water.

7.4.1.1 Phantom Experiments

The experimental results for the phantom study are shown in Figure 7.6 through Figure 7.15. Concerning infrared thermographs, only parts of representative results were given for illustrative purposes. In infrared thermographs, different colors denote different temperature ranges, from which the boundaries of the iceball (determined by the isotherm of the phantom's freezing point) at the tissue surface can be easily discriminated.

It was shown in Figure 7.6 that, for the case of a single large blood vessel without loaded nanoparticles, the temperature surrounding the vessel was much higher than that at other positions due to the heat source effect of water flow in the vessel during the freezing procedure. In addition, the tissue phantom surrounding the

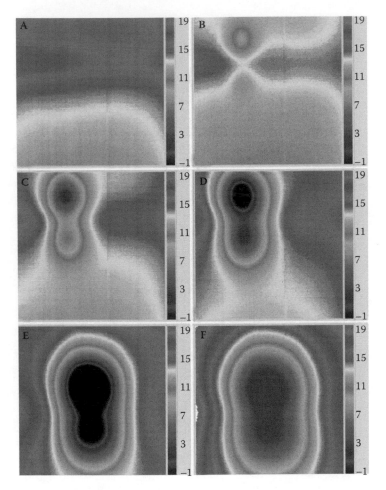

Figure 7.6 Infrared thermographs for the case of a single large blood vessel without introduced nanoparticles (phantom experiments): (A) before freezing, (B) after 5 minutes of freezing, (C) after 10 minutes of freezing, (D) after 15 minutes of freezing, (E) after 20 minutes of freezing, and (F) after 20 minutes of thawing.

vessel was relatively difficult to cool, resulting in an iceball formation that tended to deviate from the region of the large vessel. When the freezing time was long enough (20 minutes), the irregular-shaped iceball (as shown in the white area of Figure 7.1C and the black area of Figure 7.6E), which appears as a concave shape near the large vessel, finally enwrapped the large vessel. Even at this time, the simulated blood vessel was still not frozen, which can be inferred from the fact of running water flow at the outlet of the vessel.

For the case of a single large blood vessel with adjuvantly introduced nano-particles, it was found during the experiments that the water flow stopped after about 14 minutes of freezing. The corresponding infrared results for this case are shown in Figure 7.7. It can be observed from Figures 7.7A,B,C,D that before the vessel was frozen, the cooling characteristic of the phantom tissue and the shape of the isotherms are similar to those for the case of a single large vessel without introduced nanoparticles. After the vessel was frozen, as shown in Figure 7.7E, the shape of the isotherm close to the iceball is approximately elliptic, not concave. More specifically, after 20 minutes of spontaneous thawing, the elliptic shape of the isotherm in the phantom tissue became more apparent, as seen in Figure 7.7F. This indicates that the frozen blood vessel had not been thawed during the entire thawing procedure.

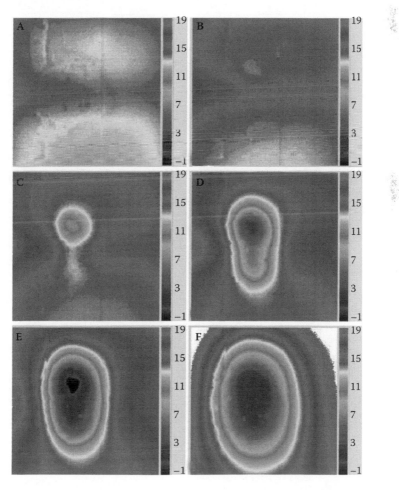

Figure 7.7 Infrared thermographs for the case of a single large blood vessel with adjuvantly introduced nanoparticles (phantom experiments): (A) before freezing, (B) after 5 minutes of freezing, (C) after 10 minutes of freezing, (D) after 15 minutes of freezing, (E) after 20 minutes of freezing, and (F) after 20 minutes of thawing.

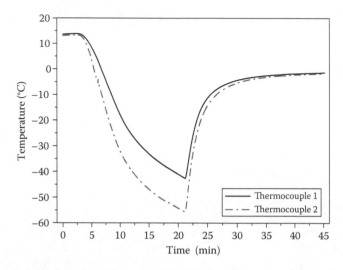

Figure 7.8   Transient temperatures at two different sites for the case of a single large blood vessel without introduced nanoparticles (phantom experiments) in which the sites of thermocouples 1 and 2 are shown in Figure 7.1A.

Figures 7.8 and 7.9 depict the transient temperatures of thermocouples 1 and 2 for the cases of a single large blood vessel without and with loaded nanoparticles, respectively. The sites of thermocouples 1 and 2 were shown in Figure 7.1. As stated above (in Section 7.2, "Experimental Procedures"), at 1 minute of

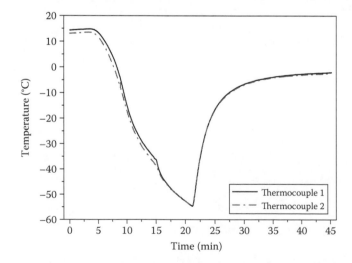

Figure 7.9   Transient temperatures at two different sites for the case of a single large blood vessel with adjuvantly introduced nanoparticles (phantom experiments) in which the sites of thermocouples 1 and 2 are shown in Figure 7.1D.

time, freezing started; and at 21 minutes, freezing stopped. It can be noted from Figures 7.8 and 7.9 that the temperature of thermocouple 1 is higher than that of thermocouple 2 during freezing, which may result from the heating effect of the blood vessel since thermocouple 1 was closer to the inlet of the water flow compared with thermocouple 2. In addition, due to the high thermal conductivity of Fe_3O_4 nanoparticles (about 40 W/m°C) [21], the temperature difference between thermocouples 1 and 2 for the case without loaded nanoparticles was larger than that for the case with nanoparticles. Moreover, in the temperature curves for the case with nanoparticles, an inflexion point appeared at about 15 minutes during freezing (i.e., a jump occurred in the cooling rate of the phantom tissue at that time). On the other hand, the temperature curves for the case without loaded nanoparticles appeared very smooth during freezing. Such a phenomenon can be more clearly seen from Figure 7.10, which may be due to the fact that the vessel was frozen after about 14 minutes of freezing (as mentioned above). On the one hand, the heating effect of the water flow would not have any further effect after the vessel was frozen. On the other hand, the thermal conductivity of the phantom tissue in the vicinity of the blood vessel would increase significantly when it is frozen. Consequently, the cooling rate of the phantom tissue would inevitably increase. The above results indicate that adjuvantly introducing nanoparticles with high thermal conductivity can significantly improve the freezing efficacy of cryoprobes.

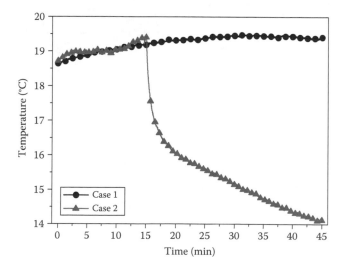

Figure 7.10   Transient temperatures of thermocouple 3 for two different cases (phantom experiments), in which case 1 denotes the case of a single large blood vessel without introduced nanoparticles, case 2 denotes a single large blood vessel with adjuvantly introduced nanoparticles, and the sites of thermocouple 3 are shown in Figures 7.1A,D for the above two cases, respectively.

Anatomically, the large blood vessels frequently are present in the form of countercurrent pairs of arteries and veins [6]. Based on this consideration, the thermal effects of parallel countercurrent vessel pairs were also investigated in the present study. Two cases were considered, without and with introduced nanoparticles. A photograph of the corresponding experimental setup is given in Figure 7.2, in which the U-type Teflon tube plays both roles of artery and vein. When the water flowed through the cubic box for the first time, it served as the arterial blood. When the water reflowed through the box, it was regarded as venous blood. The infrared thermographs for the above cases are shown in Figures 7.11 and 7.12, respectively. Similar results for the cases of a single large

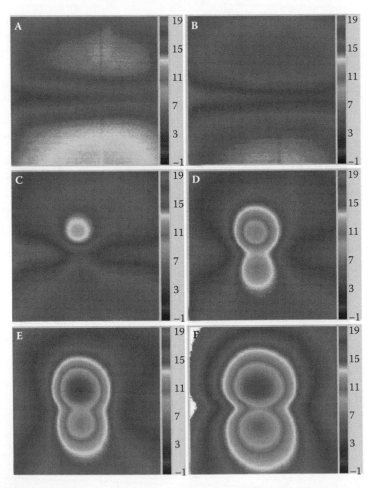

Figure 7.11   Infrared thermographs for the case of parallel countercurrent vessel pairs without introduced nanoparticles (phantom experiments): (A) before freezing, (B) after 5 minutes of freezing, (C) after 10 minutes of freezing, (D) after 15 minutes of freezing, (E) after 20 minutes of freezing, and (F) after 20 minutes of thawing.

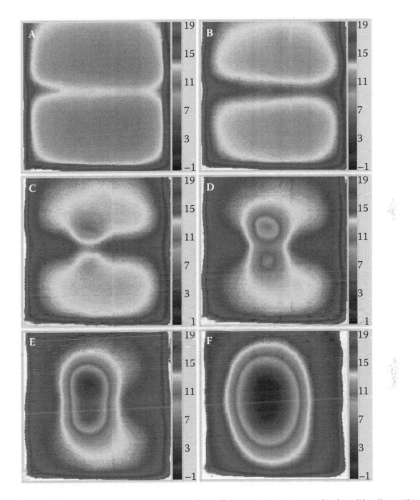

Figure 7.12   Infrared thermographs for the case of parallel countercurrent vessel pairs with adjuvantly introduced nanoparticles (phantom experiments): (A) before freezing, (B) after 5 minutes of freezing, (C) after 10 minutes of freezing, (D) after 15 minutes of freezing, (E) after 20 minutes of freezing, and (F) after 20 minutes of thawing.

vessel had been obtained. Since the water flow in countercurrent vessel pairs provides more heat energy than does a single large vessel, the tissue phantom surrounding the vessels was harder to freeze compared to the case of a single large vessel. The concave shape of the isotherm or iceball became more evident, and finally took on the shape of an "8." For the case of parallel countercurrent vessel pairs without loaded nanoparticles, the 8-shaped isotherm in the phantom tissue did not disappear since it formed about 15 minutes after freezing (as shown in Figure 7.11D), due to the heating effect resulting from uninterrupted water flow

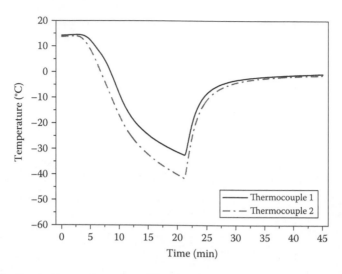

Figure 7.13   Transient temperatures at two different sites for the case of parallel countercurrent vessel pairs without introduced nanoparticles (phantom experiments) in which the sites of thermocouples 1 and 2 are shown in Figure 7.2A.

in the large vessel. For the case of parallel countercurrent vessel pairs with adjuvantly introduced nanoparticles, the 8-shaped isotherm also formed after about 15 minutes of freezing (as shown in Figure 7.12D). During the experiment, it was found that the water flow stopped after about 16.5 minutes of freezing. Then, the 8-shaped isotherm disappeared after about three more minutes of freezing since the water flow stopped (as shown in Figure 7.12E). Finally, after 20 minutes of spontaneous thawing, an elliptically shaped isotherm was formed in the phantom tissue, which is shown in Figure 7.12F. This implied that the frozen blood vessels had not been thawed during the entire thawing procedure.

Figures 7.13 and 7.14 showed the transient temperatures of thermocouples 1 and 2 for the cases of parallel countercurrent vessel pairs without and with introduced nanoparticles, respectively. The sites of thermocouples 1 and 2 are shown in Figure 7.2. As for the cases of a single large vessel presented above, freezing started at 1 minute of time, and at 21 minutes, freezing stopped. For the case without loaded nanoparticles, the temperature curves appear very smooth during freezing, as shown in Figure 7.13. For the case with nanoparticles, a jump occurs in the cooling rate of the phantom tissue, similar to that for the case of a single large blood vessel with introduced nanoparticles. The time when this jump of cooling rate occurred was at about 17.5 minutes (i.e., about 16.5 minutes after freezing started), which corresponds to the time when water flow in the large vessels stopped. Figure 7.15 depicts the transient temperatures of thermocouple 3 for the above two cases. It can also be seen from Figure 7.15 that at about 17.5 minutes, the temperature curve for the case with loaded nanoparticles has an

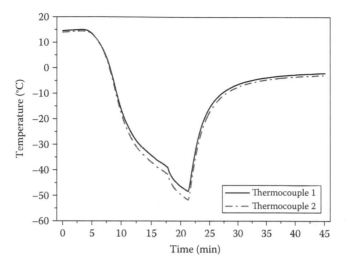

Figure 7.14   Transient temperatures at two different sites for the case of parallel countercurrent vessel pairs with adjuvantly introduced nanoparticles (phantom experiments) in which the sites of thermocouples 1 and 2 are shown in Figure 7.2D.

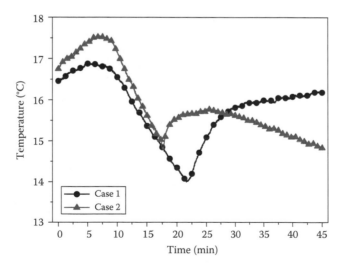

Figure 7.15   Transient temperatures of thermocouple 3 for two different cases (phantom experiments) in which case 1 denotes the case of parallel countercurrent vessel pairs without introduced nanoparticles, case 2 denotes the case of parallel countercurrent vessel pairs with adjuvantly introduced nanoparticles, and the sites of thermocouple 3 are shown in Figures 7.2A,D for the above two cases, respectively.

inflexion point. Different from the case of a single vessel with nanoparticles, the temperature for this case (shown in Figure 7.15, case 2) starts to increase after the inflexion point occurred. The reason for this phenomenon is that the location of thermocouple 3 for this case was close to the inlet of a vein, whereas for the case of a single vessel, the thermocouple was close to the inlet of an artery. At the inlet of a vein, the local temperature of the phantom tissue, which almost had not been affected by the freezing of cryoprobes, could be cooled by the water flow in the vein (flowing from the cooled section of artery). After the vessels were frozen, the cooling effect did not exist anymore, and then the temperature in the vicinity of the vein inlet could increase. Moreover, it can also be seen from Figure 7.15 that during the first several minutes after the freezing started, the temperature of the phantom tissue had slightly increased since the phantom had just been taken out of the cooling room of a refrigerator 10 minutes before the experiments, and the temperature of the phantom tissue was lower than the temperature of water flow at that time. The above results can be attributed to the higher thermal conductivity of the metallic oxide nanoparticles. These results imply that nanoparticles with high thermal conductivity can serve as effective adjuvants for enhancing the efficacy of cryosurgical treatment of tumors with embedded large blood vessels.

7.4.1.2   In Vitro Tissue Experiments

The experimental results for an *in vitro* study on porcine liver tissues are shown in Figure 7.16 through Figure 7.23. The infrared thermographs show representative results for illustrative purposes. It was shown in Figure 7.16 that, for the case of a single large blood vessel without loaded nanoparticles, the temperature of liver tissue in the vicinity of the large vessel was much higher than that at other positions due to the thermal effect of the large vessel. In addition, the iceball also had the tendency of deviating from the region of the large vessel and appears as a concave shape in the vicinity of the large vessel (as shown in the white area of Figure 7.3B and the black area of Figures 7.16B,C,D,E), similar to the results obtained from the phantom study.

For the case of a single large blood vessel with adjuvantly introduced nanoparticles, it was found during the experiment that the water flow stopped after about 12.5 minutes of freezing. The corresponding infrared results for this case are shown in Figure 7.17. It can be found from Figures 7.17B,C that before the vessel was frozen, the cooling characteristics of liver tissue and the shape of the iceball were similar to those for the case of a single large vessel without introduced nanoparticles. After the vessel was frozen, as shown in Figures 7.17D,E, the concave shape of the iceball gradually disappeared, and the iceball finally became a protuberant one (which can also be found from the optical picture, as shown in Figure 7.3D, more directly). It was indicated in Figure 7.17F that after 20 minutes of spontaneously thawing, the frozen blood vessel had not been thawed, since the shape of the unthawed iceball still appeared as protuberant, much different from that for the case of a single large vessel without introduced nanoparticles (as shown in Figure 7.16F).

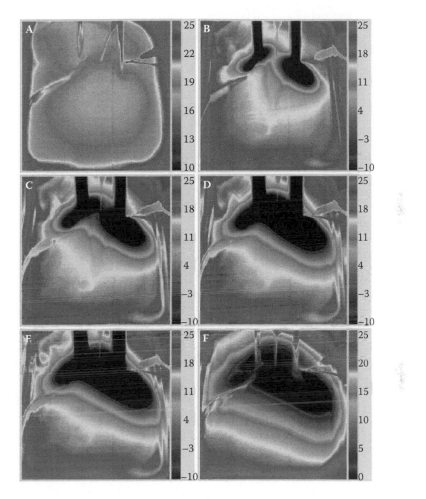

Figure 7.16   Infrared thermographs for the case of a single large blood vessel without introduced nanoparticles (*in vitro* tissue experiments): (A) before freezing, (B) after 5 minutes of freezing, (C) after 10 minutes of freezing, (D) after 15 minutes of freezing, (E) after 20 minutes of freezing, and (F) after 20 minutes of thawing.

Figures 7.18 and 7.19 depict the transient temperatures of thermocouples 1, 2, and 3 for the cases of a single large blood vessel without and with introduced nanoparticles, respectively. The sites of thermocouples 1, 2, and 3 were shown in Figure 7.3. As with the phantom study, freezing started at 1 minute of time, and at 21 minutes, freezing stopped. In Figures 7.18 and 7.19, the temperature of thermocouple 1 was the lowest one, the temperature of thermocouple 2 was the highest one, and the temperature of thermocouple 3 held the middle place. The rationale for these results was that the site of thermocouple 1 was nearest to the cryoprobes, while thermocouples 2 and 3 were inserted into the liver tissue

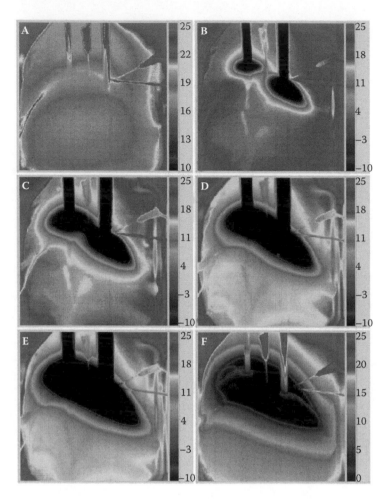

Figure 7.17   Infrared thermographs for the case of a single large blood vessel with adjuvantly introduced nanoparticles (*in vitro* tissue experiments): (A) before freezing, (B) after 5 minutes of freezing, (C) after 10 minutes of freezing, (D) after 15 minutes of freezing, (E) after 20 minutes of freezing, and (F) after 20 minutes of thawing.

close to the inlet and outlet of the large vessel, respectively. In addition, due to the high thermal conductivity of Fe_3O_4 nanoparticles (about 40 W/m°C) [21], the temperatures of thermocouples 1, 2, and 3 for the case with introduced nanoparticles (shown in Figure 7.19) were correspondingly lower than those for the case without introduced nanoparticles (shown in Figure 7.18). Moreover, for the case with introduced nanoparticles, a jump that occurred in the cooling rate of liver tissue is also shown, similar to the corresponding case of the phantom study. The time for this jump to occur was at about 13.5 minutes (12.5 minutes after freezing began), which agrees with the time when water flow in the large vessels stopped.

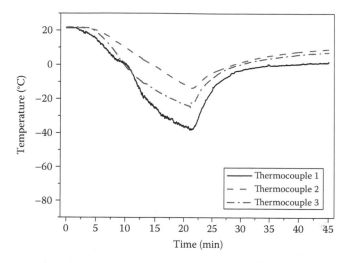

Figure 7.18 Transient temperatures at three different sites for the case of a single large blood vessel without introduced nanoparticles (*in vitro* tissue experiments), in which the sites of thermocouples 1, 2, and 3 are shown in Figure 7.3A.

As stated above, the large blood vessels frequently are present in the form of countercurrent pairs of an artery and a vein, and sometimes they are situated close together. In the phantom study, the distance of the artery and vein pairs was set at 10 mm. For the *in vitro* tissue study, the cases of parallel countercurrent

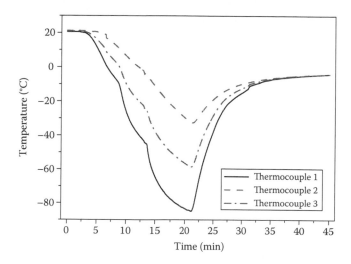

Figure 7.19 Transient temperatures at three different sites for the case of a single large blood vessel with adjuvantly introduced nanoparticles (*in vitro* tissue experiments), in which the sites of thermocouples 1, 2, and 3 are shown in Figure 7.3C.

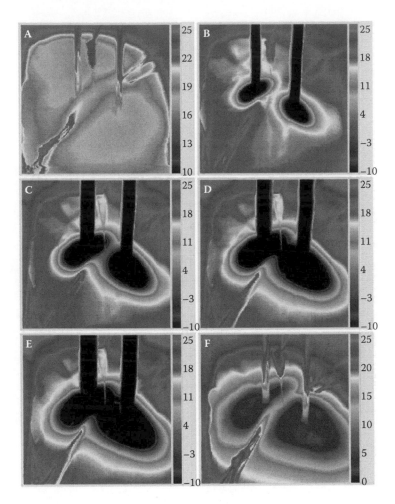

Figure 7.20   Infrared thermographs for the case of parallel countercurrent vessel pairs without intro-duced nanoparticles (*in vitro* tissue experiments): (A) before freezing, (B) after 5 minutes of freezing, (C) after 10 minutes of freezing, (D) after 15 minutes of freezing, (E) after 20 minutes of freezing, and (F) after 20 minutes of thawing.

vessel pairs were also considered, in which the separation distance of the artery and vein pairs was set more closely (about 3 mm apart). Two cases corresponding to the phantom study were considered, without and with introduced nanopar-ticles. A photograph of the experimental setup is shown in Figure 7.4, and the infrared thermographs for the above cases are shown in Figures 7.20 and 7.21, respectively. Results similar to those for the cases of a single large vessel have been obtained. Since the water flow in countercurrent vessel pairs transfers more heat energy than does that of a single large vessel, the thermal effect of large ves-sels is more evident, and the liver tissue in the vicinity of the large vessels was

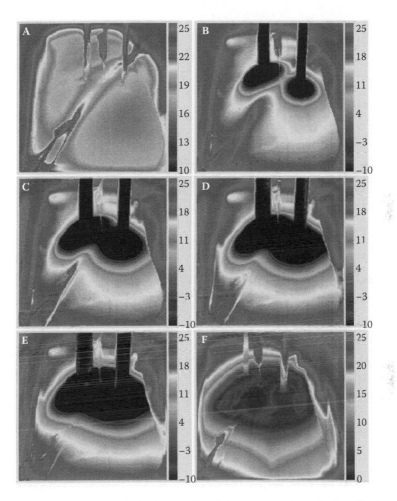

Figure 7.21   Infrared thermographs for the case of parallel countercurrent vessel pairs with adjuvantly introduced nanoparticles (*in vitro* tissue experiments): (A) before freezing, (B) after 5 minutes of freezing, (C) after 10 minutes of freezing, (D) after 15 minutes of freezing, (E) after 20 minutes of freezing, and (F) after 20 minutes of thawing.

harder to freeze compared with the case of a single vessel. For the case of parallel countercurrent vessel pairs without introduced nanoparticles, the large vessels were not frozen during the entire experimental procedure. For the case with adjuvantly introduced nanoparticles, it was found during the experiment that the water flow stopped after about 14 minutes of freezing.

Figures 7.22 and 7.23 show the transient temperatures of thermocouples 1, 2, and 3 for the cases of parallel countercurrent vessel pairs without and with introduced nanoparticles, respectively. The sites of thermocouples 1, 2, and 3 were shown in Figure 7.4. It can be found from Figure 7.22 that for the case without

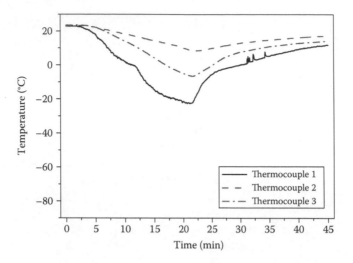

Figure 7.22 Transient temperatures at three different sites for the case of parallel countercurrent vessel pairs without introduced nanoparticles (*in vitro* tissue experiments), in which the sites of thermocouples 1, 2, and 3 are shown in Figure 7.4A.

introduced nanoparticles, the temperature curves appear to be relatively smooth. For the case with adjuvantly introduced nanoparticles, a jump in the cooling rate of liver tissue is also observed, similar to the corresponding result for the case of a single large vessel. The time for this jump to occur was at about 15 minutes (i.e.,

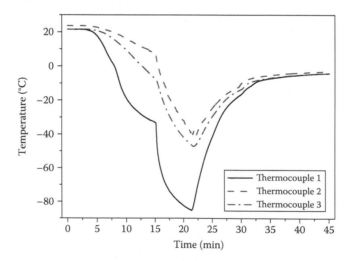

Figure 7.23 Transient temperatures at three different sites for the case of parallel countercurrent vessel pairs with adjuvantly introduced nanoparticles (*in vitro* tissue experiments), in which the sites of thermocouples 1, 2, and 3 are shown in Figure 7.4C.

about 14 minutes after freezing was started), which agrees with the time when water flow in the large vessels stopped.

The aforementioned experimental results indicated that when cryosurgical treatment is performed for tumors embedded with large vessel(s), the rapid flow of blood through the large vessel(s) will cause a heating effect on the target tissues, and such heating prevents freezing of the large vessel(s) and the surrounding tissues. Consequently, tumor cells in the vicinity of larger blood vessel(s) may survive, and the surviving tumor cells further result in the recurrence of tumors after treatment. The results for the cases with adjuvantly introduced nanoparticles demonstrate that nanocryosurgery can significantly enhance the freezing efficacy of tissues and totally freeze the tumor tissues in the vicinity of large vessel(s).

In clinics, when large vessel(s) are present at the target area, vascular inflow occlusion has been used to enhance the treatment efficiency of cryosurgery [3]. However, the need of a major surgical procedure for vascular occlusion negates minimally invasive surgery, which is one of the major advantages of cryosurgery. Therefore, nanocryosurgery is expected to serve as an attractive modality for treatment of tumors embedded with large blood vessel(s), due to its convenience in operation and excellent performance in disabling the thermal effect of large vessel(s).

7.4.2 Numerical Results

The dimensions of the overall tissue region for all vascular models (shown in Figure 7.5) were $10 \times 10 \times 20$ cm in the x, y, and z directions, respectively, and the dimensions of the tumor, which was located at the center of the tissue region, were $4 \times 4 \times 4$ cm. The vessel diameter was set as 0.8 mm for the case of validation calculation and 1 mm for other calculation cases. The constant Nusselt number was taken as $Nu = 4$ [22]. In the calculations, the cylindrical probe was approximated by a square cylinder for brevity. Although the nonuniform grid technique can be introduced to deal with the cylindrical surface of the probe, this feature was not addressed in this chapter for the sake of brevity. The two cryoprobes' active tips with $20 \times 6 \times 6$ mm size were respectively positioned in the domains of [0.034 m $\leq x \leq$ 0.04 m, 0.04 m $\leq y \leq$ 0.05 m, 0.096 m $\leq z \leq$ 0.102 m] and [0.06 m $\leq x \leq$ 0.066 m, 0.04 m $\leq y \leq$ 0.05 m, 0.096 m $\leq z \leq$ 0.102 m] for all calculation cases except the case of the validation calculation.

The typical tissue properties were applied as used in Deng and Liu [5]: $C_b = C_u = 3.6$ MJ/m^3°C, $C_f = 1.8$ MJ/m^3°C, $k_f = 2$ W/m°C, $k_u = 0.5$ W/m°C, $Q_L = 250$MJ/m^3, $T_a = 37$ °C, $T_{ml} = -8$ °C, and $T_{mu} = -1$ °C. The temperature of the cryoprobe tip was assumed to be constant ($T_w = -196$ °C). Both the blood perfusion and metabolic rate are very different for normal and tumor tissues [23], and were respectively taken as

$$\omega_b = \begin{cases} 0.0005 \text{ ml/s/ml}, & x, y \notin \Omega_t \\ 0.002 \text{ ml/s/ml}, & x, y \in \Omega_t \end{cases}, \quad Q_m = \begin{cases} 4200 \text{ W/m}^3, & x, y \notin \Omega_t \\ 42000 \text{ W/m}^3, & x, y \in \Omega_t \end{cases}$$

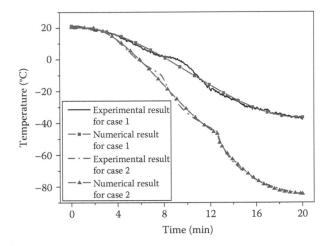

Figure 7.24   Comparison of experimental and numerical results for *in vitro* porcine liver tissues, in which case 1 denotes the case of a single large blood vessel without introduced nanoparticles, and case 2 denotes the case of a single large blood vessel with adjuvantly introduced nanoparticles.

where Ω_t denotes the tumor domain. Since the value of the thermal conductivity of the mixtures of tissue and Fe_3O_4 nanoparticles was still not available, the thermal conductivities of frozen and unfrozen tissues injected with nanoparticle suspensions were respectively taken as $k_u = 0.8$ W/m°C and $k_f = 2.8$ W/m°C based on the data used in Deng and Liu [12], both of which are a little higher than that of the normal situations. The tissue domain injected with nanoparticle suspensions was the same as the tumor region for all involved calculations.

In order to validate the theoretical model, a comparison between the experimental results for *in vitro* porcine liver tissues and the corresponding numerical results is shown in Figure 7.24. The experimental results were taken from the *in vitro* tissue experiments for the case of a single large blood vessel (as shown in Figures 7.18 and 7.19, the temperature curves of thermocouple 1 during the freezing procedure), and the numerical results were obtained from calculations using the SATT model. In the calculations, the configuration of the blood vessel and the cryoprobes and other conditions were the same as those for the *in vitro* tissue experiments. From the temperature curves presented in Figure 7.24, it can be seen that the calculated values fit fairly well with the experimental data, both for the case without introduced nanoparticles and for the case with adjuvantly introduced nanoparticles.

Although the experimental results presented in this study have demonstrated the feasibility of using nanocryosurgery to completely freeze tumor tissues embedded with a large blood vessel, numerical study on *in vivo* tissue was further performed to more realistically disclose the thermal effect of large blood vessels during an actual cryosurgery, since *in vitro* experiments do differ from real

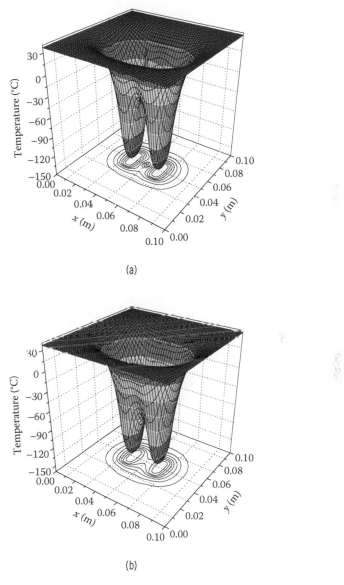

(a)

(b)

Figure 7.25   Temperature distributions at cross section $z = 0.094$ m at $t = 1200$ s for the SATT model, in which (a) is for the case without introduced nanoparticles and (b) is for the case with adjuvantly introduced nanoparticles.

cryosurgical treatment in the human body. Figure 7.25 depicts the temperature distributions at the cross section $z = 0.094$ m at $t = 1200$ s for the SATT model, in which the center line of the single artery was at $x = 0.05$ m, $y = 0.05$ m. It can be seen from Figure 7.25 that the tissue in the vicinity of the large vessel was significantly warmer than that in other regions. It can also be noted in Figure 7.25a that for the case without introduced nanoparticles, there is a temperature peak at the position of the large vessel, and the value of this peak temperature was obviously larger than the freezing point of liver tissue. It indicates that the vessel had not been frozen in the cryosurgical procedure. In fact, as observed in clinics [3,10], it is very difficult to freeze a large blood vessel by cryosurgical treatment, even when using multiple cryoprobes. For the case with introduced nanoparticles, the temperature at the position of the blood vessel (i.e., $x = 0.05$ m, $y = 0.05$ m) was obviously lower than the freezing point of tissue. It indicates that the vessel had been frozen, and that using adjuvantly introduced nanoparticles to totally freeze tumor tissue embedded with a large vessel is feasible. In addition, it should be observed that the temperature in the vicinity of a large vessel was still higher than that in other tissue regions, even after the blood flow was occluded by freezing. This phenomenon may have resulted from the thermal effect of the large blood vessel before it was frozen. Such a phenomenon should be addressed when designing a treatment plan for nanocryosurgery of a tumor embedded with large vessels in order to ensure that the whole tumor can be frozen to a temperature lower than the lethal temperature (which is usually significantly lower than the freezing point of tissue, and ranges from –20°C to –70°C for different tumor tissues) [24]. Figure 7.26 depicts the boundaries of the iceball formed at the cross section $z = 0.094$ m at $t = 1200$ s for the SATT model. It is clearly shown in this figure that the large vessel had not been frozen for the case without introduced nanoparticles, while the vessel had been frozen for the case with introduced nanoparticles. In addition, due to the high thermal conductivity of nanoparticles, the size of the iceball for the case with introduced nanoparticles was slightly larger than that for the case without introduced nanoparticles. Similar results had also been observed in the experimental results presented earlier, as shown in Figures 7.3B,D.

Figure 7.27 shows the temperature distributions at the cross section $z = 0.094$ m at $t = 1200$ s for the CVTT model, in which the center lines of the

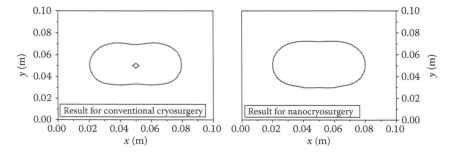

Figure 7.26   Boundaries of iceball formed at cross section $z = 0.094$ m at $t = 1200$ s for the SATT model, in which conventional cryosurgery denotes the case without introduced nanoparticles, and nanocryosurgery denotes the case with adjuvantly introduced nanoparticles.

artery and the vein were at $x = 0.05$ m, $y = 0.05$ m and $x = 0.05$ m, $y = 0.048$ m, respectively. Similar results to the cases using the SATT model are obtained. The temperature peak at the position of the large vessels for the case without introduced nanoparticles was more evident than that for the corresponding case using the SATT model, since the blood flow in countercurrent vessel pairs provides more heat energy than does a single large vessel. Figure 7.28 depicts the

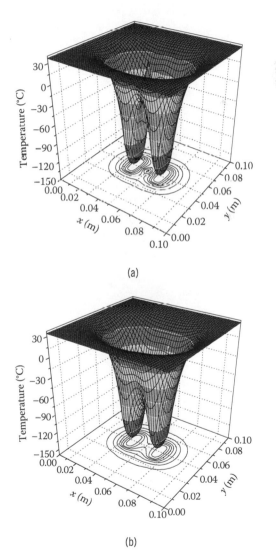

(a)

(b)

Figure 7.27   Temperature distributions at cross section $z = 0.094$ m at $t = 1200$ s for the CVTT model, in which (a) is for the case without introduced nanoparticles and (b) is for the case with adjuvantly introduced nanoparticles.

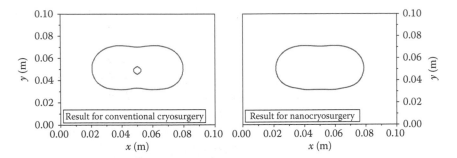

Figure 7.28 Boundaries of iceball formed at cross section $z = 0.094$ m at $t = 1200$ s for the CVTT model, in which conventional cryosurgery denotes the case without introduced nanoparticles, and nano-cryosurgery denotes the case with adjuvantly introduced nanoparticles.

boundaries of the iceball formed at the cross section $z = 0.094$ m at $t = 1200$ s for the CVTT model. It is also demonstrated that the large vessels had not been frozen for the case without introduced nanoparticles and had been frozen for the case with nanoparticles. Figure 7.29 gives a comparison of numerical results for the SATT and CVTT models, in which the curves depict the temperature distribution at $t = 1200$ s in the artery at $x = 0.05$ m, $y = 0.05$ m. It is clearly shown that for the case with introduced nanoparticles, the large blood vessel(s) had been

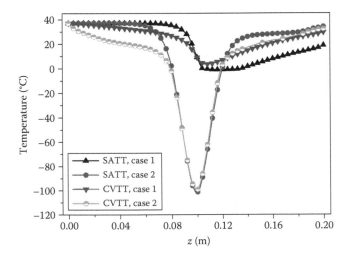

Figure 7.29 Comparison of numerical results for both SATT and CVTT models, in which case 1 denotes the case without introduced nanoparticles, and case 2 denotes the case with adjuvantly introduced nanoparticles.

frozen for both the SATT and CVTT models, whereas the large vessel(s) could not have been frozen for the case without introduced nanoparticles even when using the SATT model (in which only one large vessel was present).

The foregoing numerical results indicate that the heating effect created by blood flow in large vessel(s) can produce steep temperature gradients and lead to inadequate cooling of the surrounding tumor tissues. Therefore, they may seriously contribute to the failed killing of tumors during cryosurgery. These results also imply that highly conductive nanoparticles can serve as effective adjuvants for enhancing the efficacy of cryosurgical treatment of tumors embedded with large vessels, and this feature enhances the possibility to totally destroy tumor tissues surrounding large vessel(s) by nanocryosurgery. However, it should be pointed out that when performing nanocryosurgery for treatment of a tumor embedded with some major vessel(s) that has important functions for tissues or organs at other regions, it must be seriously considered whether freezing such vessel(s) will result in severe postoperative complications.

7.5 CONCLUSIONS

In this study, both experimental and numerical investigations were performed to probe into the thermal effects of a single large blood vessel and of parallel countercurrent vessel pairs on tissue temperature responses during conventional cryosurgery and nanocryosurgery. The results demonstrate that large blood vessel(s) embedded in tumor tissues have significant heating effects on its (their) surrounding tumor tissues and thus may result in insufficient freezing, which contributes to the nonkilling of part of the tumor cells. When nanoparticles with high thermal conductivity were introduced into the target area, it was found that the freezing efficacy can be significantly enhanced, and consequently the tumor in the vicinity of large vessel(s) and the vessel(s) can be effectively frozen. In conclusion, both experimental and numerical results indicate that nanocryosurgery can be used to completely destroy tumors embedded with large blood vessel(s), and nanocryosurgery is expected to provide cryosurgeons with more choices in performing a highly efficient, minimally invasive treatment for tumor killing.

ACKNOWLEDGMENTS

This work was partially supported by the National Natural Science Foundation of China under Grants 50576104 and 50575219 and the Initial Funding for the Gainer of Excellent Doctoral Dissertation Award of the Chinese Academy of Sciences. The authors appreciate the very constructive suggestions and help from Dr. John Abraham.

NOMENCLATURE

C:	volumetric heat capacity [J/m³°C]
D:	diameter of vessel [m]
h:	convective heat transfer coefficient [W/m²°C]
k:	thermal conductivity [W/m°C]
Nu:	Nusselt number
p:	perimeter of vessel [m]
Q_l:	volumetric latent heat [J/m³]
Q_m:	metabolic heat generation [W/m³]
S:	cross-sectional area of vessel [m²]
t:	time [seconds]
T:	temperature [°C]
T_a:	artery temperature [°C]
T_b:	mean blood temperature [°C]
T_{ml}:	lower phase-transition temperature of tissue [°C]
T_{mu}:	upper phase-transition temperature of tissue [°C]
T_w:	temperature at probe tip [°C]
T_{wb}:	temperature of the vessel wall [°C]
v:	velocity of blood flow [m/s]
x,y,z:	Cartesian coordinate [m]
X:	location [m]

GREEK SYMBOLS

ω_b:	blood perfusion [ml/s/ml]
Ω:	computation domain

SUBSCRIPTS

f:	frozen tissue
u:	unfrozen tissue

SUPERSCRIPTS

$\sim$:	effective value

REFERENCES

1. B. Rubinsky, Cryosurgery, *Annual Review of Biomedical Engineering*, vol. 2, pp. 157–187, 2000.
2. J. K. Seifert and D. L. Morris, Indicators of recurrence following cryotherapy for hepatic metastases from colorectal cancer, *British Journal of Surgery*, vol. 86, pp. 234–240, 1999.

3. T. Mala, L. Frich, L. Aurdal, O. P. Clausen, B. Edwin, O. Screide, and I. P. Gladhaug, Hepatic vascular inflow occlusion enhances tissue destruction during cryoablation of porcine liver, *Journal of Surgical Research*, vol. 115, pp. 265–271, 2003.

4. Y. T. Zhang, J. Liu, and Y. X. Zhou, Pilot study on cryogenic heat transfer in biological tissues embedded with large blood vessels, *Forschung im Ingenieurwesen (Engineering Research)*, vol. 67, pp. 188–197, 2002.

5. Z. S. Deng and J. Liu, Numerical study on the effects of large blood vessels on three-dimensional tissue temperature profiles during cryosurgery, *Numerical Heat Transfer, Part A: Applications*, vol. 49, pp. 47–67, 2006.

6. Z. S. Deng, J. Liu, and H. W. Wang, Disclosure of the significant thermal effects of large blood vessels during cryosurgery through infrared temperature mapping, *International Journal of Thermal Sciences*, vol. 47, pp. 530–545, 2008.

7. J. C. Chato, Heat transfer to blood vessels, *ASME Journal of Biomechanical Engineering*, vol. 102, pp. 110–118, 1980.

8. J. Crezee and J. J. W. Lagendijk, Temperature uniformity during hyperthermia: the impact of large vessels, *Physics in Medicine and Biology*, vol. 37, pp. 1321–1337, 1992.

9. W. K. Berger and J. Poledna, New strategies for the placement of cryoprobes in malignant tumors of the liver for reducing the probability of recurrences after hepatic cryosurgery, *International Journal of Colorectal Disease*, vol. 16, pp. 331–339, 2001.

10. W. Jungraithmayr, M. Szarzynski, H. Neeff, J. Haberstroh, G. Kirste, A. Schmitt-Graeff, E. H. Farthmann, and S. Eggstein, Significance of total vascular exclusion for hepatic cryotherapy: an experimental study, *Journal of Surgical Research*, vol. 116, pp. 32–41, 2004.

11. T. H. Yu, J. Liu, and Y. X. Zhou, Selective freezing of target biological tissues after injection of solutions with specific thermal properties, *Cryobiology*, vol. 50, pp. 174–182, 2005.

12. Z. S. Deng and J. Liu, Numerical simulation of selective freezing of target biological tissues following injection of solutions with specific thermal properties, *Cryobiology*, vol. 50, pp. 183–192, 2005.

13. R. K. Visaria, R. J. Griffin, B. W. Williams, E. S. Ebbini, G. F. Paciotti, C. W. Song, and J. C. Bischof, Enhancement of tumor thermal therapy using gold nanoparticle-assisted tumor necrosis factor-α delivery, *Molecular Cancer Therapeutics*, vol. 5, pp. 1014–1020, 2006.

14. S. Mornet, S. Vasseur, F. Grasset, and E. Duguet, Magnetic nanoparticle design for medical diagnosis and therapy, *Journal of Materials Chemistry*, vol. 14, pp. 2161–2175, 2004.

15. G. F. Paciotti, L. Myer, D. Weinreich, D. Goia, N. Pavel, R. E. McLaughlin, and L. Tamarkin, Colloidal gold: a novel nanoparticle vector for tumor directed drug delivery, *Drug Delivery*, vol. 11, pp. 169–183, 2004.

16. I. Brigger, C. Dubernet, and P. Couvreur, Nanoparticles in cancer therapy and diagnosis, *Advanced Drug Delivery Reviews*, vol. 54, pp. 631–651, 2002.

17. J. Liu, J. F. Yan, and Z. S. Deng, Nano-cryosurgery: a basic way to enhance freezing treatment of tumor, *ASME International Mechanical Engineering Congress and RD&D Expo*, Seattle, Washington, November 11–15, 2007.

18. J. Liu, Y. X. Zhou, T. H. Yu, L. Gui, Z. S. Deng, and Y. G. Lv, Minimally invasive probe system capable of performing both cryosurgery and hyperthermia treatment on target tumor in deep tissues, *Minimally Invasive Therapy and Allied Technologies*, vol. 13, pp. 47–57, 2004.

19. Z. P. Chen and R. B. Roemer, The effects of large blood vessels on temperature distributions during simulated hyperthermia, *ASME Journal of Biomechanical Engineering*, vol. 114, pp. 473–481, 1992.

20. Z. S. Deng and J. Liu, Numerical simulation of 3-D freezing and heating problems for combined cryosurgery and hyperthermia therapy, *Numerical Heat Transfer, Part A: Applications*, vol. 46, pp. 587–611, 2004.

21. Y. G. Lv, Z. S. Deng, and J. Liu, Study on the induced heating effects of embedded micro/nano particles on human body subject to external medical electromagnetic field, *IEEE Transactions on NanoBioscience*, vol. 4, pp. 284–294, 2005.

22. H. W. Huang, C. L. Chan, and R. B. Roemer, Analytical solutions of Pennes bio-heat transfer equation with a blood-vessel, *ASME Journal of Biomechanical Engineering*, vol. 116, pp. 208–212, 1994.

23. J. Liu, L. Zhu, and L. X. Xu, Studies on the three-dimensional temperature transients in the canine prostate during transurethral microwave thermal therapy, *ASME Journal of Biomechanical Engineering*, vol. 122, pp. 372–379, 2000.

24. A. A. Gage and J. Baust, Mechanism of tissue injury in cryosurgery, *Cryobiology*, vol. 37, pp. 171–186, 1998.

8

Whole-Body Human Thermal Models

E. H. Wissler

CONTENTS

8.1 INTRODUCTION

The expression "human thermal model" can be defined in various ways. A broad definition includes any relationship between one or more bodily temperatures and environmental and metabolic variables, such as ambient temperature, humidity, and intensity of exercise. Algorithms included under a broad definition are often part of schemes for evaluating thermal comfort under various conditions. Such models usually ignore the geometry and composition of the human body, and treat physiological variables, such as skin blood flow, shivering, and sweating, in a highly empirical manner. Narrower definitions of human thermal model limit use of the term to models in which the temperature field is computed for a reasonably faithful representation of the human form and relevant physiological factors. Although we will mention briefly models included under a broad definition, the principal focus of this chapter will be on more rigorously defined models.

The usefulness of any mathematical model depends on many factors, the most important of which is the accuracy and completeness of underlying fundamental equations. While the accuracy and speed of computational methods used to obtain numerical values of various quantities are important, computational elegance cannot make up for the shortcomings of a fundamentally flawed model. With a few exceptions [1–3], human thermal models have been developed by engineers whose knowledge of, and interest in, human physiology appears to be limited. In this chapter, we will review in detail physiological phenomena that are important to human thermal regulation.

Human thermal models serve several useful purposes. Perhaps the most important purpose is to incorporate diverse physiological and physical phenomena into an internally consistent representation of human thermal regulation. The difficulty of accomplishing that goal was described by Stolwijk and Hardy in 1966 [4]. They wrote in describing their model,

In fact, it can be argued that no completely satisfactory model of a (physiological) subsystem such as temperature regulation can be formulated by itself and that inclusion of all thermal and non-thermal physiological data is essential for a solution. While recognizing this difficulty as basic in the study of physiology generally, the separate analyses involved in each subsystem and component are essential in each case and are therefore justified within limits for the present study.

An important example of interaction between systems is provided by the effect of exercise on skin blood flow, which is discussed in some detail later.

Another purpose of models is to predict human behavior under potentially life-threatening conditions. For example, expected survival time during accidental immersion in cold water is an important factor in managing search and rescue missions, both in terms of the response time required for a successful outcome and the maximum time a search should be continued. Laboratory studies involving human subjects provide useful information about initial cooling rates during immersion in cold water, but such studies are never carried to life-threatening limits, and predicting survival time involves significant extrapolation of the resulting data. A good human thermal model can be very useful for that purpose, although it also inevitably involves important assumptions, such as the conditions under which a severely hypothermic individual no longer shivers.

A third purpose is to apply results obtained using human thermal manikins to human performance. For example, protective garments worn by military personnel invariably impose considerable thermal stress on the wearer. Instead of testing each garment on 10 human subjects under various conditions, it is preferable to measure the properties of the garment on a manikin and use those results to predict human behavior under various field conditions.

There are also important medical applications of human thermal models, although not much has been done in that regard. The greatest interest to date has been in preventing hypothermia during surgery, and commercial systems have been developed for that purpose. On the other hand, patients are routinely cooled during open-heart surgery to alleviate undesirable effects of cardiopulmonary bypass. There is also considerable current interest in possible beneficial effects of cooling victims of heart attack and stroke. It is reasonable to expect additional medical applications of thermal models.

The final purpose to be mentioned is the prediction of human comfort under various conditions. Although relating individual perceptions of comfort to physical variables, such as skin and central temperatures, sweat rate, and skin wettedness, is a rather imprecise art, there is considerable interest in such applications. Automotive engineers appear to be particularly interested in establishing comfortable conditions for passengers.

All previously developed human thermal models define the regulation of skin blood flow, sweating, and shivering in terms of a central temperature (presumably representative of hypothalamic temperature) and mean skin temperature. Relevant control functions are usually defined to achieve good agreement between computed and observed temperatures under various conditions, but that process does not necessarily yield results consistent with results obtained in other physiological studies. An important objective of this chapter is to demonstrate that control functions derived from independent physiological studies yield a model with good thermal properties.

8.2 EVOLUTION OF HUMAN THERMAL MODELS

Before we delve into the details of model formulation, it is worthwhile to review briefly the history of human thermal models. Two of the earliest models are still in use in modified form by those whose principal concern is thermal comfort under moderately stressful conditions. The evolution of human thermal models outlined in this section reflects the author's prejudices, and others might construct a rather different version.

Probably the first model was proposed by A. C. Burton in 1934 [5]. His very simple model represented the human body as a single homogeneous cylinder with uniform metabolic heat generation. It suggested that the steady-state temperature profile is parabolic in agreement with the experimental observations of Bazett and McGlone [6]. Subsequently, Burton and Bazett [7] used the transient solution of the heat conduction equation for the same system to interpret data obtained with their bath calorimeter.

Since the results of experimental studies are nearly always reported as central and mean skin temperatures, it followed quite naturally that a two-node model would be developed. Machle and Hatch [8] described a "core and shell model" in their extensive 1947 review paper summarizing the results of research conducted during World War II at the Armored Medical Research Laboratory in Fort Knox, Kentucky, and at the Pierce Laboratory at Yale University. Application of that model was not discussed and it was probably not very useful, because it neglected heat transfer between core and shell, assuming instead that core and shell temperatures were linearly related.

A significant milepost in the development of human thermal models occurred in 1948 with the publication of Pennes's classic paper [9] on the effect of blood flow on tissue temperature. Pennes advanced the notion that heat transfer between blood in small vessels and adjacent tissue is proportional to the product of the perfusion rate and the difference between blood and tissue temperatures. Although computed temperature profiles in the forearm had a slightly different shape from profiles measured by drawing a fine thermocouple through the forearm, the difference was eventually attributed to the way data for arms of different size were analyzed [10]. Most current detailed models employ the Pennes model (often referred to as the bioheat equation) to describe heat transfer in perfused tissue. Implicit in the Pennes model is the assumption that no heat transfer occurs between blood in small arteries and veins connected to capillaries and surrounding tissue. That assumption has been criticized and analyzed in considerable detail (for a summary of early work, see Charney [11]), with the resulting conclusion that Pennes's original model probably overestimates the rate of heat transfer between blood and tissue. Multiplication of the Pennes expression by a factor that depends on the perfusion rate provides a more accurate result [12–14].

Wissler [15] applied concepts developed in Pennes's paper to develop the first steady-state, multielement, human thermal model. His model consisted of six homogeneous cylindrical elements representing the head, trunk, two arms, and two legs, which were connected by circulating blood. The thermal energy balance for blood made allowance for the effect of countercurrent heat transfer between arterial and venous blood. When reasonable values were assigned to metabolic and perfusion rates in the six elements and allowance was made for respiratory heat loss, acceptable agreement between computed and measured temperatures was obtained. That model did not consider the effect of thermal state on blood flow, sweat secretion, or shivering.

Information gained from studies conducted at the Pierce Laboratory during the next 5 years contributed greatly to our understanding of human thermoregulation. Transient changes in central (usually rectal) and mean skin temperatures were recorded during exposure of seated, lightly clad, male subjects to air temperatures ranging from 13°C to 48°C. Rates of metabolic heat generation and sweat secretion were determined by partitional calorimetry. The resulting data are still invaluable for testing human thermal models.

In 1966, Stolwijk and Hardy [4] significantly advanced the art of human thermal modeling with their publication of a theoretical study in which the concepts of feedback control were applied to human thermoregulation. Although those concepts were not original with Stolwijk and Hardy, the thoroughness of their analysis enhanced the validity of human thermal modeling.

The 1966 Stolwijk and Hardy model consisted of three cylindrical elements representing the head, trunk, and extremities. That model was implemented on an analog computer, which undoubtedly limited the number of elements and amount of detail it could contain. A total of seven regions represented the head core (brain), trunk muscle, trunk core (viscera), extremity core, and a 2 mm thick layer of skin on each element. The radius and length of each cylinder were defined so that it had a mass and surface area appropriate to the anatomical region represented. Heat transfer by conduction occurred between adjacent regions in proportion to the difference in regional temperatures. Blood contained in a central pool exchanged heat with tissue located in each region. The principal contribution of that model was that it incorporated control functions for skin blood flow, sweating, and shivering into a physically reasonable model. Subsequent models have all employed the approach introduced in that paper.

Two notable models evolved from the 1966 Stolwijk and Hardy model. In 1970, Stolwijk [1] developed a six-element model that ran on an early digital computer. Each element was subdivided into four regions representing a central core surrounded by muscle, subcutaneous fat, and skin. The regions were perfused with blood drawn from a central pool. Heat transfer by conduction between adjacent regions occurred at a rate proportional to the temperature difference between them. Using physiological control functions for skin blood flow, sweating, and shivering based on the best available information, Stolwijk and Hardy significantly advanced the art of human thermal modeling. Their model was used in the design and operation of the Portable Life Support System for the Apollo missions, and it is still employed in several thermal comfort models. For example, the Berkeley multinode comfort model [16] is based on the 25-node Stolwijk model, although it has been augmented in many ways.

Also in 1970, Gagge et al. [2] developed a simple two-node model for the purpose of evaluating thermal stress imposed by a given environment. Their objective was to develop an effective temperature scale that would allow engineers and environmental scientists to compare thermal environments on the basis of energy exchange. That simple model still finds application, although one must question the virtue of extreme simplicity when powerful computational facilities are readily available. Moreover, a recent study by Jay et al. [17] reaffirmed that using core and mean skin temperatures to estimate the internal energy content of the human body is quite inaccurate [18]. However, for the record, we mention an example of that approach provided by Bruse's individual thermal comfort model [19], which is based on the two-node Gagge model.

The summary presented above is by no means complete. Other variants of human thermal models evolved from the Stolwijk and Hardy model, but their existence was often transient, and limitations of space and time preclude including them in this document.

8.3 MODEL STRUCTURE

Factors that affect human thermal regulation are both physical and physiological in nature. Physical factors include the geometry and composition of the body. Heat transfer from exposed skin or clothing is also a purely physical phenomenon, as is evaporation of sweat. Ambient conditions, such as fluid velocity, temperature, humidity, and incident radiant flux, are also physical factors. In general, the effect of physical factors is well understood and amenable to rational analysis, although applications to human modeling are often rudimentary. For example, modern computational fluid dynamic (CFD) techniques should permit accurate computation of local heat and mass transfer coefficients for the human body, but that is seldom done. Another physical factor normally treated rather approximately is heat and mass transfer in clothing.

Under moderate environmental conditions, human beings can achieve acceptable temperature control by regulating three physiological variables: blood flow to skin, sweating, and shivering. While those variables respond primarily to thermal stimuli, usually defined in terms of a central temperature assumed to be the temperature of the hypothalamus, mean skin temperature, and local skin temperature, they are also affected by nonthermal factors. An example is that circulatory responses to thermal conditions may be modulated by exercise, because the primary purpose of the circulatory system is to satisfy the metabolic requirements of tissue, especially the requirements of critical organs and active muscle.

8.3.1 Geometry

Nearly all human thermal models approximate individual elements of anatomy as circular cylinders, although several represent the head as a sphere. An early arrangement is shown in Figure 8.1. That one-dimensional model developed in 1964 [20] assumed that physical properties and temperature are functions of radial position and time and required a large central computer for execution. As more powerful computers became commonplace, additional cylindrical elements were added to models, and time-dependent physical properties and temperature were allowed to vary with both r and θ. Axial conduction is still generally neglected. For example, Fiala et al. [21,22] developed a 15-element model,

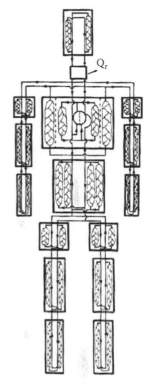

Figure 8.1 A typical representation of the human geometry.

and Qi [23] developed a 54-element model. The author has recently developed a new 21-element model in which the head is represented by two elements, the trunk is represented by three elements, and each arm and leg is represented by four elements. All of the models mentioned allow a reasonably good representation of major organs, muscles, arteries, and veins.

Allowing physical properties and temperature to vary with angular position removes a serious limitation of early one-dimensional models, which could not represent nonuniform boundary conditions on a given element. For instance, conditions on the anterior surfaces of an individual seated in an air-conditioned automobile are markedly different from conditions on the posterior surface, and there is no rational basis for defining "average conditions" in a one-dimensional model.

One might expect that allowing temperature to vary with θ greatly complicates numerical analysis, but that is not necessarily true. According to Fiala et al. [21] two considerations suggest that conduction in the θ direction can be neglected without seriously affecting accuracy. Within the very thin skin region, the product of the temperature gradient parallel to the surface and the area through which heat is conducted in the θ direction is much smaller than the corresponding product for heat transfer normal to the skin and, therefore, is negligible. In deeper regions, temperature gradients are small, heat transport by convection is dominant, and tissue temperature is determined largely by the rates of metabolic heat generation and perfusion by blood. While those arguments are intuitively appealing, they appear to be untested.

Several recent papers describe in somewhat nonspecific terms models in which finite element methods are used to represent accurately the human form [24–26]. Those models appear to describe faithfully external features of human geometry, but it is unclear how accurately they represent internal structural features.

8.3.2 Composition

The temperature field within a region depends on its composition, because thermal conductivity, thermal diffusivity, rate of heat generation, and perfusion rate are all functions of tissue type. One of the more important tissue components is subcutaneous fat, which has a relatively low thermal conductivity and low perfusion rate, and is located directly under skin where it can limit heat transfer to cool environments.

The local thickness of subcutaneous fat has been determined for more than 50 years by measuring the skinfold thickness with a standard calipers. A correlation relating percent body fat determined by underwater weighing to skinfold thicknesses measured at four sites was developed in 1974 by Durnin and Womersley [27], and is still widely used. However, when local subcutaneous fat thickness derived from skinfold measurement at the site has been compared with thickness determined from magnetic resonance imaging (MRI) or X-ray images, the skinfold-derived value has usually been found to be quite inaccurate, often underestimating the actual fat thickness by 50% [28,29].

In one of the more comprehensive studies, Hayes et al. [30] compared subcutaneous fat thickness determined from MRI images with corresponding thicknesses derived from skinfold measurements at 89 sites on 20 male and 20 female subjects. They found that skinfold measurement significantly underestimated the subcutaneous fat thickness. They also reported that the distribution of subcutaneous fat varied markedly as a function of percentage of body fat and gender, which is often not taken into consideration.

Since that report is not generally available, selected data are summarized in Table 8.1. Groups are defined in terms of mean subcutaneous fat thickness (*SCF*) as determined by MRI measurement. For females: in Group 1, 12.1 mm < *SCF*; in Group 2, 12.1 mm ≤ *SCF* < 15.2 mm; and in Group 3, 15.2 mm ≤ *SCF*. For males: in Group 1, 4.8 mm < *SCF*; in Group 2, 4.8 mm ≤ *SCF* < 9.0 mm; and in Group 3, 9.0 mm ≤ *SCF*. The body segments are as follows: 1 = upper trunk; 2 = lower trunk; 3 = head; 4, 5, and 6 = proximal, medial, and distal segments of the legs; and 7, 8, and 9 = proximal, medial, and distal segments of the arms. The

Table 8.1   Ratio of Local Subcutaneous Fat Thickness to Mean Fat Thickness Determined by MRI

Females									
Body Segment	1	2	3	4	5	6	7	8	9
Group 1	0.64	1.56	0.92	1.45	0.92	0.72	0.76	0.63	0.64
Group 2	0.72	1.70	0.62	1.36	0.85	0.61	0.81	0.55	0.47
Group 3	0.79	1.67	0.72	1.36	0.84	0.60	0.75	0.55	0.45
Males									
Body Segment	1	2	3	4	5	6	7	8	9
Group 1	0.77	1.94	0.64	1.53	0.99	1.08	0.39	0.13	0.01
Group 2	0.96	1.67	0.66	1.23	0.84	0.78	0.78	0.25	0.09
Group 3	1.07	1.63	0.68	1.05	0.73	0.50	0.75	0.55	0.27

Source: Hayes et al. [30].

following relationships define "true mean" SCF in terms of the mean skinfold thickness (SkF_4) measured by calipers at four sites. For females,

$$SCF = 0.445\ SkF_4 - 1.013 \tag{8.1}$$

and for males,

$$SCF = 0.721\ SkF_4 + 7.300 \tag{8.2}$$

8.4 PHYSIOLOGICAL CONTROL FUNCTIONS

Human thermal models are often defined in terms of passive and active systems. The "passive system" refers to the physical system in which the transient temperature field is defined by the bioheat equation with appropriate boundary and initial conditions, while the "active system" defines control functions for blood flow (especially to skin and muscle), sweat secretion, and shivering. As we noted previously, definition of the passive system ranges from very simple to very sophisticated. Although a simple system may provide an acceptable description of resting individuals in a warm environment, a detailed description is usually required for cold environments when appreciable temperature gradients exist within the body. We will assume that readers are quite knowledgeable about methods available to construct the passive system and will focus attention on the active system.

8.4.1 Regulation of Blood Flow to Muscle

Since convection by circulating blood is an effective transport mechanism for heat transfer, the manner in which local perfusion rates are computed is an important attribute of human thermal models. Highly variable blood flow to muscle and skin is especially important. Definition of suitable control functions for the circulatory system is complicated by the fact that blood transports vital chemical species, as well as heat, and both functions must be performed adequately. The heart is a two-stage positive displacement pump that supplies a vascular system in which the resistance of various branches is regulated in a manner that assures adequate flow to vital organs, such as the heart and brain. Consequently, thermoregulatory control of blood flow cannot be treated as though it were an independent system. That is discussed in great detail by Rowell [31,32].

The perfusion rate of active muscle affects the temperature increase owing to enhanced metabolic heat generation during exercise. Typically, the steady-state temperature of active muscle is 1°C higher than the local arterial blood temperature. Although the fundamental mechanisms that determine muscle blood flow remain obscure, empirical data firmly establish that the perfusion rate of active muscle increases promptly as the local rate of oxygen consumption ($\dot{V}_{O_2}$) increases. Values of $\dot{V}_{O_2}$ above the resting rate are determined by the rate at which external work is done, the mechanical efficiency of the body (typically about 25%), and the relative involvement of various muscles.

The perfusion rate (q) is related to the arteriovenous oxygen difference ($\Delta O_{2,av}$) by the relationship

$$q = \frac{\dot{V}_{O_2}}{\Delta O_{2,av}} \tag{8.3}$$

where $\Delta O_{2,av}$ in resting muscle is about 5 ml O_2/100 ml of blood, and increases very rapidly to approximately 13 to 17 ml O_2/100 ml blood during exercise. Proctor et al. [33] observed that the reduction of femoral venous oxygen content occurs at surprisingly low exercise levels (for example, at 20 W), and Nielsen et al. [34] showed that $\Delta O_{2,av}$ in active muscle is not a strong function of temperature.

An important question is whether blood flow to inactive muscle varies with local tissue temperature. That question was especially pertinent a few years ago, when venous occlusion plethysmography (VOP) was the predominant method for measuring skin blood flow in the human forearm during rest and exercise in a warm environment. That technique involves measuring the volume of a section of forearm when venous outflow is blocked by applying appropriate pressure at the wrist and below the elbow. Pressure applied at the wrist blocks both arterial and venous blood flow, while pressure applied below the elbow blocks only venous outflow from the forearm. Several studies [35–37] established conclusively that blood flow to inactive muscle does not increase with increasing temperature. On the other hand, other studies strongly suggest that the perfusion rate of inactive muscle decreases when muscle is cooled, although the precise nature of thermally induced vasoconstriction in muscle remains obscure.

8.4.2 Regulation of Skin Blood Flow

Extensive investigation of skin blood flow has been motivated by several considerations. One is that the rate of blood flow to maximally dilated skin is the order of 7 l/min, which is an appreciable fraction of maximal cardiac output. Moreover, since cutaneous vasculature is rather compliant, a significant transfer of blood from the central venous pool to skin occurs during heat stress, which may adversely affect cardiac performance. Skin is also the only organ in which blood flow can be measured noninvasively.

Skin blood flow is an important factor in human thermoregulation. Most models define skin blood flow simply as a function of central and mean skin temperatures. While those temperatures are undoubtedly important determinants of skin blood flow, more than 50 studies carried out during the past half century have established that local skin temperature and several nonthermal factors are also important. The results of those studies can be summarized as follows:

1. Two branches of the sympathetic nervous system affect efferent neural mechanisms that control cutaneous vasculature of the head, trunk, and limbs. One branch is an active vasodilator system, and the other branch is an active vasoconstrictor system.
2. Modulation of skin blood flow owing to cold stress, moderate heat stress, and exercise can be attributed to variable vasoconstrictor tone. As skin and central temperatures increase, active vasodilation becomes effective, and the large increase in skin blood flow seen during severe heat stress is caused primarily by active vasodilation.
3. Active vasoconstriction responds reflexively to mean skin temperature.
4. Local skin temperature affects local cutaneous vasoconstrictor tone.
5. Active vasodilation increases linearly with increasing central temperature above a threshold temperature that is modulated by mean skin temperature, exercise, and posture.

8.4.2.1 Measurement of Skin Blood Flow

Prior to 1980 the only quantitative measure of skin blood flow was provided by forearm blood flow (FBF) measurement using VOP. Around 1980, laser-Doppler (LD) techniques that detect the velocity of red blood cells as they pass through cutaneous capillaries were developed for sensing skin blood flow in a small area. Choosing a laser of appropriate wavelength allows incident light to be almost completely absorbed or reflected within the skin so that blood flow within underlying muscle is not detected [38]. When Johnson et al. [39] compared skin blood flow measured using an LD instrument with flow measured by VOP, they found that LD flow rates were linearly related to flow rates measured plethysmographi-cally, although the relationship was specific to each subject. LD instruments offer the advantages that they can be used on any skin area, they respond rapidly to changing flow rate, and the temperature of skin in the area of measurement can be closely controlled.

Since LD measurements do not provide absolute values for skin blood flow (SkBF), they are usually reported as a percentage of some reference value, usually either a thermally neutral initial value or a maximal value recorded at a local skin temperature of 42°C. In addition, LD measurements are usually reported as cutaneous vascular conductance (CVC), which is the ratio of the flow rate to the mean arterial pressure.

The use of chemical agents to block centrally mediated changes in skin blood flow is a time-honored technique. When combined with LD measurement of SkBF, it allows investigators to differentiate clearly responses of the active vaso-dilator system from responses of the vasoconstrictor system. Kellogg et al. [40,41] very successfully employed bretylium blockade of active vasoconstriction to iso-late the effect of active vasodilation on SkBF during exercise.

8.4.3 Algorithm for Computing Skin Blood Flow

Experimental results from studies reported during the past half century can be represented reasonably well by the following relationship:

$$q_s = q_{s,r} \times AVD \times CVCM \times CVCL \times CVCE \qquad (8.4)$$

where the quantity q_s is the local cutaneous perfusion rate (typically reported as ml blood/[100 ml tissue min]), and $q_{s,r}$ is the perfusion rate under reference con-ditions, which we define as $T_s = 34°C$, $\bar{T}_s = 34.5°C$, and $T_c <$ the threshold central temperature $(T_{c,th})$ for active vasodilation. AVD defines the centrally mediated drive for active vasodilation, $CVCM$ accounts for the reflex effect of $\bar{T}_s$ on CVC, $CVCL$ defines the locally mediated effect of T_s on CVC, and $CVCE$ accounts for the direct effect of dynamic exercise on vascular conductance and the increase in mean arterial pressure (MAP) that occurs during exercise. By definition, each of the functions has a value of unity at the reference condition.

With the exception of active vasodilation (AVD), definitions of the component functions are based on CVC data. Forearm blood flow data are generally consis-tent with CVC data when reasonable assumptions are made about muscle blood flow. If we assume that plethysmography detects only skin and muscle blood flows, we have

$$FBF = X_s\, q_s + X_m\, q_m \qquad (8.5)$$

in which X_s and X_m are the volume fractions of skin and muscle, respectively, in the forearm; and q_s and q_m are the respective perfusion rates. Cooper et al. [42]

reported that $X_s = 0.086$ and $X_m = 0.636$ in five forearms obtained postmortem from different subjects.

8.4.3.1 Active Vasodilation: Thermal Factors

Defining the separate and collective contributions of central temperature and $\bar{T}_s$ to regulation of skin blood flow has been very challenging, largely because it is difficult to change one temperature without affecting the other. Central temperature can be increased either by raising the skin temperature, or by increasing metabolic heat production through exercise. A complicating factor associated with the second approach is that moderate to heavy exercise causes a decrement in skin blood flow under certain conditions.

Several papers published from 1974 through 1975 [43,44] established that FBF increases linearly with T_c and $\bar{T}_s$ over certain temperature ranges. If FBF varies linearly with T_c, q_s also varies linearly with T_c when q_m is constant. Accordingly, we define AVD as follows: when $T_c < T_{c,th}$, $AVD = 1.0$, and when $T_{c,th} < T_c$,

$$AVD = 1.0 + \alpha\,(T_c - T_{c,th}) \tag{8.6}$$

in which α is a gain constant. $T_{c,th}$ depends on $\bar{T}_s$ and several nonthermal factors. If we let $T_{c,th} = T^o_{c,th} + \Delta T_{c,th}$ in which $T^o_{c,th}$ is the threshold central temperature for a resting, supine individual when $\bar{T}_s = 33°C$, $\Delta T_{c,th}/\Delta \bar{T}_s \approx -0.05$.

8.4.3.2 Active Vasodilation: The Effect of Exercise

Since exercise and heat both cause cardiac output to increase and modify the distribution of blood volume, it is reasonable to assume that exercise affects circulatory responses to heat. The fact that splanchnic and renal blood flow and blood flow to inactive muscle decrease during exercise also suggests that skin blood flow might decrease during exercise. However, that possibility must be tempered by the fact that skin blood flow plays an important role in thermoregulation, while splanchnic and renal and inactive muscle blood flows do not. Numerous early studies stimulated by the possibility that exercise affects skin blood flow yielded somewhat contradictory results, although a reasonably consistent picture can now be constructed.

Studies of the effect of exercise on SkBF conducted before 1990 focused on the vasodilator system, that is, on AVD in Equation (8.4). In that context, exercise could modulate SkBF either by modifying the slope of the T_c–SkBF relationship or by changing $T_{c,th}$. Johnson [45] was the strongest proponent of the notion that nonthermal factors, such as exercise and posture, affect SkBF by modulating T_{cth}. Early support for that concept came primarily from studies conducted with an elevated $\bar{T}_s$ of 38°C to 38.5°C established by means of a liquid-perfused suit. Experiments usually consisted of comparing the values of $T_{c,th}$ during rest with values observed during exercise of moderate intensity. $T_{c,th}$ was defined as the

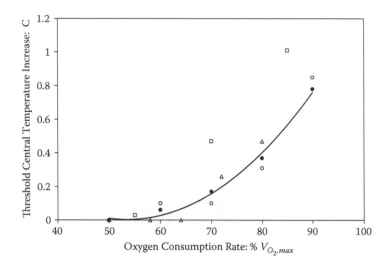

Figure 8.2 $\Delta T_{c,th}$ as a function of $\dot{V}_{O_2}$ when $\bar{T}_s \leq 33\,^{\circ}\mathrm{C}$. Filled and open circles identify plethysmographic and LD data, respectively, of Smolander et al. [48]; open triangles identify data of Taylor et al., [47]; and open squares identify data from Table 2 of Kenny et al. [49]. The graph of Equation (8.8) is also shown.

value of T_{es} at which FBF increased sharply with increasing T_c. In a typical study, Johnson and Park [46] concluded that $T_{c,th}$ increases 0.28°C with exercise, and 0.11°C with the change from supine to upright posture.

Contradictory studies that showed no effect of exercise on $T_{c,th}$ were conducted in cool room air at lower mean skin temperature. While it was not apparent at the time, it now appears that the difference between results obtained by different investigators can be attributed to the difference in $\bar{T}_s$. A lack of awareness of the importance of skin temperature during the 1970s is suggested by the fact that $\bar{T}_s$ was often not reported when exercise was performed in air.

The murky picture of the effect of exercise on SkBF began to clear with publication of a study by Taylor et al. in 1988 [47]. Those investigators evaluated the effect of five different intensities of supine cycling in 21.1°C air on the cutaneous vascular responses of four men. Data from their study indicated that exercise had no effect on CVC at low intensities of exercise, but exercise at intensities above 125 W, which corresponds to $\dot{V}_{O_2}$ of approximately 45% of $\dot{V}_{O_2,max}$, caused a decrement in CVC at a given value of T_{es}. Taylor et al. attributed the decrement in CVC, which was proportional to the workload, to an increment in $T_{c,th}$. Values of $\Delta T_{c,th}$ derived from their data are plotted in Figure 8.2.

In a subsequent study, Smolander et al. [48] employed VOP to measure FBF in six men during 15 minutes of cycling exercise in 25°C air. Measurements were made at five workloads varying from 50% to 90% of $\dot{V}_{O_2,max}$. Mean skin temperatures close to 33.5°C were independent of exercise intensity. The principal finding of this study was that SkBF during dynamic work was significantly attenuated when a subject's oxygen uptake exceeded 80% of his maximum oxygen consumption rate ($\dot{V}_{O_2,max}$). LD measurement of SkBF and VOP measurement of FBF provided comparable indications of $T_{c,th}$ for active vasodilation.

Smolander et al. [48] also attributed the decrement in FBF at moderate workloads to an increase in $T_{c,th}$. If we assume that $T_{c,th}$ for $\dot{V}_{O_2} = 0.5 \ \dot{V}_{O_2,max}$ is the normal threshold temperature for enhanced SkBF, then threshold temperature increments ($\Delta T_{c,th}$) for higher workloads can be computed from values reported by Smolander et al. The values of $\Delta T_{c,th}$ computed in that way are 0.00, 0.06, 0.17, 0.37, and 0.78°C for $\dot{V}_{O_2} = 0.5$, 0.6, 0.7, 0.8, and 0.9 $\dot{V}_{O_2,max}$, respectively. Those values are also plotted in Figure 8.2.

Another paper that helped to clarify the effect of exercise on SkBF was published by Kellogg et al. [40], who (as mentioned in Section 8.4.2.1) employed bretylium blockade of active vasoconstriction [41] to isolate the effect of active vasodilation on SkBF during exercise. Figure 3 in that paper shows clearly that exercise delays the onset of active vasodilation by raising $T_{c,th}$. For the conditions of their study, $T_{c,th}$ during exercise was 0.28°C higher than during rest. An additional observation was that the initiation of exercise causes a reduction in CVC before active vasodilation is initiated.

The studies by Taylor et al. [47] and Smolander et al. [48] established that SkBF depends on the intensity of exercise under conditions that impose moderate to high stress on the circulatory system. We have chosen to define the relationship between intensity of exercise and SkBF in terms of the relative rate of oxygen consumption,

$$\dot{V}_{O_2,r} = \frac{\dot{V}_{O_2}}{\dot{V}_{O_2,max}} \tag{8.7}$$

Equation (8.8), derived from the data of Smolander et al. and Taylor et al., defines the increase in $\Delta T_{c,th}$ owing to exercise in the absence of heat stress.

$$\Delta T_{c,th} = 1.41 \ \Delta V^2 - 0.194 \ \Delta V + 0.01°C \tag{8.8}$$

where $\Delta V = (\dot{V}_{O_2,r} - \dot{V}_{O_2,crit})/(1.0 - \dot{V}_{O_2,crit})$. For the cool mean skin temperatures employed by Taylor et al. and Smolander et al., $\dot{V}_{O_2,crit} = 0.5$. The relationship defined by Equation (8.8) is plotted in Figure 8.2.

Although Equation (8.8) accounts for the effect of exercise on $\Delta T_{c,th}$ at moderate skin temperatures, it fails to account for the effect of moderately heavy exercise on $\Delta T_{c,th}$ at high skin temperatures. For example, Equation (8.8) yields

$\Delta T_{c,th} = 0°C$ when $\bar{T}_s = 38°C$ and $\dot{V}_{O_2,r} = 0.5$, which is inconsistent with the threshold increment of 0.28°C observed under those conditions by Johnson and Park [46] and by Kellogg et al. [40]. A possible solution for that dilemma is to assume that $\dot{V}_{O_2,crit}$ decreases with increasing $\bar{T}_s$, which is not unreasonable, because cutaneous blood flow and volume both increase when the skin temperature is high. We will assume that $\dot{V}_{O_2,crit}$ decreases linearly from $\dot{V}_{O_2,crit} = 0.5$ for $\bar{T}_s \leq 33°C$ to $\dot{V}_{O_2,crit} = 0$ for 38°C $\leq \bar{T}_s$. Then, if $\bar{T}_s = 38°C$ and $\dot{V}_{O_2,r} \approx 0.5$, as in the studies of Johnson and Park, and Kellogg et al., the exercise-mediated $\Delta T_{c,th}$ predicted by Equation (8.8) is 0.26°C, which is reasonably close to the observed value of 0.28°C.

Data on which Equations (8.7) and (8.8) are based were derived from studies in which the onset of active cutaneous vasodilation was observed during a period of rising central temperature. A very important complementary study by Kellogg et al. [50] investigated the effect of exercise on established active vasodilation. Conditions of that study were similar to those in the first study by Kellogg et al. [40], except that moderately heavy exercise did not commence until after active vasodilation had been firmly established by 35 to 40 minutes of passive heating. In both studies, CVC was determined at two sites, one treated with bretylium tosylate to block vasoconstriction and the other untreated. Contrary to implications of previous observations, exercise in the second study reduced CVC at the untreated site, and not at the bretylium-treated site, which implied that exercise modulated cutaneous vasoconstriction, but did not affect active vasodilation.

Several mathematical solutions for the paradox presented in the preceding paragraph are possible. One is to assume that the central thermoregulatory center processes afferent signals by integrating the rate of change, instead of subtracting $T_{c,th}$ from T_c. That is, we can assume that

$$AVD = 1.0 + \alpha \int_{t_o}^{t} \frac{dT_c}{dt'} dt' \tag{8.9}$$

in which t_o is the instant at which $T_c = T_{c,th}$. If we assume that integration continues as long as $T_{c,th} < T_c$, modifying $T_{c,th}$ after active vasodilation has begun has no effect on SkBF. Another possibility is that $\Delta T_{c,th}$ owing to exercise is transient in nature and diminishes as active vasodilation progresses. That possibility is completely unexplored.

8.4.3.3 Active Vasodilation: Other Factors

Extrapolation of relationships representative of moderate conditions of heat stress and exercise to more severe conditions may require cardiac output that approaches, or exceeds, the capacity of the heart. Consequently, it is necessary to limit SkBF to physically realizable values. Several studies have shown that during moderate exercise, the slope of the T_c–FBF curve decreases significantly for central temperatures above 38°C, with the slope above $T_{es} = 38$°C being roughly one-half the slope at lower central temperatures [51,52]. Kellogg et al. [53] established that the reduced slope can be attributed to reduced active vasodilation, which in our model amounts to attenuating AVD. Therefore, if 38°C < T_c and 0.4 < $\dot{V}_{O_2,r}$ < 0.9, AVD defined by Equation (8.6) is modified as follows:

$$AVD = AVD* - 0.5\ (T_c - 38.0) \qquad (8.10)$$

where AVD* is the value computed using Equation (8.6).

Moreover, according to Smolander et al. [48], the slope of the T_c–FBF curve is greatly reduced over its entire range at exercise intensities above 0.9 $\dot{V}_{O_2,max}$. Therefore, when 0.9 < $\dot{V}_{O_2,r}$, AVD is computed as follows:

$$AVD = 0.25\ (T_c - T_{c,o} - \Delta T_{c,th}) \qquad (8.11)$$

8.4.3.4 Cutaneous Vasoconstriction (CVCM): The Reflex Effect of Mean Skin Temperature

Studies by Charkoudian and Johnson [54], Stephens et al. [55], and Kenney et al. [56] established that, in addition to shifting $T_{c,th}$, $\bar{T}_s$ has a reflex effect on cutaneous vasoconstriction. Since $\bar{T}_s$ in those studies was lowered from an initially comfortable level, T_c remained below $T_{c,th}$, and centrally mediated vasodilation was not a factor. Local skin temperature at the measurement site was maintained constant in each study.

Results from that study are plotted in Figure 8.3. All data have been normalized so that CVC is unity at $\bar{T}_s = 32$°C. A shortcoming of those data is that there are no experimental data to define an upper limit for the ratio CVC/CVC_{32}°C, although we know that SkBF has an upper limit [57]. The reason for the lack of data is probably that it is impossible to raise $\bar{T}_s$ to 42°C without simultaneously increasing T_c, causing active vasodilation. There is also a paucity of data for $\bar{T}_s$ below 30°C. The curve shown in Figure 8.3 is a graph of the function

$$CVCM = \frac{1.422 + \tanh\left[\,0.275\,(\bar{T}_s - 32.0)\right]}{2.018} \qquad (8.12)$$

8.4.3.5 Effect of T_s on Skin Blood Flow (SkBF) (CVCL)

The least ambiguous aspect of SkBF regulation is the effect of T_s on SkBF, which can be studied under comfortable conditions that do not involve active vasodilation, and for which the effect of $\bar{T}_s$ is fairly constant. Direct LD measurements of SkBF were made by Charkoudian et al. [58] in their study of the effect of female reproductive hormones on local control of SkBF. Those data are plotted

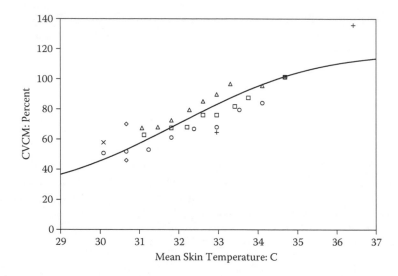

Figure 8.3   Normalized CVC plotted as a function of $\overline{T}_s$. (Data were obtained from: open circles, CVC from Stephens et al. [55] Figure 1 (saline); diamonds, CVC from Stephens et al., Figure 3a (saline); pluses, CVC from Stephens et al., Figure 5 (saline); and triangles, CVC from Charkoudian and Johnson [54], Figure 2.)

in Figure 8.4. Forearm blood flow data reported by Barcroft and Edholm [35,59], Brown et al. [60], and Wenger et al. [61] agree qualitatively with the *CVC* data of Charkoudian et al. in that they show a strong withdrawal of vasoconstrictor tone at local skin temperatures above 32°C.

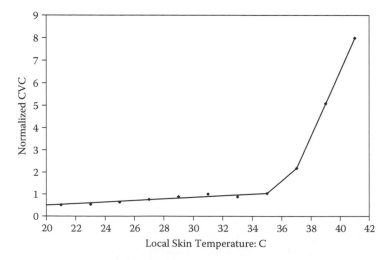

Figure 8.4   Values of relative *CVC* normalized to the value of unity at $T_s = 34$°C. (*Note*: Also shown is the graph of *CVCL*(T_s).) (From Charkoudian et al. [58].)

8.4.3.6 Effect of Exercise on Cutaneous Vasoconstriction (CVCE)

Data from several studies suggest that exercise causes a significant decrease in CVC. Kellogg et al. [40,50] observed that strenuous exercise reduces CVC by increasing cutaneous vasoconstriction in both normothermia and hyperthermia. Taylor et al. [62] observed that moderate leg exercise causes a reduction in blood flow of a locally heated arm. In that study FBF at four temperatures, 36, 38, 40, and 42°C, was measured by VOP. While there was some variation with T_s, the mean reduction was the order of 3 ml/(100 ml min).

Taylor et al. [63] evaluated the effect of absolute and relative work, and the kind of exercise on CVC. They found that the CVC decrement was proportional to absolute workload, and not relative workload. In addition, they found that isometric exercise had no effect. Their data are represented reasonably well by the following relationship:

$$CVCE = 1.0 - \frac{0.071}{110} W \qquad (8.13)$$

in which W is the workload in watts.

The decrease in CVC with exercise is offset to some extent by an increase in MAP. We will assume that MAP increases during leg exercise according to the relationship

$$MAP = 85 + 23 \, \dot{V}_{O_2,r} \quad Torr \qquad (8.14)$$

in which $\dot{V}_{O_2,r}$ is the fraction of the maximum oxygen consumption rate.

8.4.3.7 Combination of AVD, CVCL, CVCM, and CVCE to Define q_s

The four factors described above, AVD, CVCL, CVCM, and CVCE, were derived from experiments that provide little guidance about the manner in which they should be combined to form a comprehensive model of SkBF regulation. For example, experimental investigations carried out at a fixed elevated $\overline{T}_s$ provide no information about the combined effect of $\overline{T}_s$ and T_c. Although several authors have concluded that responses owing to central and cutaneous temperatures combine additively, a careful analysis of relevant data suggest that a multiplicative combination is more appropriate.

We note that since the quadruple product, $AVD \times CVCM \times CVCL \times CVC$, has a value of unity for the reference state, $T_c < 37.0°C$, $\overline{T}_s = 34.5°C$, $T_s = 34°C$, and $\dot{V}_{O_2,r} < 0.5$, it must be multiplied by $q_{s,r}$. Our estimate of $q_{s,r}$ is 14.7 ml/(100 cc of skin min), which is consistent with an FBF reference value of 3.0 ml/(100 cc min). Hence, we have

$$q_s = 14.7 \times AVD \times CVCM \times CVCL \times CVCE \frac{ml\ blood}{100\ cc\ of\ tissue\ min} \qquad (8.15)$$

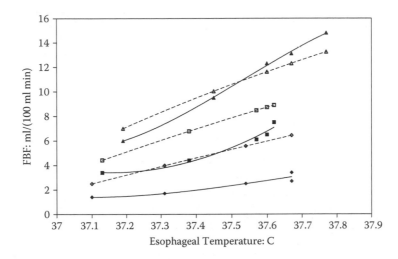

Figure 8.5   FBF measured by Nadel et al. [52] for exercise at 40% of $\dot{V}_{O_2,max}$ and $\bar{T}_s = 32.0°C$ (circles), 33.7°C (squares), and 35.2°C (triangles). Filled symbols identify measured values, and open symbols identify computed values.

A study that supports our hypothesis concerning the combined effect of the four factors was performed by Nadel et al. [52]. They used VOP to measure FBF in three relatively fit young men who performed moderate and heavy cycling exercise at three room temperatures: 20°C, 26°C, and 36°C. Corresponding mean skin temperatures were approximately 32.0°C, 33.5°C, and 35.5°C. The esophageal temperature varied from 37.1°C to 37.7°C during exercise at 40% $\dot{V}_{O_2,max}$, and from 37.1°C to 38.8°C during exercise at 70% $\dot{V}_{O_2,max}$. Forearm skin temperature was not measured. Data from that study are plotted in Figures 8.5 and 8.6. Corresponding values computed using Equations (8.1) and (8.19) and assuming that $T_s = \bar{T}_s$ are also plotted. R^2 for the computed and measured values is 0.95. Two important conclusions can be drawn from the data plotted in Figures 8.5 and 8.6. One is that the slope of the T_{es}–FBF relationship increases with increasing $\bar{T}_s$, and the other is that FBF is suppressed by moderate exercise.

8.4.3.8 Comparison of Computed and Measured Forearm Blood Flow (FBF) Data

Although the definitions of CVCM, CVCL, and CVCE are based on relatively recent CVC data, it would be imprudent to ignore earlier FBF data. FBF measured by VOP includes blood flow to both skin and muscle. While several studies have established rather conclusively that forearm q_m does not increase with bodily temperature, the possibility remains that it decreases under hypothermic conditions, and FBF data tend to support that possibility.

In Figure 8.7, values of FBF computed using the algorithm described above are compared with measured values reported by various investigators. That comparison certainly lends credence to the validity of our approach.

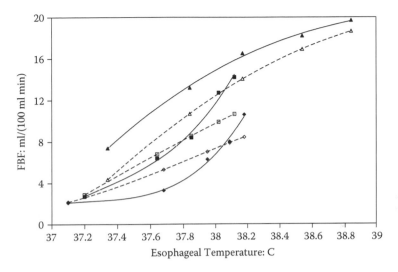

Figure 8.6   FBF measured by Nadel et al. [52] for exercise at 70% of $\dot{V}_{O_{2,max}}$ and $\overline{T}_s = 32.0°C$ (circles), 33.6°C (squares), and 35.5°C (triangles). Filled symbols identify measured values, and open symbols identify computed values.

8.4.4 Regulation of Sweating

Sweating is essential for survival in a warm environment. Various studies have established that the neutral range of skin temperature for resting individuals is approximately 33°C to 35°C. In other words, 35°C is the maximum mean skin

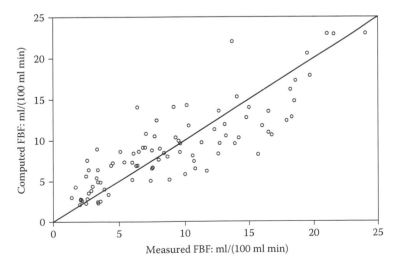

Figure 8.7   Comparison of computed and measured FBF.

temperature for which heat generated in metabolically active organs at their normal temperatures can be transported to the skin. If 80 W of metabolically generated heat are transferred from skin to air by convection and radiation, and the combined heat transfer coefficient is 10 W/(m²°C), the mean skin-to-air temperature difference is roughly 4°C. It follows that resting individuals must sweat when the air temperature is above 31°C.

Basic attributes of sweating are summarized below:

1. Local sweat rate (SR) is affected by T_c, $\bar{T}_s$, and T_s.
2. Sweating at a given thermal state appears to be greater during exercise than during rest.
3. Local sweat rate varies over the surface of the body. In general, the slope of the T_c–SR relationship is greater on surfaces where sweating begins early as T_c increases.
4. Fitness and heat acclimation both facilitate sweating.
5. Prolonged wetting of skin reduces sweat secretion—a process known as hidromeiosis.

Although sweating is absolutely essential to survival in a hot environment, it is actually a rather crude thermoregulatory mechanism. Sweat is secreted copiously during severe heat stress, but the rate of evaporative cooling is often limited by the rate of diffusion of water from skin to air, rather than by the rate of sweat secretion. In a humid environment, sweating may have little effect on thermal response.

8.4.4.1 Measurement of Sweat Rate

Continuous weighing of a subject, with suitable allowance for respiratory loss of mass, is the preferred technique for determining the transient whole-body rate of sweat evaporation. Local rates of sweat secretion are usually determined by measuring the rate of evaporation under a capsule ventilated with a steady stream of dry air with resistance hygrometry used to measure the humidity of the effluent air stream [64]. While that technique is useful for assessing the effect on sweating of various factors, such as heat acclimation, it is difficult to determine the whole-body rate of sweat secretion from local values [65].

Sweat rates measured on several skin areas in a single study indicate that the slope of the T_c–SR curve and $T_{c,th}$ for sweating vary from region to region. Figure 8.8 shows how the local sweat rate at six sites increased with time during exhaustive exercise (90% of $\dot{V}_{O_2,max}$) in 30°C air [66]. Since the esophageal temperature during those trials increased at the rate of 0.2°C/min, the sweating sensitivity for the chest was roughly 0.1 mg/(min cm²°C). Sweating began almost simultaneously on the abdomen, chest, and back. Sweating on the forearm began about 2.25 minutes later, which corresponds to a 0.45°C higher $T_{c,th}$.

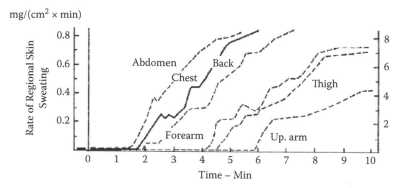

Figure 8.8   Change in sweating over different skin areas at the start of heavy exercise in 30°C room air. The solid line is an average of five observations and the other five curves are single observations plotted in time and magnitude relative to the chest curve. (Saltin et al. [66], Figure 7. With permission.)

Central threshold temperatures for sweating on the thigh and upper arm were 0.6°C and 0.8°C higher, respectively, than $T_{c,th}$ for the chest. Moreover, as Nadel et al. [67] noted, there appears to be a tendency for lower sweating sensitivity on areas that have higher $T_{c,th}$.

8.4.4.2 An Algorithm for Computing the Sweat Rate

The discussion in this section focuses on the dependence of sweating on thermal factors, acclimation to heat, fitness, and exercise. Two papers by Nadel and his colleagues [67,68] defined the principal characteristics of sweating. In the following discussion of those papers, central temperature refers to the esophageal temperature, and sweat rate is defined as grams of sweat produced per square meter of skin per second. Since the latent heat of vaporization of water at 32°C is 2425 J/gm, evaporation at the rate of 0.1 gm/(m² s) removes heat at the rate of 242.5 W/m². For a person whose body surface area is 1.8 m², an evaporation rate of 0.1 gm/(m² s) corresponds to a whole-body evaporation rate of 0.65 kgm/hr.

The first paper by Nadel et al. [68] described experiments in which they continuously measured the weight of supine, seminude subjects supported on 5 cm fish netting suspended from an aluminum frame on a sensitive beam balance. A subject's anterior skin temperature was rapidly varied by controlling the radiant heat flux from lamps located overhead. Esophageal temperature was increased by imposing brief periods of exercise between observation intervals. This study yielded a canonical relationship between sweat rate and relevant body temperatures. That relationship has the following form:

$$SR = \left[\alpha\,(T_c - T_{c,o}) + \beta\,(\overline{T}_s - \overline{T}_{s,o}) + \gamma\left(\frac{d\overline{T}_s}{dt} + r_o \right) \right] \exp\left(\frac{T_s - \overline{T}_{s,o}}{\delta} \right) \qquad (8.16)$$

in which the term that involves $\frac{d\bar{T}_s}{dt}$ is included only if $\frac{d\bar{T}_s}{dt} < -r_o °C/min$. A specific example for whole-body sweating (derived by Nadel et al. from mean data for six resting subjects with $T_{ty} < 37.5°C$) is

$$
SR = \begin{bmatrix} 0.08\,(T_c - T_{c,o}) + 0.01\,(\bar{T}_s - T_{s,o}) \\[2mm] + 0.05\left(\dfrac{d\bar{T}_s}{dt} + 0.2 \right) \end{bmatrix} \exp\left(\frac{T_s - 34.0}{10.0} \right) \frac{gm}{m^2\,s} \qquad (8.17)
$$

The second paper by Nadel et al. [67] refined and extended concepts presented in the first paper. Three relatively fit male subjects performed 10-minute bouts of cycling exercise at 80% of $\dot{V}_{O_2,max}$. Ambient conditions were an air temperature of 26°C and a relative humidity of 40%. A sensitive beam balance was used to measure whole-body weight loss. Data from two sweat collection capsules placed on various skin areas established that sweat secretion was initiated on different areas at different esophageal temperatures. In general, early sweating was associated with high local sensitivity (i.e., with a larger slope of the T_{es}–SR relationship).

Nadel et al. interpreted Equation (8.17) as the product of a central drive that increases linearly with central and mean skin temperatures, and a peripheral modifying factor that increases exponentially with local skin temperature. They also concluded that the increase in central drive per degree increase of T_{es} is approximately 10 times the increase per degree of $\bar{T}_s$. In addition, they concluded that the natural logarithm of the peripheral factor increases 0.1 per degree increase of local skin temperature.

Nadel et al. [67] correlated their data for the steady-state rate of sweating on the chest with the following equation:

$$
SR = 0.21\,[\,(T_{es} - 36.6) + 0.02\,(\bar{T}_s - 34)\,]\exp\left(\frac{T_s - 34}{10} \right) - 0.033\,\frac{gm}{m^2\,s} \qquad (8.18)
$$

They hypothesized that the last term $[0.033\ gm/(m^2\ s)]$ represents the minimum rate of glandular secretion for which sweat is expelled from sweat ducts; in other words, they assumed that sweat secreted from a gland at less than $0.033\ gm/(m^2\ s)]$ is reabsorbed through the surface of the duct and does not reach the skin surface.

Other research studies clarified certain aspects of sweating without changing appreciably basic tenets presented above. Wurster and McCook [69], Saltin et al. [66], and Wyss et al. [43] also observed that sweating is strongly attenuated by falling mean skin temperature. Wyss et al. [43] measured sweat rate under a capsule on the forearm during direct heating and cooling of four male subjects. Skin temperature was rapidly varied by controlling the temperature of water in a tube suit worn by the subjects, as shown in Figures 8.1 and 8.2 of their article [43]. Data for their subjects were well represented by the following equation:

$$SR = \alpha \, (T_{es} - 36.5) + \beta \, (\bar{T}_s - 33.0) + \delta \frac{d\bar{T}_s}{dt} + \gamma \qquad (8.19)$$

in which α = 0.12 (0.002) gm/(m²s°C), β = 0.013 (0.0005) g/(m²s°C), and γ = -0.11 g/(m² s). The parameter, δ, varies depending on the sign of $\frac{d\bar{T}_s}{dt}$. When $\frac{d\bar{T}_s}{dt}$ is positive, δ = 0.026 (0.005); and when $\frac{d\bar{T}_s}{dt}$ is negative, δ = 0.015 (0.002). Values within parentheses are standard deviations. It is not surprising that parameters derived from this study are different from those reported by Nadel et al. [67], because sweat rate on the forearm was measured in this study while whole-body sweat rate was measured by Nadel et al., and it is well known that sweat secretion is not uniform over the skin surface.

We will adopt Equation (8.16) as the basic relationship defining the local rate of sweating in terms of T_c, $\bar{T}_s$, $\frac{d\bar{T}_s}{dt}$, and T_s. Application of Equation (8.16) requires specification of seven parameters, which calls for some rather arbitrary decisions. First, we assume that $\beta \sim (0.05 \text{ to } 0.1) \, \alpha$, which is consistent with results from several studies [66,67]. Accepting the values reported by Nadel et al. [68] for γ, δ, and r_o yields the following relationship between sweat rate and thermal factors:

$$SR = \left\{ \alpha [\, T_c - T_{c,o} + 0.1 \, (\bar{T}_s - \bar{T}_{s,o})] + 0.05 \left(\frac{d\bar{T}_s}{dt} + 0.2 \right) \right\} \exp\left(\frac{T_s - T_{s,o}}{10} \right) \qquad (8.20)$$

If we also assume that $T_{s,o} = \bar{T}_{s,o}$, Equation (8.20) then contains three arbitrary parameters: the sweating sensitivity, α, and two reference temperatures, $T_{c,o}$ and $\bar{T}_{s,o}$. The reference temperature $\bar{T}_{s,o}$ can be specified arbitrarily, because it appears in combination with α and $T_{c,o}$, at least during heating, which is our primary concern. Setting $\bar{T}_{s,o}$ = 34.0°C leaves two parameters to be specified, α and $T_{c,o}$. Results from several experimental studies are summarized in Table 8.2.

Unfortunately, the wide range of values in Table 8.2 makes assigning appropriate values to α and $T_{c,o}$ rather difficult. For young resting individuals who are neither especially fit nor heat acclimated, we will use Equation (8.11) from the paper by Saltin et al. [66], expressed as follows:

$$SR = \begin{bmatrix} 0.046 \, (T_{es} - 36.4) + 0.00285 \, (\bar{T}_s - 34.0) \\[2mm] + 0.02 \frac{d\bar{T}_s}{dt} \end{bmatrix} \exp\left(\frac{\bar{T}_s - 34.0}{10} \right) \frac{gm}{m^2 \, s} \qquad (8.21)$$

The parameters shown in Equation (8.21) might appear to be odd choices because α = 0.046 gm/(m²s°C) is certainly on the low end of the values in Table 8.2.

Table 8.2 Threshold Central Temperature for Sweating and Sweat Sensitivity Determined in Various Studies

First Author	Activity	Skin Area	$T_{c,th}$ (°C)	Sensitivity (gm/(m²s°C))
Nadel et al. [68]	Rest	Thigh	37.2	0.06
Mack et al. [70]	Rest	Chest	36.8	0.30
Kondo et al. [71]	Rest	Forearm	36.8	0.19
Kondo et al. [71]	Exercise	Forearm	36.8	0.12
Nadel et al. [68]	Exercise	Chest	36.6	0.21
Nadel et al. [72]	Exercise	Chest	37.4	0.21
Roberts et al. [73]	Exercise	Chest	37.5	0.178
Pandolf et al. [74]	Exercise	Upper arm	37.0	0.083
Cotter et al. [75]	Exercise	Mean of eight sites	36.9	0.58
Patterson et al. [76]	Exercise	Mean of five sites	37.0	0.7
Nadel et al. [68]	Rest	Whole body	36.7	0.08
Hardy [77]	Rest	Whole body	36.6	0.07
Stolwijk [78]	Rest	Whole body	37.0	0.038
Saltin et al. [66]	Exercise	Whole body	36.4	0.046

However, those values were chosen because they are based on whole-body data collected for both resting and exercising subjects over a considerable period of time at the Pierce Foundation Laboratory. By comparison, data published in several papers by Nadel and colleagues indicate that local sweat rates measured with aspirated capsules tend to be higher than whole-body rates measured under similar conditions [66,67].

8.4.4.3 *Effect of Acclimation to Heat and Fitness on Sweating*

Since the purpose of sweating is to facilitate heat removal from skin when convective and radiative heat transfer are inadequate, the rate of sweat secretion is determined primarily by thermal factors. However, that does not preclude the possibility that nonthermal factors also influence sweating. We will consider four possible nonthermal factors: acclimation to heat, fitness, exercise, and skin wettedness.

Various studies have shown that enhanced sweating is the principal effect of acclimation to heat, which can be realized either by reducing the threshold central temperature for sweating or by increasing the slope of the T_{es}–SR relationship. In their third paper, Nadel et al. [72] investigated the separate effects of physical training and heat acclimation on sweating. Six relatively unfit men underwent physical training by cycling 1 hour per day for 10 consecutive days in a cool environment ($T_a = 22°C$). Physical training was followed by a similar period of acclimation to heat effected by exercise in a warm, dry environment. The sweat sensitivity of each subject was determined by measuring the rate of

Table 8.3 Effect of Training and Heat Acclimation on Sweating Parameters

	Pretraining	Posttraining	Postacclimation
Slope: gm/(m²s°C)	0.16	0.22	0.24
Threshold temperature: °C	37.5	37.4	37.3

Source: Roberts et al. [73].

sweat secretion on the chest as a function of esophageal and local skin temperatures. The authors concluded that physical training and heat acclimation both
enhance the rate of sweat secretion at given values of esophageal and mean skin
temperature, although they do so by entirely different mechanisms. Physical
training increases the slope of the T_{es}–SR relationship; that is, it increases the
value of α in Equation (8.16). Heat acclimation, on the other hand, lowers the
threshold central temperature for initiating the central drive for sweating. These
studies were conducted with each subject serving as his own control, and while a
majority of the subjects responded as described above, some exhibited little
response, or even responded negatively.

Results from a subsequent study by Roberts et al. [73] were consistent with the
conclusions of Nadel et al. [72]. Roberts et al. also employed a two-step procedure
(10 days of exercise training followed by 10 days of heat acclimation) to assess
the effect of training and acclimation on chest sweating. Mean values of the slope
and threshold temperature are shown in Table 8.3. Although those values suggest
that the slope increases with training and heat acclimation, differences between
the three conditions were not statistically significant.

Pandolf et al. [74] studied the effect of 10 days of heat acclimation on the relationship between rectal temperature and whole-body sweat rate for two groups of
subjects, one young and the other middle-aged. Their results are summarized in
Table 8.4. The difference in slopes and threshold temperatures between this study
and the study of Roberts et al. [73] can probably be attributed to the fact that
Roberts et al. measured local sweating on the chest, while Pandolf et al. measured
whole-body sweating. Otherwise, results from the two studies are consistent in
that they show a small increase in slope and a 0.3°C decrease in threshold temperature with acclimation.

Table 8.4 Effect of Heat Acclimation on Sweating Parameters

	Slope: gm/(m²s°C)		Threshold Temperature: °C	
	Day 1	Day 10	Day 1	Day 10
Young	0.083	0.098	36.99	36.67
Middle-aged	0.078	0.105	36.79	36.64

Source: Pandolf et al. [74].

Similar results were reported by Armstrong and Kenney [79], who studied the effect of age and heat acclimation on the mean body temperature (defined as $0.8\ T_{re} + 0.2\ \overline{T}_{sk}$)–sweating relationship during 90 minutes of passive heating. In both groups, 9 days of acclimation lowered the threshold mean body temperature about 0.4°C.

8.4.4.4 Effect of Exercise on Sweating

The effect of exercise on sweating is mildly controversial. Van Beaumont and Bullard [80] were the first to investigate carefully whether exercise directly affects sweating. They used resistance hygrometry to measure the rate of sweating under 7 cm² capsules located on the forearm and calf. Subjects exercised at 160 W on a cycle ergometer after an initial, 1-hour period of rest in the test chamber. When the chamber temperature was 37.5°C, light sweating occurred before exercise began. Under those conditions, definite increases in sweat rates on both the resting forearm and the working leg were observed within 2 seconds of the onset of exercise, before any change in central or skin temperature occurred. When the chamber temperature was 30°C, preexercise sweating was absent and sweating began after 30 to 80 seconds of exercise.

In a subsequent investigation, Gisolfi and Robinson [81] also observed rapid changes in the rate of sweat secretion during intermittent dynamic exercise when subjects had a warm core (T_c above 38°C), a warm skin ($\overline{T}_{sk}$ above 35.5°C), or both. Robinson [82] and Robinson et al. [83] reported that, for a given thermal state defined by central temperature, mean skin temperature, and body heat load, the rate of sweating is higher during exercise than during rest.

Similar results were reported by Yanagimoto et al. [84], whose subjects cycled for 80 seconds at 30, 50, and 70% of $\dot{V}_{O_2,max}$ in a chamber where the temperature was 35°C and the relative humidity was 50%. Those conditions caused light sweating during the 1-hour preexercise period. Increased sweating on the chest, thigh, and forearm was observed to begin during the first 2 to 16 seconds of exercise. The latent period decreased and the rate of sweating increased with increasing intensity of exercise.

Yamazaki et al. [85] measured esophageal and mean skin temperatures and forearm sweat rates as six male subjects performed cycling exercise that varied sinusoidally in intensity from 10% to 60% of $\dot{V}_{O_2,max}$. Four periods (1, 3, 4, and 8 seconds) were employed. Since sweat rate varied sinusoidally during the 1- and 3-second oscillations even though there was little variation in esophageal or mean skin temperature, Yamazaki et al. [85] concluded that exercise modulates sweat rate.

Tam et al. [86] studied two subjects, one normal and the other paraplegic, during passive heating and during exercise. During exercise in a warm environment, a sharp increase in whole-body sweat rate (evaluated as a weighted mean of four local sweat rates) [65] was observed before there was a measurable increase in aural or mean skin temperature. The initial sharp increase in sweat rate was followed by a slower increase with increasing central temperature. The slope during the second phase was similar to the slope for passive heating, but an apparent threshold central temperature obtained by extrapolating the tympanic temperature–SR curve to zero sweat rate was about 1°C lower than the threshold tympanic temperature for rest.

Although a considerable body of evidence exists indicating that sweating is enhanced during dynamic exercise, results from other studies fail to support that concept. For example, Johnson and Park [46] observed that T_{es}–forearm sweating relationships for supine rest and upright exercise were virtually identical. In both cases, subjects wore a liquid-perfused suit that maintained a mean skin temperature of 38.0 to 38.5°C. One difference between those experiments and the experiments of Van Beaumont and Bullard [80] and Gisolfi and Robinson [81] is that light sweating was present in the latter two studies, which exhibited an enhancement of sweating when work commenced, while that apparently was not true of the study by Johnson and Park.

Saltin et al. [66] performed a series of experiments with one subject to define more clearly how exercise affects the onset of sweating. Their subject cycled at 90% of $\dot{V}_{O_2,max}$ in a chamber where the temperature was 30°C and the humidity was low. No sweating was observed during the first minute of exercise, but by the end of the second minute, sweating occurred on at least 50% of the skin.

8.4.4.5 Hidromeiosis

Another nonthermal effect is hidromeiosis, which defines the progressive suppression of sweating when the skin is wet. An early description of hidromeiosis was published by Taylor and Buettner [87], who presented previously unpublished data collected in 1932, 1942, and 1946. Although their analysis is somewhat limited by the assumption that the thermal drive for sweating is proportional to the mean skin temperature, their data indicated clearly that sweat secretion depends on environmental factors, such as humidity, wind speed, and air pressure, which they called the "environmental effect." They concluded that any environmental factor that facilitates evaporation of sweat promotes sweat secretion at a given mean skin temperature. The presence of water on the skin, whether caused by intense sweating or by immersion, inhibits sweat secretion [88,89].

Brown and Sargent [90] also conducted an extensive study of hidromeiosis, from which emerged the following six characteristics:

1. In an environment of constant high ambient temperature, a person exhibits initially an outburst of sweating that reaches a peak in 1 to 2 hours, after which the rate progressively declines.
2. The rate of decline is steeper in moist heat than in dry heat.
3. The degree of sweat depression appears to be related to the magnitude of the initial maximum rate, suggesting that a threshold rate of sweating must be reached to initiate hidromeiosis.

4. Hidromeiosis does not appear to be an adaptive process, since frequently the rectal temperature rises in the face of the declining sweat rate.
5. Dehydration may accelerate hidromeiosis.
6. The duration of exposure to thermal stress, rather than the intensity of work, appears to be a factor in the occurrence of hidromeiosis.

Brown and Sargent [90] concluded that there is a threshold rate of sweating below which hidromeiosis does not occur, and that rate is sufficient to maintain a fully wetted skin. Since hidromeiosis usually becomes apparent only after 1 or 2 hours of intense sweating, it is not normally a factor in experiments of shorter duration.

Two very interesting early studies conducted by Gerking and Robinson [91,92] were consistent with the first four points of Brown and Sargent [90]. Male subjects in the Gerking-Robinson studies walked for 6 hours with a 5-minute break for weighing at the end of each hour. The predominant metabolic rate was 220 W/m², although that was reduced to 150 W/m² for several subjects. Experiments were carried out under two different ambient conditions: humid, and hot and dry. Humid conditions involved an air temperature of 31.9°C to 38°C and relative humidity of 95% to 51%. In the hot and dry conditions, air temperature ranged from 40.0°C to 50.1°C with a relative humidity of 38% to 18%. Subjects either wore shorts, shoes, and socks, or a poplin tropical uniform, shoes, and socks.

In 50 experiments, the average rate of sweating during the first 2 hours was 1400 gm/hr. The rate of sweat secretion during the sixth hour varied from to 10% to 80% of the initial rate, depending on environmental conditions, with high humidity resulting in a pronounced decline in the rate of sweating. Subjects were able to maintain a steady sweating rate of 780 gm/hr in moderate conditions.

The Robinson-Gerking studies [91,92] are unique in that they evaluated the effect of clothing on sweat secretion. As expected, clothing inhibited evaporation of sweat, and excess sweat accumulated on the skin and in the clothing. Consequently, the rate of sweat secretion by clothed subjects decreased continuously after the second hour, and by the fourth hour, the rate of evaporation was no longer sufficient to prevent skin and rectal temperatures from increasing. Sweat deficiency during the last 3 hours of work was partially alleviated by evaporation of sweat that had accumulated in the cotton uniform during the first 3 hours (as much as 800 gm in one case).

Nadel and Stolwijk [93] studied hidromeiosis as one part of the series of investigations discussed above. In the hidromeiosis study they reduced the amount of sweat accumulated on skin either by wiping the skin with a towel, or by increasing the wind speed to facilitate evaporation. In both cases, they observed that drying the skin promptly increased the rate of sweat secretion. Qualitatively, their observations were similar to those of Brown and Sargent [90].

Candas et al. [94] investigated the effect on hidromeiosis of acclimation to humid heat. Their study took place over 10 consecutive days during which eight subjects clad only in shorts reclined for 2 hours and 45 minutes on a web bed in a humid environment. The initial rate of sweat secretion for seven subjects increased linearly with days of acclimation; the eighth subject showed no effect of acclimation. Candas et al. found that the rate of decline of sweat secretion owing to hidromeiosis was proportional to the rate of sweat secretion during the prehidromeiosis period. Their data are well represented by the following relationship:

$$SR = SR_o \left[1 - 0.009\,(t - t_o)\right] \tag{8.22}$$

where SR_o is the maximum rate of sweat secretion before the onset of hidromeiosis at t_o. Time is measured in minutes. The degree of sweat secretion attenuation observed by Candas et al. is several times larger than was observed by Robinson and Gerking [92]. Based on their observations of sweat secretion (deduced from the rate of sweat drippage), Candas et al. concluded that hidromeiosis does not occur on a given region until it is fully wet.

From the limited point of view of thermoregulation of nude subjects, hidromeiosis is unimportant, because it does not occur until an area is fully wet, and then the rate of evaporation is determined by environmental conditions. Nevertheless, hidromeiosis cannot be completely neglected for two reasons. One is that the rate of sweat secretion affects one's level of hydration when adequate water replacement does not occur, and appreciable dehydration has thermoregulatory consequences. Another reason is that excess sweat accumulates in garments and changes their physical properties. For example, as noted above, Robinson and Gerking noted that one subject accumulated 824 grams of sweat in his poplin uniform during the first 3 hours of walking at a metabolic rate of 220 W/m² under hot and dry conditions. During the last 3 hours of the 6-hour experiment, his sweat rate was below the rate of evaporation from his clothing, and one-third of the previously accumulated sweat evaporated.

8.4.4.6 *Comparison of Computed and Measured Whole-Body Sweat Rates*

Values computed using Equation (8.21) with suitable modification to account for the effect of training, heat acclimation, and exercise are compared with 42 experimentally observed values in Figure 8.9. $T_{es,o}$ was assigned a value of 36.7°C for resting nonacclimated individuals. We assumed that acclimation plus training reduce $T_{es,o}$ by 0.3°C, and exercise reduces $T_{es,o}$ by 0.9°C; the two effects are additive. Represented are three combinations of acclimation and exercise: unacclimated rest, unacclimated exercise, and acclimated exercise. Although agreement between computed and measured values might be improved by modifying the factors given above, which are supported by experimental data, those factors provide acceptable agreement between computed and measured values.

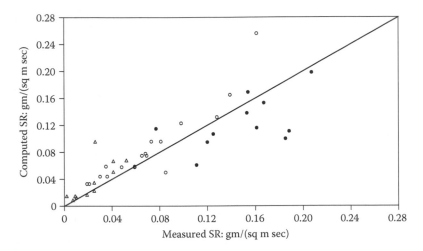

Figure 8.9   Comparison of sweat rates computed using Equation (8.21) with corresponding measured values. Filled circles denote acclimated exercising subjects, open circles denote unacclimated exercising subjects, and open triangles denote unacclimated resting subjects.

8.4.5 Shivering

During exposure to extreme cold, metabolic heat generation owing to shivering can approach five times the resting metabolic rate [95]. While shivering helps to prevent hypothermia under most conditions, the net benefit may be small during immersion in cold water, because peripheral blood flow also increases. Shivering responds to signals generated both centrally and peripherally.

8.4.5.1 An Algorithm for Computing the Rate of Shivering Metabolism

Our basic representation of shivering is derived from the study of Hayward et al. [96], who measured the oxygen consumption rate together with tympanic and rectal temperatures in two groups of subjects during immersion in 10°C water. One group had a mean skinfold thickness of 7.8 mm, while the mean skinfold thickness of the other group was 10.4 mm. That study is particularly interesting because it demonstrates that shivering, unlike skin blood flow and sweating, exhibits a rather clear dependence on the rate of change of mean skin temperature.

Two other studies suggest that shivering is also stimulated by rapidly falling central temperature. Piantadosi et al. [97] attributed enhanced shivering to rapid central cooling (defined in terms of changing rectal temperature) in subjects who breathed hyperbaric heliox at pressures corresponding to depths of 1400 and 1800 ft sw. In the other study, Nadel et al. [98] induced a rapid 0.6°C decrease in tympanic temperature by having subjects ingest 500 gm of ice cream during a 10-minute interval. They attributed the ensuing sharp, brief interval of enhanced shivering to the transient decrease in central temperature. However, we will not include this factor in the shivering algorithm because there are not enough experimental data to define the effect, and it is unlikely to be important in most situations to which thermal models are applied.

Consequently, we assume that the rate of shivering can be represented as the sum of two parts:

$$M_{sh} = M_{sh,1} + M_{sh,2} \qquad (8.23)$$

$M_{sh,1}$, which accounts for the effect of rapidly decreasing mean skin temperature, is defined as follows:

$$\frac{dM_{sh,1}}{dt} = F_1\left(\frac{d\overline{T}_s}{dt}\right) - \beta_1 M_{sh,1} \qquad (8.24)$$

where $F_1 = 0$ if $-1.5°C/min < \frac{d\overline{T}_s}{dt}$; $F_1 = A_1(-\frac{d\overline{T}_s}{dt} - 1.5)$ if $-3.5°C/min < \frac{d\overline{T}_s}{dt} < -1.5°C$; and $F_1 = 2.0\,A_1$ if $\frac{d\overline{T}_s}{dt} < -3.5°C/min$. According to this model, slowly decreasing $\overline{T}_s$ does not stimulate shivering, and a maximum stimulus occurs when $\overline{T}_s$ decreases more rapidly than $-3.5°C/min$. Values of the two parameters are $A_1 = 120$ W/°C and $\beta_1 = 0.5$ m^{-1}.

Under most conditions, the largest contribution to shivering is provided by $M_{sh,2}$, which is defined as follows:

$$\frac{dM_{sh,2}}{dt} = \beta_2\,[F_2(T_c, \overline{T}_s) - M_{sh,2}] \qquad (8.25)$$

in which $F_2(T_c, \overline{T}_s)$ is the rate of shivering normally observed under steady-state conditions. The value of β_2 is 0.177 m^{-1}. Specific relationships for F_2 can be derived from papers by Hayward et al. [96] and Tikuisis and Giesbrecht [99]. We use the function defined by Tikuisis and Giesbrecht with slightly modified parameters.

$$F_2 = \frac{155.5\,(36.6 - T_c) - [1.57\,(28.0 - \overline{T}_s) - 47.0]\,(30.0 - \overline{T}_s)}{\sqrt{BF}} \qquad (8.26)$$

where BF is the percentage of body fat.

Benzinger [100] concluded from his extensive early study of shivering that shivering metabolism decreases when the mean skin temperature is below 20°C. Although several of his hypotheses were refuted by subsequent investigators, a recent study by Tikuisis and Giesbrecht [99] seems to support his hypothesis concerning the effect of skin temperature on shivering. Accordingly, when $\overline{T}_s < 20°C$, we reduce the shivering rate computed according to Equation (8.26) by adding the multiplicative factor, $1.0 - 0.004\,(20.0 - \overline{T}_s)^2$.

8.5 MODEL PERFORMANCE

As we mentioned earlier, an important objective of this chapter is to investigate whether physiological control functions determined in their own right allow an acceptable description of human thermal regulation when incorporated into a human thermal model. Before we consider specific cases, it is worthwhile to discuss several factors relevant to comparison of computed and measured values. We start with the following quotation attributed to W. I. Beveridge (1908), which appears in the last paragraph of Rowell's book on human circulation: "No one believes an hypothesis except the originator, but everyone believes an experiment except the experimentor" [31].

Probably the greatest frustration encountered by a physical scientist engaged in human thermal modeling is the biological variability of individuals. Physiological experiments usually involve a small number of subjects whose physical characteristics may vary considerably. For example, the physical characteristics of eight subjects in the cold-immersion experiments performed by Hayward et al. [96] are summarized in Table 8.5.

The mean rectal temperature of the first three less fat subjects decreased to 34.0°C during 45 minutes of immersion in 10°C water, while the mean rectal temperature of the other five fatter subjects remained above 34.5°C during 60 minutes of immersion at the same temperature. Clearly, subcutaneous fat thickness affects an individual's response to severe cold, but can we assume that the response of an individual with mean physical characteristics is the same as the mean response of a group with diverse characteristics?

In general, models easily account for two variable physical factors, body mass and subcutaneous fat thickness, although, as we saw earlier, fat thickness determined using calipers may be quite inaccurate. Degree of acclimation to heat and

Table 8.5 Physical Characteristics of Subjects in the Cold-Immersion Study of Hayward et al.

Subject	Age (years)	Weight (kgm)	Height (cm)	Skinfold (mm)[a]
1	21	71.4	188.4	7.4
2	21	75.9	179.1	7.8
3	22	80.9	181.3	8.2
4	21	64.5	174.2	9.0
5	20	92.7	193.0	9.8
6	20	76.4	181.6	10.3
7	22	74.1	178.5	11.0
8	26	76.8	185.4	12.0

[a] Mean of thicknesses at four sites.
Source: Hayward et al. [96].

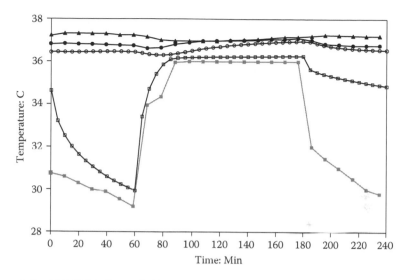

Figure 8.10 Measured and computed temperatures for seminude subjects resting for 60 minutes at 18°C, 120 minutes at 42°C, and 60 minutes at 18°C. Plotted are the measured tympanic (filled circles), rectal (filled triangles), and mean skin temperatures, and the computed esophageal (open circles) and mean skin temperatures (open triangles).

physical fitness can also be important factors under particular conditions, but they are not usually reported. In addition, factors such as the threshold temperatures for active vasodilation and the onset of sweating vary considerably between individuals and from day to day for a given individual.

Another important consideration is lack of precision in measured values. For example, the central temperature for thermoregulatory control is usually approximated by the rectal, esophageal, or tympanic temperature, none of which provides a consistently accurate approximation for the temperature of the inaccessible hypothalamus. Measured central temperatures may differ from one to another by as much as 1°C during transient periods. The disparity is illustrated in Figure 8.10 and discussed more fully following Figure 8.11.

Mean skin temperature typically determined as the weighted mean of six to ten skin temperatures is only an approximation for the true mean skin temperature. In addition to the problem of definition, we note that simply taping a thermocouple on the skin alters the temperature of the skin. Moreover, there is very little experimental evidence pertaining to the integrated cutaneous input when conditions vary considerably over the surface of the body, for example during partial immersion in cold water.

In summary, differences between computed and measured values should not automatically be attributed to deficiencies in the model; valid reasons for the differences may be attributable to the manner in which the experiment was performed, or the results are interpreted.

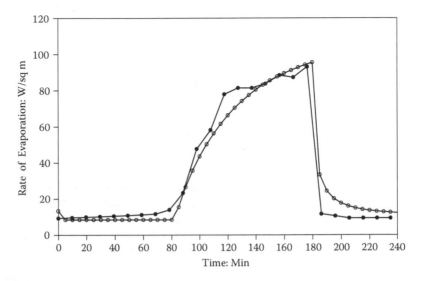

Figure 8.11 Measured (filled circles) and computed (open circles) rates of evaporative cooling for the conditions shown in Figure 8.10.

We will compare computed and measured values for four cases. The first case involves extended exposure of resting subjects to a hot environment, the second case involves exercise at three intensities in cool and warm environments, the third case involves passive exposure to cold air, and the fourth case involves immersion in cold water. In each case, subjects were seminude, which eliminates uncertainty associated with modeling clothing and provides a clearer evaluation of the physiological model. These particular cases were chosen to test the model because we have experimentally measured values for one or two central temperatures, a mean skin temperature, the metabolic rate, and the rate of evaporative cooling (when sweating is an important factor).

8.5.1 Passive Exposure to Heat

The first case is one of many reported by Hardy and Stolwijk [77,78]. Their experiments carried out in the partitional calorimeter of the Pierce Foundation Laboratory at Yale University were fairly unique in that the rate of sweat evaporation was measured continuously using a sensitive beam balance to record subject weight. Seminude subjects sat quietly in a chair during the experiment, with minimal air movement in the chamber. The rate of oxygen uptake, rectal temperature, tympanic temperature, and mean skin temperature were also measured.

Plotted in Figures 8.10 and 8.11 are results for a 4-hour experiment in which three subjects spent the first hour in a chamber where $T_{air} = 18°C$, the next 2 hours in a chamber where $T_{air} = 42°C$, and the fourth hour back in the first chamber. Computed values of the esophageal temperature, mean skin temperature, and rate of evaporative cooling are plotted.

Several points are worth mentioning about the results of this case. First, a significant difference exists between measured rectal and tympanic temperatures. The difference is largest during relatively steady conditions at the beginning and end of the experiment, when the rectal temperature is approximately 0.5°C higher than the tympanic temperature. That difference is fairly typical of steady-state rectal and tympanic temperatures. During the initial heating period, the rectal temperature decreased for 20 minutes (presumably owing to an increase in peripheral circulation), while the tympanic temperature increased. Because the tympanic temperature increased more rapidly than the rectal temperature immediately after subjects moved into the warm room, the difference between the two temperatures was negligible during much of the heating period. It should be noted that part of the increase in tympanic temperature was probably due to the influence of ambient air temperature on tympanic temperature. Several studies have established that tympanic temperature is affected by ambient temperature, although the cause is not clear. One likely cause is that venous drainage from the face affects temperature in the ear through countercurrent exchange of heat with blood in the carotid artery. Since the actual change in arterial blood temperature was probably intermediate between changes in the rectal and tympanic temperatures, the computed change in esophageal temperature may be slightly too large.

Agreement between measured and computed mean skin temperatures is reasonable given the problems associated with such measurements. The reason for the sizeable temperature disparity during the third phase of the experiment is unknown. Agreement between measured and computed rates of evaporative cooling is excellent.

8.5.2 Exercise at Three Intensities and Two Temperatures

Experiments simulated in the second case were also carried out in the Pierce Foundation partitional calorimeter [101]. This case differs from the previous case in that this study involved a seminude subject who performed cycling exercise at three intensities (25%, 50%, and 75% of $\dot{V}_{O_2,max}$). Measured results for this case are particularly interesting because they are for one subject, instead of being mean values for five or six subjects. Hence, we are simulating the behavior of a subject for whom all of the physical parameters are known.

Rectal, active muscle, and mean skin temperatures were measured together with the metabolic and evaporative cooling rates. Measured and computed results are plotted in Figures 8.12 and 8.13 for 10°C ambient conditions, and in Figures 8.14 and 8.15 for 30°C ambient conditions. Results for 20°C were consistent with those two conditions.

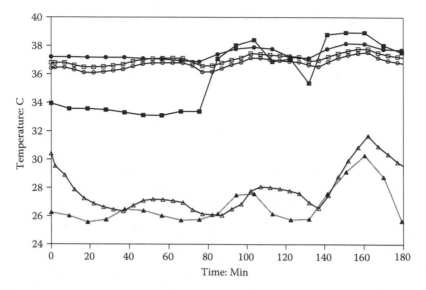

Figure 8.12 Measured and computed temperatures for a seminude subject during three periods of cycling exercise in 10°C air. Plotted are the measured rectal (filled circles), active muscle (filled squares), and mean skin (filled triangles) temperatures, and computed esophageal (open circles), muscle (open squares), and mean skin (open triangles) temperatures.

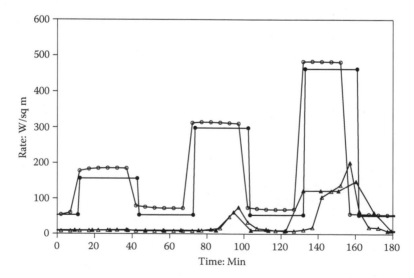

Figure 8.13 Measured and computed metabolic (circles) and evaporative cooling (triangles) rates during cycling exercise in 10°C air. Filled symbols denote measured values and open symbols denote computed values.

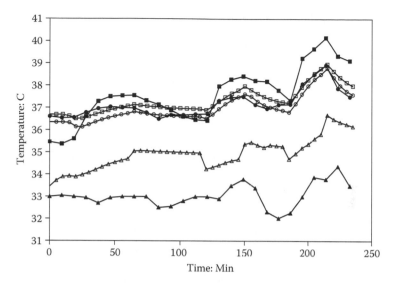

Figure 8.14   Measured and computed temperatures for seminude subjects during three periods of cycling exercise in 30°C air. Symbols have the same meaning as in Figure 8.12.

Agreement between measured and computed values for this rather demanding case is reasonable. The greatest discrepancies are between measured and computed mean skin temperatures in the 30°C chamber, and between measured and computed active muscle temperatures in both chambers. The discrepancy between computed and measured mean skin temperatures could be caused by

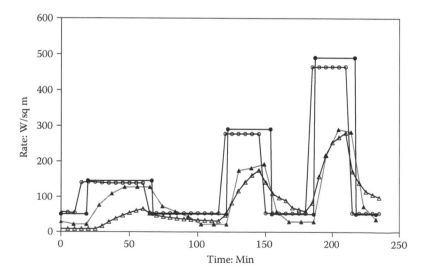

Figure 8.15   Measured and computed metabolic and evaporative cooling rates during cycling exercise in 30°C air. Symbols have the same meaning as in Figure 8.13.

several factors, including uncertainty in the heat and mass transfer coefficients, improper distribution of sweat secretion, and errors in the computed local blood flow rates. The smaller than expected increase in computed muscle temperature might be attributable to our assumption that the entire mass of leg muscle participates equally in the work of cycling, which is probably not true, especially at lower intensities of exercise.

8.5.3 Passive Exposure to Cold Air

The last two cases involve exposure to conditions sufficiently cold to invoke shivering. Experimental data for the third case were obtained from four studies [102–105] performed at the Centre de Reserches du Service de Sante des Armees in Lyon, France. In those experiments, recumbent, seminude subjects were exposed for 2 hours to 1°C, 5°C, and 10°C air. Computed results are compared with measured values in Figures 8.16, 8.17, and 8.18.

Although there was some disparity between results from different studies, two rather interesting conclusions emerged from the observations. One is that shivering and strong vasoconstriction are able to prevent a large decrease in central temperature during several hours of cold-air exposure. The second, somewhat surprising conclusion is that a smaller decrease in central temperature occurs during exposure to 1°C air than during exposure to 10°C air, as is illustrated by the temperatures plotted in Figure 8.16. While that behavior can be attributed in part to the higher metabolic rate observed during exposure to 1°C air, as shown

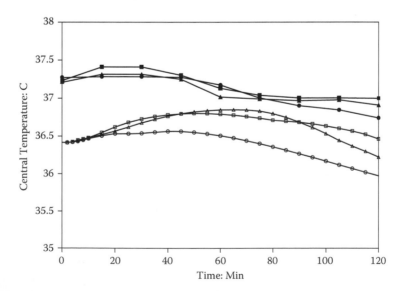

Figure 8.16 Measured rectal temperature [103] (filled symbols) and computed esophageal temperature (open symbols) during exposure of seminude subjects to cold air. Ambient temperatures were 1°C (squares), 5°C (triangles), and 10°C (circles).

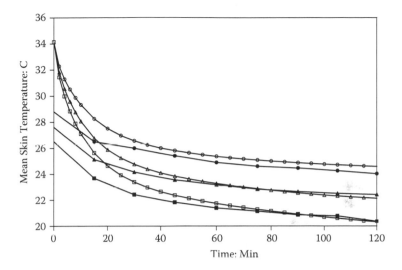

Figure 8.17   Measured [103] and computed mean skin temperature for seminude subjects during exposure to cold air. Symbols have the same significance as in Figure 8.16.

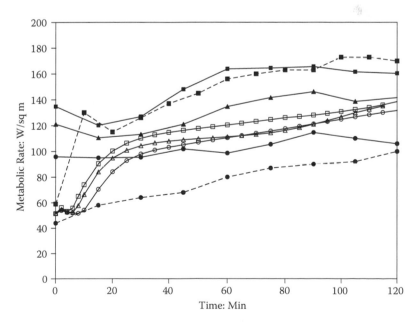

Figure 8.18   Measured [103] and computed mean metabolic rate for seminude subjects during exposure to cold air. Symbols have the same significance as in Figure 8.16. Measured values on the dashed lines are mean values for nine subjects [102,105].

in Figure 8.18, enhanced vasoconstriction induced by rapidly decreasing skin temperature during exposure to 1°C air also seems to be involved.

It is difficult to compare computed and measured mean skin temperatures during the first few minutes of cooling because the measured values reported at time = 0 were apparently recorded after the cold exposure began. Nevertheless, it is encouraging to note that computed mean skin temperatures during the last half of the experiment are close to measured values.

The sizeable discrepancy between computed and measured metabolic rates might indicate that the shivering control function used in the model is inadequate, which is not totally unreasonable because it was derived from experimental studies in which subjects were exposed to rather cold water, while these experiments involved less severe exposure to cold air. The experimental data plotted in Figure 8.18 suggest that shivering began almost simultaneously with exposure to cold air, but the data in Table 8.2 of the paper from which those data were taken [104] reported that shivering began after cold exposures of 6.1 minutes, 15.7 minutes, and 17.5 minutes for air temperatures of 1°C, 5°C, and 10°C, respectively. In addition, a graph of the mean metabolic rates for nine subjects during exposure to 1°C and 10°C air, represented by the broken lines in Figure 8.18, suggests that shivering develops more slowly in 10°C air than the data plotted in Figure 8.18 indicate. Although determination of shivering metabolic rate appears to be straightforward, that may not be completely true. Interested readers may want to read a discussion by Buskirk et al. [106] of possible artifacts that can affect such measurements during mild cold exposure.

8.5.4 Immersion in Cold Water

The fourth case provides a more severe test of the model's ability to simulate responses to cold exposure. In these experiments, eight young men whose physical characteristics are shown in Table 8.5 were immersed to the neck in 10°C water. One criterion for termination of the experiment was a rectal temperature of 35°C. The three subjects with the lowest mean skinfold thicknesses reached that condition after 45 minutes of immersion, while five fatter subjects tolerated 60 minutes of immersion. Measured rectal temperatures and computed esophageal temperatures for the two groups are plotted in Figure 8.19. Measured and computed metabolic rates are plotted in Figure 8.20.

The computed change in esophageal temperature for the thin subjects (−1.5°C) is significantly less than the measured change in tympanic temperature (−2.8°C). On the other hand, the computed change in esophageal temperature for the fatter subjects (−1.8°C) is close to the measured change in tympanic temperature (−1.9°C). While the second result is encouraging, it may be somewhat fortuitous because of differences that can occur between tympanic and esophageal temperatures during periods of rapid change. It is disturbing that the model does not appear to account properly for the effect of subcutaneous fat thickness on response during immersion in cold water. Another important difference

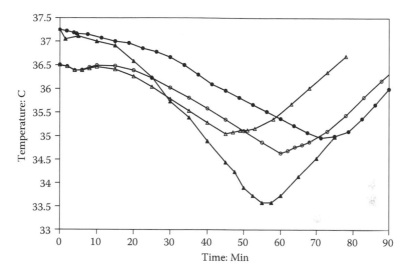

Figure 8.19 Measured tympanic temperature (filled symbols) and computed esophageal temperature (open symbols) during immersion in 10°C water. Circles denote heavy subjects (mean skinfold thickness = 10.4 mm) and triangles denote thin subjects (mean skinfold thickness = 7.8 mm).

between computed and measured central temperatures for both groups is that the model does not account for the "after-drop" in central temperature, which is often observed in cold-immersion experiments. Unfortunately, the physical and physiological reasons for after-drop are unclear, and there is no obvious reason for the difference between simulated and observed responses.

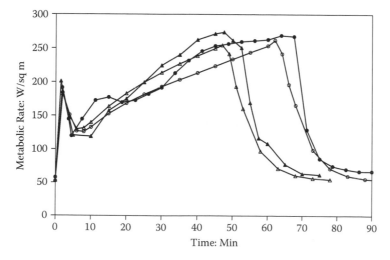

Figure 8.20 Measured and computed metabolic rates during immersion in 10°C water. Symbols have the same significance as in Figure 8.19.

A final comment about results for the third and fourth cases is in order. Although a strong objective of this development has been to establish that the physiological control functions for skin blood flow, sweating, and shivering determined without regard to thermoregulation per se yield a model with acceptable thermoregulatory characteristics, it was necessary for the two cold-exposure cases to make some ad hoc adjustments. Although those adjustments are not supported by independent physiological experiments, neither do they violate such observations.

We found that the model responded properly to cold-air exposure only after it was modified by assuming that rapidly falling mean skin temperature causes vasoconstriction in inactive muscle and venoconstriction in the arms and legs (which increases arteriovenous countercurrent heat transfer). Unfortunately, changes required to conserve bodily heat during exposure to cold air also prevented cooling during immersion in cold water, and, therefore, it was necessary to assume that those changes are reversed when the skin temperature falls below 20°C. Hence, vasoconstriction in inactive muscle and venoconstriction were stronger during exposure to 1°C air than during exposure to either 10°C air or 10°C water. Although independent experimental evidence supporting the second assumption was reported by Cannon and Keatinge [107], the validity of both responses warrants further experimental investigation.

8.6 CONCLUSIONS

To develop a useful human thermal model, one must be able to formulate physical concepts in mathematical terms and have a good grasp of scientific literature dealing with thermal and circulatory physiology. Unlike physical scientists, physiologists do not express their findings in quantitative terms that are easily incorporated into complex models. For example, there are only two or three early papers that propose a quantitative relationship between skin blood flow and central and mean skin temperatures, and those relationships have been ignored by subsequent investigators. Consequently, there are no sources to which modelers can turn for quantitative relationships between physiological variables. The principal purpose of this chapter is to present relationships that were derived from an extensive examination of the relevant literature.

The relationships presented in this chapter seem to work well when incorporated into a human thermal model. Although that model is not described, it is similar to other models developed during the last 30 or 40 years. The important thing is that the model provides reasonable agreement between computed and measured thermal responses to conditions involving rest and exercise during exposure to hot and cold ambient conditions.

Challenges remain for those who are fascinated by the functioning of the human body, as I have been for 50 years. It seems to me that one of the biggest challenges is to discover how the body responds to distributed stimuli. For example, if a soldier walks immersed to the waist in cold water, do skin blood flow and shivering respond to mean skin temperature computed for the entire body, or do they respond preferentially to cold stimuli coming from wet skin? Answers to such questions will allow the development of better models, and well-designed human thermal models can play an important role in answering such questions.

REFERENCES

1. J. A. J. Stolwijk, A mathematical model of physiological temperature regulation in man, *NASA-9-9531*, September 1970.
2. P. A. Gagge, J. A. J. Stolwijk, and Y. Nishi, An effective temperature scale based on a simple model of human physiological regulatory response, *ASHRAE Trans.*, vol. 70 (Part 1), pp. 247–262, 1970.
3. K. K. Kraning and R. R. Gonzalez, A mechanistic computer simulation of human work in heat that accounts for physical and physiological effects of clothing, aerobic fitness, and progressive dehydration, *J. Therm. Biol.*, vol. 22, pp. 331–342, 1977.
4. J. A. J. Stolwijk and J. D. Hardy, Temperature regulation in man: a theoretical study, *Pfluegers Archiv*, vol. 291, pp. 126–162, 1966.
5. A. C. Burton, The application of the theory of heat flow to the study of energy metabolism, *J. Nutrition*, vol. 7, pp. 497–533, 1934.
6. H. C. Bazett and B. McGlone, Temperature gradients in the tissues in man, *Am. J. Physiol.*, vol. 82, pp. 415–451, 1927.
7. A. C. Burton and H. C. Bazett, A study of the average temperature of the tissues, of the exchanges of heat and vasomotor responses in man by means of a bath calorimeter, *Am. J. Physiol.*, vol. 117, pp. 36–54, 1936.
8. W. Machle and T. F. Hatch, Heat: man's exchanges and physiological responses, *Physiol. Rev.*, vol. 27, pp. 200–227, 1947.
9. H. H. Pennes, Analysis of tissue and arterial blood temperature in the resting human forearm, *J. Appl. Physiol.*, vol. 1, pp. 93–122, 1948.
10. E. H. Wissler, Pennes' 1948 paper revisited, *J. Appl. Physiol.*, vol. 85, pp. 35–41, 1998.
11. C. K. Charney, Mathematical models of bioheat transfer, in *Advances in Heat Transfer*, vol. 22, pp. 19–155, Y. I. Cho, J. P. Hartnett, and T. F. Irvine Jr., editors, New York: Academic Press, 1992.
12. H. Brinck and J. Werner, Efficiency function: improvement of classic bioheat approach, *J. Appl. Physiol.*, vol. 77, pp. 1617–1622, 1994.
13. S. Weinbaum, L. X. Xu, L. Zhu, and A. Ekpene, A new fundamental bioheat equation for muscle tissue, part I: blood perfusion term, *ASME J. Biomech. Eng.*, vol. 119, pp. 278–288, 1997.
14. L. Zhu, L. X. Xu, Q. He, and S. Weinbaum, A new fundamental bioheat equation for muscle tissue: part II: temperature of SAV vessels, *ASME J. Biomech. Eng.*, vol. 124, pp. 121–132, 2002.
15. E. H. Wissler, Steady-state temperature distribution in man, *J. Appl. Physiol.*, vol. 16, pp. 734–740, 1961.
16. H. Zhang, C. Huizenga, Z. Hui, E. Arens, and T. Yu, Considering individual physiological differences in a human thermal model, *J. Therm. Biol.*, vol. 26, pp. 401–408, 2001.

17. O. Jay, F. D. Reardon, P. Webb, M. B. DuCharme, T. Ramsay, L. Nettlefold, and G. P. Kenny, Estimating changes in mean body temperature for humans during exercise using core and skin temperatures is inaccurate even with a correction factor, *J. Appl. Physiol.*, vol. 103, pp. 443–451, 2007.

18. A. L. Vallerand, G. Savourey, and J. H. M. Bittel, Determination of heat depth in the cold: partitional calorimetry vs. conventional methods, *J. Appl. Physiol.*, vol. 72, pp. 1380–1385, 1992.

19. M. Bruse, ITCM: a simple dynamic 2-node model of the human thermo-regulatory system and its application in a multi-agent system, *Ann. Meteorol.*, vol. 41, pp. 398–401, 2005.

20. E. H. Wissler, A mathematical model of the human thermal system, *Bull. Math. Biophysic.*, vol. 26, pp. 147–166, 1964.

21. D. Fiala, K. J. Lomas, and M. Stohrer, A computer model of human thermoregulation for a wide range of environmental conditions: the passive system, *J. Appl. Physiol.*, vol. 87, pp. 1957–1972, 1999.

22. D. Fiala, K. J. Lomas, and M. Stohrer, Computer prediction of human thermoregulatory and temperature responses to a wide range of environmental conditions, *Int. J. Biometeorol.*, vol. 45, pp. 143–159, 2001.

23. Y. Qi, A new two-dimensional human thermal model and study of heat transfer in living tissue, unpublished doctoral dissertation, University of Texas at Austin, 1994.

24. J. P. Rugh and D. Bharathan, Predicting human thermal comfort in automobiles, Paper 2005-01-2008, *Presented at the Vehicle Thermal Management Systems Conference and Exhibition*, Toronto, May 2005.

25. N. Gao, J. Niu, and H. Zhzng, Coupling CFD and human body thermo-regulation model for the assessment of personalized ventilation, *HVAC&R Research*, vol. 12, pp. 497–518, 2006.

26. Q. Xu and X. Luo, Dynamic thermal comfort numerical simulation model on 3D garment CAD, *Appl. Math. Computation*, vol. 182, pp. 106–118, 2006.

27. G. V. A. Durnin and J. Womersley, Body fat assessed from total body density and its estimation from skinfold thickness: measurement on 481 men and women aged 16 to 72 years, *Brit. J. Nutrition*, vol. 32, pp. 77–97, 1974.

28. J. H. Himes, A. F. Roche, and R. M. Siervogel, Compressibility of skinfolds and measurement of subcutaneous fatness, *Am. J. Clinical Nutrition*, vol. 32, pp. 1734–1740, 1979.

29. P. A. Hayes, P. J. Sowood, A. Belyavin, J. B. Cohen, and F. W. Smith, Subcutaneous fat thickness measured by magnetic resonance imaging, ultrasound, and calipers, *Med. Sci. Sports Exer.*, vol. 20, pp. 303–309, 1988.

30. P. A. Hayes, J. B. Cohen, and P. J. Sowood, Subcutaneous fat distribution of adult males and females measured by nuclear magnetic resonances, in *IAM Report No. 655*, Farnborough, UK: RAF Institute of Aviation Medicine, 1987.

31. L. B. Rowell, *Human circulation regulation during physical stress*, New York: Oxford University Press, 1986.

32. L. B. Rowell, Human cardiovascular adjustments to exercise and thermal stress, *Physiol. Rev.*, vol. 54, pp. 75–159, 1974.

33. D. N. Proctor, S. C. Newcomer, D. W. Koch, K. U. Le, D. A. MacLean, and U. A. Leuenberger, Leg blood flow during submaximal cycle ergometry is not reduced in older normally active men, *J. Appl. Physiol.*, vol. 94, pp. 1859–1869, 2003.

34. B. Nielsen, G. Savard, E. A. Richter, M. Hargreaves, and B. Saltin, Muscle blood flow and muscle metabolism during exercise and heat stress, *J. Appl. Physiol.*, vol. 69, pp. 1040–1046, 1990.

35. H. Barcroft and O. G. Edholm, The effect of temperature on blood flow and deep temperature in the human forearm, *J. Physiol.*, vol. 102, pp. 5–20, 1943.

36. H. Barcroft, K. D. Bock, H. Hensel, and A. H. Kitchin, Die Muskeldurchblutung des Menschen bei indirekter Erwärmung und Abkühlung. *Pflüg. Arch. ges. Physiol.*, vol. 261, pp. 199–210, 1955.

37. J.-M. R. Detry, G. L. Brengelmann, L. B. Rowell, and C. Wyss, Skin and muscle components of forearm blood flow in directly heated man, *J. Appl. Physiol.*, vol. 32, pp. 506–511, 1972.

38. J. L. Saumet, D. L. Kellogg Jr., W. F. Taylor, and J. M. Johnson, Cutaneous laser-Doppler flowmetry: influence of underlying muscle blood flow, *J. Appl. Physiol.*, vol. 65, pp. 478–481, 1988.

39. J. M. Johnson, W. F. Taylor, A. P. Shepherd, and M. K. Park, Laser-Doppler measurement of skin blood flow: comparison with plethysmography, *J. Appl. Physiol.*, vol. 56, pp. 786–803, 1984.

40. D. L. Kellogg, J. M. Johnson, and W. A. Kosiba, Control of internal temperature threshold for active vasodilation by dynamic exercise, *J. Appl. Physiol.*, vol. 71, pp. 2476–2482, 1991.

41. D. L. Kellogg, J. M. Johnson, and W. A. Kosiba, Selective abolition of adrenergic vasoconstrictor responses in skin by local iontophoresis of bretylium, *Am. J. Physiol.*, vol. 257 (*Heart Circ. Physiol.*, vol. 26), pp. H11599–H1606, 1989.

42. K. E. Cooper, O. G. Edholm, and R. F. Mottram, The blood flow in skin and muscle of the human forearm, *J Physiol.*, vol. 128, pp. 258–267, 1955.

43. C. R. Wyss, G. L. Brengelmann, J. M. Johnson, L. B. Rowell, and M. Niederberger, Control of skin blood flow, sweating, and heart rate: role of skin vs. core temperature, *J. Appl. Physiol.*, vol. 36, pp. 726–733, 1974.

44. C. R. Wyss, G. L. Brengelmann, J. M. Johnson, L. B. Rowell, and D. Silverstein, Altered control of skin blood flow at high skin and core temperatures, *J. Appl. Physiol.*, vol. 38, pp. 839–845, 1975.

45. J. M. Johnson, Nonthermoregulatory control of human skin blood flow, *J. Appl. Physiol.*, vol. 61, pp. 1613–1622, 1986.

46. J. M. Johnson and M. K. Park, Effect of upright exercise on threshold for cutaneous vasodilation and sweating, *J. Appl. Physiol.: Resp. Env. Exercise Physiol.*, vol. 50, pp. 814–818, 1981.

47. W. F. Taylor, J. M. Johnson, W. A. Kosiba, and C. M. Kwan, Graded cutaneous vascular responses to dynamic leg exercise, *J. Appl. Physiol.*, vol. 64, pp. 1803–1809, 1988.

48. J. Smolander, J. Saalo, and O. Korhonen, Effect of work load on cutaneous vascular response to exercise, *J. Appl. Physiol.*, vol. 71, pp. 1614–1619, 1991.

49. G. P. Kenny, J. Periard, W. S. Journeay, R. J. Sigal, and F. D. Reardon, Cutaneous active vasodilation in humans during passive heating postexercise, *J. Appl. Physiol.*, vol. 95, pp. 1025–1031, 2003.

50. D. L. Kellogg, J. M. Johnson, and W. A. Kosiba, Competition between cutaneous active vasoconstriction and active vasodilation during exercise in humans, *Am. J. Physiol.*, vol. 261 (*Heart Circ. Physiol.*, vol. 30), pp. H1184–H1189, 1991.

51. G. L. Brengelmann, J. M. Johnson, L. Hermansen, and L. B. Rowell, Altered control of skin blood flow during exercise at high internal temperatures, *J. Appl. Physiol.*, vol. 43, pp. 790–794, 1977.

52. E. R. Nadel, E. Cafarelli, M. F. Roberts, and C. B. Wenger, Circulatory regulation during exercise in different ambient temperatures, *J. Appl. Physiol: Resp. Env. Physiol.*, vol. 46, pp. 430–437, 1979.

53. D. L. Kellogg, J. M. Johnson, W. L. Kenney, P. E. Pergola, and W. A. Kosiba, Mechanisms of control of skin blood flow during prolonged exercise in humans, *Am. J. Physiol.*, vol. 265 (*Heart Circ. Physiol.* 34), pp. H562–H568, 1993.

54. N. Charkoudian and J. M. Johnson, Reflex control of cutaneous vaso-constrictor system is reset by exogenous female reproductive hormones, *J. Appl. Physiol.*, vol. 87, pp. 381–385, 1999.

55. D. P. Stephens, K. Aoki, W. A. Kosiba, and J. M. Johnson, Non-norandrenergic mechanism of reflex cutaneous vasoconstriction in men, *Am. J. Physiol. Heart Circ. Physiol.*, vol. 280, pp. H1496–H1504, 2001.

56. W. L. Kenney and C. G. Armstrong, Reflex peripheral vasoconstriction is diminished in older men, *J. Appl. Physiol.*, vol. 80, pp. 512–515, 1996.

57. W. F. Taylor, J. M. Johnson, D. O'Leary, and M. K. Park, Effect of high local temperature on reflex cutaneous vasodilation, *J. Appl. Physiol.: Resp. Env. Exercise Physiol.*, vol. 57, pp. 191–196, 1984.

58. N. Charkoudian, D. P. Stephens, K. C. Pirkle, W. A. Kosiba, and J. M. Johnson, Influence of female reproductive hormones on local thermal control of skin blood flow, *J. Appl. Physiol.*, vol. 87, pp. 1719–1723, 1999.

59. H. Barcroft and O. G. Edholm, Temperature and blood flow in the human forearm, *J. Physiol.*, vol. 104, pp. 366–376, 1946.

60. G. M. Brown, J. D. Hatcher, and J. Page, Temperature and blood flow in the forearm of the Eskimo, *J. Appl. Physiol.*, vol. 5, pp. 410–420, 1953.

61. C. B. Wenger, L. A. Stephenson, and M. A. Durkin, Effect of nerve block on response of forearm blood flow to local temperature, *J. Appl. Physiol.*, vol. 61, pp. 227–232, 1986.

62. W. F. Taylor, J. M. Johnson, D. S. O'Leary, and M. K. Park, Modification of the cutaneous vascular response to exercise by local skin temperature, *J. Appl. Physiol.: Resp. Env. Exercise Physiol.*, vol. 57, pp. 1878–1884, 1984.

63. W. F. Taylor, J. M. Johnson, and W. A. Kosiba, Roles of absolute and relative load in skin vasoconstriction responses to exercise, *J. Appl. Physiol.*, vol. 69, pp. 1131–1136, 1990.

64. R. W. Bullard, Continuous recording of sweat rate by resistance hygrometry, *J. Appl. Physiol.*, vol. 17, pp. 735–737, 1962.

65. H.-S. Tam, R. C. Darling, J. A. Downey, and H.-Y. Cheh, Relationship between evaporation rate of sweat and mean sweating rate, *J. Appl. Physiol.*, vol. 41, pp. 777–780, 1976.

66. B. Saltin, A. P. Gagge, U. Bergh, and J. A. J. Stolwijk, Body temperatures and sweating during exhaustive exercise, *J. Appl. Physiol.*, vol. 32, pp. 635–643, 1972.

67. E. R. Nadel, J. W. Mitchell, B. Saltin, and J. A. J. Stolwijk, Peripheral modifications to central drive for sweating, *J. Appl. Physiol.*, vol. 31, pp. 828–822, 1971.

68. E. R. Nadel, R. W. Bullard, and J. A. J. Stolwijk, Importance of skin temperature in the regulation of sweating, *J. Appl. Physiol.*, vol. 31., pp. 80–87, 1971.

69. R. D. Wurster and R. D. McCook, Influence of rate of change in skin temperature on sweating, *J. Appl. Physiol.*, vol. 25, pp. 237–240, 1969.

70. G. W. Mack, L. M. Shannon, and E. R. Nadel, Influence of β-andrenergic blockade on the control of sweating in humans, *J. Appl. Physiol.*, vol. 61, pp. 1701–1705, 1986.

71. N. Kondo, M. Shibasaki, K. Aoki, S. Koga, Y. Inoue, and C. G. Crandall, Function of human eccrine sweat glands during dynamic exercise and passive rest, *J. Appl. Physiol.*, vol. 90, pp. 1877–1881, 2001.

72. E. R. Nadel, K. B. Pandolf, M. F. Roberts, and J. A. J. Stolwijk, Mechanism of thermal acclimation to exercise and heat, *J. Appl. Physiol.*, vol. 37, pp. 515–520, 1974.

73. M. F. Roberts, C. B. Wenger, J. A. J. Stolwijk, and E. R. Nadel, Skin blood flow and sweating changes following exercise training and heat acclimation, *J. Appl. Physiol.*, vol. 43, pp. 133–137, 1977.

74. K. B. Pandolf, B. S. Cadarette, M. N. Sawka, A. J. Young, R. P. Francesconi, and R. R. Gonzalez, Thermoregulatory responses of middle-aged and young men during dry-heat acclimation, *J. Appl. Physiol.*, vol. 65, pp. 65–71, 1988.

75. J. D. Cotter, M. J. Patterson, and N. A. S. Taylor, Sweat distribution before and after repeated heat exposure, *Eur. J. Appl. Physiol.*, vol. 76, pp. 181–186, 1997.

76. M. J. Patterson, J. M. Stocks, and N. A. S. Taylor, Humid heat acclimation does not elicit a preferential sweat redistribution toward the limbs, *Am. J. Physiol. Reg. Integ. Comp. Physiol.*, vol. 286, pp. R512–R518, 2004.

77. J. D. Hardy and J. A. J. Stolwijk, Partitional calorimetric studies of man during exposures to thermal transients, *J. Appl. Physiol.*, vol. 21., pp. 1799–1806, 1966.

78. J. A. J. Stolwijk and J. D. Hardy, Partitional calorimetric studies of responses of man in thermal transients, *J. Appl. Physiol.*, vol. 21, pp. 967–977, 1966.

79. C. G. Armstrong and W. L. Kenney, Effects of age and acclimation on responses to passive heat exposure, *J. Appl. Physiol.*, vol. 75, pp. 2162–2167, 1993.

80. J. Van Beaumont and R. W. Bullard, Sweating: its rapid response to muscular work, *Science, New Series*, vol. 141, pp. 643–646, 1963.

81. C. Gisolfi and S. Robinson, Central and peripheral stimuli regulating sweating during intermittent work in men, *J. Appl. Physiol.*, vol. 29, pp. 761 768, 1970.

82. S. Robinson, Temperature regulation in exercise, *Pediatrics*, vol. 32, pp. 691–702, 1963.

83. S. Robinson, F. R. Meyer, J. L. Newton, C. H. Ts'ao, and L. O. Holgersen, Relations between sweating, cutaneous blood flow, and body temperature in work, *J. Appl. Physiol.*, vol. 20, pp. 575–582, 1965.

84. S. Yanagimoto, T. Kuwahara, Y. Zhzng, S. Koga, Y. Inoue, and N. Kondo, Intensity-dependent thermoregulatory responses at the onset of dynamic exercise in mildly heated individuals, *Am. J. Physiol. Reg. Integ. Comp. Physiol.*, vol. 285, pp. R200–R207, 2003.

85. F. Yamazaki, R. Sone, and H. Ikegami, Responses of sweating and body temperature to sinusoidal exercise, *J. Appl. Physiol.*, vol. 76, pp. 2541–2545, 1994.

86. H.-S. Tam, R. C. Darling, H.-Y. Cheh, and J. A. Downey, Sweating response: a means of evaluating the set-point theory during exercise, *J. Appl. Physiol.*, vol. 45, pp. 451–458, 1978.

87. C. L. Taylor and K. Buettner, Influences of evaporative forces upon skin temperature dependency of human perspiration, *J. Appl. Physiol.*, vol. 6, pp. 113–123, 1953.

88. D. F. Brebner and D. McK. Kerslake, The time course of the decline in sweating produced by wetting the skin, *J. Physiol. London*, vol. 175, pp. 295–302, 1964.

89. D. F. Brebner and D. McK. Kerslake, The effects of soaking the skin in water at various temperatures on the subsequent ability to sweat, *J. Physiol. London*, vol. 194, pp. 1–11, 1968.

90. W. K. Brown and F. Sargent II, Hidromeiosis, *Arch. Environ. Health*, vol. 11, pp. 442–453, 1965.

91. S. D. Gerking and S. Robinson, Declines in the rates of sweating of men working in severe heat, *Am. J. Physiol.*, vol. 147, pp. 370–378, 1946.

92. S. Robinson and S. D. Gerking, Thermal balance of men working in severe heat, *Am. J. Physiol.*, vol. 149, pp. 476–488, 1947.

93. E. R. Nadel and J. A. J. Stolwijk, Effect of skin wettedness on sweat gland response, *J. Appl. Physiol.*, vol. 35, pp. 689–694, 1973.

94. V. Candas, J. P. Libert, and J. J. Vogt, Effect of hidromeiosis on sweat drippage during acclimation to humid heat, *Eur. J. Appl. Physiol.*, vol. 44, pp. 123–133, 1980.

95. D. A. Eyolfson, P. Tikuisis, X. Xu, G. Weseen, and G. G. Giesbrecht, Measurement and prediction of peak shivering intensity in humans, *Eur. J. Appl. Physiol.*, vol. 84, pp. 100–106, 2001.

96. J. S. Hayward, J. D. Eckerson, and M. L. Collis, Thermoregulatory heat production in man: prediction equation based on skin and core temperatures, *J. Appl. Physiol.*, vol. 42, pp. 377–384, 1977.

97. C. A. Piantadosi, E. D. Thalmann, and W. H. Spaur, Metabolic response to respiratory heat loss-induced core cooling, *J. Appl. Physiol.*, vol. 50, pp. 829–834, 1981.

98. E. R. Nadel, S. M. Horvath, C. A. Dawson, and A. Tucker, Sensitivity to central and peripheral stimulation in man, *J. Appl. Physiol.*, vol. 29, pp. 603–609, 1970.

99. P. Tikuisis and G. G. Giesbrecht, Prediction of shivering heat production from core and mean skin temperatures, *Eur. J. Appl. Physiol.*, vol. 79, pp. 221–229, 1999.

100. T. H. Benzinger, Heat regulation: homeostasis of central temperature in man, *Physiol. Rev.*, vol. 49, pp. 671–759, 1969.

101. B. Saltin, A. P. Gagge, and J. A. J. Stolwijk, Body temperatures and sweating during thermal transients caused by exercise, *J. Appl. Physiol.*, vol. 28, pp. 318–327, 1970.

102. J. H. M. Bittel, Heat debt as an index for cold adaptation in men, *J. Appl. Physiol.*, vol. 62, pp. 1627–1634, 1987.

103. J. H. M. Bittel, C. Nonotte-Varley, G. H. Livecchi-Gonnot, G. L. M. J. Savourey, and A. M. Hanniquet, Physical fitness and thermoregulatory reactions in a cold environment in men, *J. Appl. Physiol.*, vol. 65, pp. 1984–1989, 1988.

104. G. Savourey and J. Bittel, Thermoregulatory changes in cold induced by physical training in humans, *Eur. J. Appl. Physiol.*, vol. 78, pp. 379–384, 1998.

105. A. L. Vallerand, G. Savourey, and J. H. M. Bittel, Determination of heat debt in the cold: partitional calorimetry vs. conventional methods, *J. Appl. Physiol.*, vol. 72, pp. 1380–1385, 1992.

106. E. R. Buskirk, R. H. Thompson, and G. D. Whedon, Metabolic response to cold air in men and women in relation to total body fat content, *J. Appl. Physiol.*, vol. 18, pp. 603–612, 1963.

107. P. Cannon and W. R. Keatinge, The metabolic rate and heat loss of fat and thin men in heat balance in cold and warm water, *J. Physiol.*, vol. 154, pp. 329–344, 1960.

9

Computational Infrastructure for the Real-Time Patient-Specific Treatment of Cancer

K. R. Diller, J. T. Oden, C. Bajaj, J. C. Browne,
J. Hazle, I. Babuška, J. Bass, L. Bidaut,
L. Demkowicz, A. Elliott, Y. Feng, D. Fuentes,
S. Goswami, A. Hawkins, S. Khoshnevis,
B. Kwon, S. Prudhomme, and R. J. Stafford

CONTENTS

9.1 INTRODUCTION

The computational control system under development at the University of Texas at Austin combines a numerical implementation of the Pennes equation of bioheat transfer with the precise timing and orchestration of the problems of calibration, optimal control, data transfer, registration, finite element mesh refinement, cellular damage prediction, and laser control. The ultimate goal of this research is to provide the medical community a predictive computational tool that may be used by a surgeon during a minimally invasive hyper- or hypothermia treatment of a cancer-infected tissue. The tool controls the thermal source, provides a prediction of the entire outcome of the treatment, and, using intraoperative data, updates itself to increase the accuracy of the prediction. A current working snapshot of the entire control system is provided within this chapter. Current results demonstrate the importance of modeling the heterogeneity within the patient-specific biological domain to the accuracy of the computational solution. Through inversion of the constitutive nonlinearities, results also reinforce the experimentally observed phenomena of decreased perfusion in the damage region and hyperperfusion surrounding the damage region.

Minimally invasive treatments of cancer are key to improving posttreatment quality of life. Thermal therapies delivered under various treatment modalities are a form of minimally invasive cancer treatment that has the potential to become an effective option to eradicate the disease, maintain the functionality of infected organs, and minimize complications and relapse. However, the ability to control the energy deposition to prevent damage to adjacent healthy tissue is a limiting factor in all forms of thermal therapies [1], including cryotherapy, microwave, radio frequency, ultrasound, and laser. The combination of image guidance with computational prediction has the potential to allow unprecedented control over the bioheat transfer. Image guidance facilitates real-time treatment monitoring through temperature feedback during treatment delivery [2,3], and high-performance numerical implementations of mathematical bioheat transfer models can use the current-time thermal-imaging data to predict the outcome of the treatment minutes in advance [4].

The cyberinfrastructure under development at the University of Texas at Austin is an example of a dynamic data-driven feedback control system wherein the digitized bioheat transfer models control the heat transfer while simultaneously using real-time imaging data to update the accuracy of the prediction. The aim of the control system is to provide the medical community with a real-time computational tool for visualization of the predicted temperature and damage fields, allowing for patient-specific optimized therapy guidance. The current control system is applicable to tissues that are stationary during imaging and uses an interstitial laser fiber as the thermal source. The purpose of this chapter is to provide a working snapshot of the current cyberinfrastructure.

9.2 CONTROL SYSTEM IMPLEMENTATION

Figure 9.1 provides an outline of the data flow between the major software modules of the control system. Communication connecting the software modules at an actual laboratory at the M. D. Anderson Cancer Center in Houston, Texas, to the computing and visualization center in Austin, Texas, is currently handled via batch *secure file transfer protocol* (sftp) over a commercial GigE Internet connection. The Level Set Boundary-Interior-Exterior (LBIE) Mesher* generates a patient-specific finite element mesh of the biological domain using preoperative magnetic resonance imaging (MRI) data. Prior to treatment, the location of optical fiber and laser power are optimized to control heat shock protein (HSP) expression, eliminate and sensitize cancer cells, and minimize damage to healthy cells. During treatment, intraoperative MRI data are used to register the computational domain with the biological domain, and real-time thermal-imaging MR thermal imaging (MRTI) data drive the calibrations aligning the parameters of the biohcat transfer model to the patient's biological tissue values. As new thermal-imaging data are acquired intermittently, the computational prediction is compared to the measurements of the real-time thermal images and the differences seen are used to update the computations of the optimal laser parameters as well as goal-oriented mesh adaptation [5], where appropriate. The image acquisition by the computers in Austin implicitly controls the power wattage output of the laser in Houston. The software infrastructure is built from the Petsc [6] parallel computing paradigm and the Toolkit for Advanced Optimization (TAO; Argonne National Laboratory, Argonne, Illinois) [7] parallel optimization library. Advanced Visual Systems (AVS; Waltham, Massachusetts) [8] is used in conjunction with a virtual network computing (VNC) server for remote visualization. AVS coroutines are used to manage and coordinate the simultaneous visualization of the MRI anatomical image, the MRTI thermal image, and finite element data sets.

---

* Software available at http://cvcweb.ices.utexas.edu/cvc.

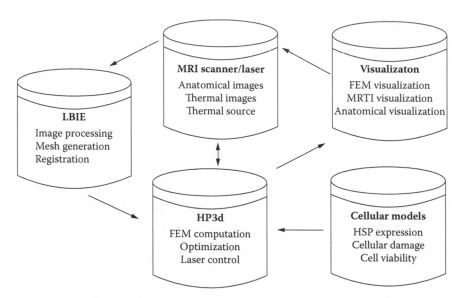

Figure 9.1   The computational infrastructure has five main modules: heating and imaging, finite element mesh generation, finite element computation, cellular damage models, and visualization. The data flow between the control system modules is illustrated.

A high-level abstraction of the HP3d* finite element solver is shown in Figure 9.2. The finite element solver is run at the Texas Advanced Computing Center (TACC), located in Austin. As illustrated in Figure 9.2, the problems of real-time calibration, optimal control, and goal-oriented error estimation are solved in parallel by separate groups of processors. Periodically during treatment, the groupwise optimization solutions and error estimates are collected on the control task. A skeleton of the entire finite element method (FEM) mesh of the biological domain is stored on the control task. Using the collective error estimates, a mesh refinement strategy is computed on the control task, and both the collective optimization solutions and refinement strategy are broadcast to the individual computational groups. The data server shown in Figure 9.2 reads in the thermal images from disk, filters the thermal images to remove noise, broadcasts the thermal images to each computational group as needed, and transmits the laser power to Houston. The timing of the laser power control is implicit through the image acquisition by the HP3d data server in Austin. As new thermal images are written to disk physically in Houston, they are transferred to disk at TACC. When the data server detects that the full set of thermal images for a time instance is available, the power to be used for the next time interval is sent to Houston.

---

* Software available at http://dddas.ices.utexas.edu.

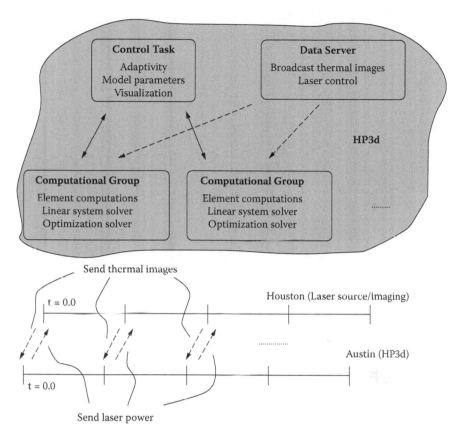

Figure 9.2   The finite element computations are performed on a parallel computing architecture using multiple groups of computed tasks to simultaneously solve disjoint numerical problems of the control system. A control task is used to gather and broadcast the individual solutions of the computational groups. A data server broadcasts filtered thermal images to individual computational groups as requested. The imaging implicitly controls the laser power output. As a new thermal image is acquired by HP3d in Austin, the power wattage for the next time interval is transmitted to the laser.

9.3 IMAGING TO FINITE ELEMENT MESH PIPELINE

The imaging-to-modeling software system for anatomical MRI data employs both image-processing and geometry-processing functionalities to produce a suitable linear or higher-order meshed model of the anatomy. Figure 9.3 describes the data flow layout. The major algorithmic components of each of the processing units is described in this section. The reader must note that the modules are selectively used depending on the nature and quality of the imaging data.

9.3.1 Image Processing

The input raw imaging data are often of poor quality, which makes it difficult to build a quality meshed model of the anatomy under investigation. In order to

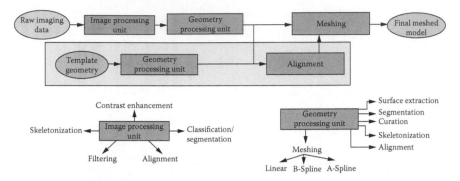

Figure 9.3   Data flow of imaging-to-meshing software system: There are two major processing units: image processing and geometry processing. The image-processing unit consists of modules for *contrast enhancement, classification and segmentation, filtering, skeletonization*, and *alignment*. The geometry-processing unit consists of *surface extraction, curation, segmentation, skeletonization, alignment*, and *meshing*; meshing is further subdivided into linear and higher-order boundary element and finite element mesh generation components. The three-dimensional (3D) anatomical magnetic resonance imaging (MRI) data are first passed through the image-processing unit for improvement of image quality, and are then processed by the geometry-processing unit for extraction of a clean geometry annotated with the present features. Finally, the clean geometry is converted to a linear or higher-order mesh. Occasionally, to deal with incomplete or low-quality imaging data, a twin data-processing pipeline is employed where a template geometry is processed to extract vital geometric information.

improve the image quality, we have developed a suite of image-processing functionalities that facilitate further processing. The modules encapsulated in the image-processing units are as follows:

1. *Contrast enhancement*: Improves the contrast of the image to help extraction of the domain of interest [9].
2. *Filtering*: Removes the noise by modifying the input image using bilateral filtering coupled with an anisotropic geometric diffusion **partial differential equation** (PDE) [10].

$$\partial_t \phi - \|\nabla \phi\| \, div\left(D^\sigma \frac{\nabla \phi}{\|\nabla \phi\|} \right) = 0$$

3. *Segmentation*: Segments an image into anatomically separate regions of interest using a fast marching method [11]; each region can then be extracted from the raw image [12].
4. *Image skeletonization*: Extracts lower-dimensional features from the image by analyzing the critical points of the imaging data [13,14].

5. *Flexible alignment*: Performs affine transformation to best fit an image of a biological system onto a different instance of the same [15].

9.3.2 Geometry Processing

After passing the input raw image through the modules of the image-processing unit, an improved image is obtained on which the geometry-processing routines are applied. The modules encapsulated in this unit are dedicated to better understanding the features of the model in order to improve the topological and geometric qualities that help in producing a correct meshed model of the domains of interest.

1. *Surface extraction*: Geometry extraction from the imaging data is done either by using contouring [16,17] or by reconstructing a piecewise triangulated or higher-order surface model from the boundary voxels of the segmented regions [18–20].

2. *Curation and filtering*: The initial surface model extracted from the imaging data has topological anomalies, namely, small components, spurious noisy features, and the like. The algorithms developed to cure the model of such anomalies include point cloud regularization [21], and volumetric primal and dual space feature quantification [22,23].

3. *Segmentation*: Geometric segmentation of an initial model often leads to better understanding of the quality of the model in terms of its topology N-d geometry. We have developed a geometry segmentation module based on the distance function induced by the geometry under investigation [22].

4. *Skeletonization*: The skeletal feature of a model provides a lower-dimensional description of a geometry that is helpful in building further meshed models of the anatomy, as was utilized in Zhang et al. [24]. We have developed a skeletonization algorithm that extracts a polylinear or polygonal skeletal structure from the geometric model [25] and further helps in annotating the shape into tubular or flat regions.

5. *Meshing*: The task of meshing is primarily divided into two parts— boundary element and finite element meshing. Each part has three subparts depending on if the resulting mesh is linear or higher order. For boundary element meshing, we have three options, namely, triangle or quadrilateral meshing [26], B-spline meshing, and A-spline meshing [27]. Similarly, for finite element meshing, we also have developed three different meshing modules, namely, tetrahedral or hexahedral meshing [26,28], solid nonuniform rational B-spline (NURBS) meshing [24], and A-spline meshing [29].

9.4 GOVERNING EQUATIONS

The governing state equations of the control system are built around the Pennes bioheat transfer model.

Find the spatially and temporally varying temperature field $u(\beta,\mathbf{x},t)$ such that

$$\rho c_p \frac{\partial u}{\partial t} - \nabla \cdot (k(u,\mathbf{x},\beta)\nabla u) + \omega(u,\mathbf{x},\beta)c_{blood}(u-u_a) = Q_{laser}(\beta,\mathbf{x},t) \quad \text{in } \Omega$$

given the Cauchy and Neumann boundary conditions

$$-k(u,\mathbf{x},\beta)\nabla u \cdot \mathbf{n} = h(u-u_\infty) \qquad \text{on } \partial\Omega_C$$

$$-k(u,\mathbf{x},\beta)\nabla u \cdot \mathbf{n} = \mathcal{G} \qquad \text{on } \partial\Omega_N$$

and the initial condition

$$u(\mathbf{x},0) = u^0 \qquad \text{in } \Omega$$

where c_p and c_{blood} are the specific heats of the tissue and blood, respectively; u_a is the arterial temperature; ρ is the density of the tissue; and h is the coefficient of cooling. The constitutive equations for the thermal conductivity, $k[\frac{J}{s \cdot m \cdot K}]$, and blood perfusion, $\omega[\frac{kg}{s\,m^3}]$, assume a nonlinear form, as shown in Figure 9.4.

$$k(u,\mathbf{x},\beta) = k_0(\mathbf{x}) + k_1 \operatorname{atan}(\tilde{k}_2(u-\hat{k}_3))$$

$$\omega(u,\mathbf{x},\beta) = \omega_0(\mathbf{x}) + \omega_1 \operatorname{atan}(\tilde{\omega}_2(u-\hat{\omega}_3))$$

Note that $\omega_0(\mathbf{x})$ is allowed to vary over the spatial dimension, as the blood perfusion within the necrotic core of a cancerous tumor or within a damaged tissue is expected to be significantly lower than that within the surrounding healthy tissue. $k_0(\mathbf{x})$ is also allowed to vary over the spatial dimension to capture the biological tissue heterogeneity. The isotropic laser source term, Q_{laser}, is of the form

$$Q_{laser}(\beta,\mathbf{x},t) = 3P(t)\mu_a\mu_{tr} \frac{\exp(-\mu_{eff}\|\mathbf{x}-\mathbf{x}_0\|)}{4\pi\|\mathbf{x}-\mathbf{x}_0\|} \qquad \begin{aligned} \mu_{tr} &= \mu_a + \mu_s(1-g) \\ \mu_{eff} &= \sqrt{3\mu_a\mu_{tr}} \end{aligned}$$

where μ_a and μ_s are laser coefficients related to laser wavelength and give the probability of tissue absorption and scattering of photons, respectively. $P(t)$ is the laser power wattage as a function of time. For the defined bioheat transfer model, let β denote all the model parameters:

$$\beta \equiv (k_0(\mathbf{x}), k_1, \tilde{k}_3, \hat{k}_3, \omega_0(\mathbf{x}), \omega_1, \tilde{\omega}_3, \hat{\omega}_3, P(t), \mu_a, \mu_s, \mathbf{x}_0)$$

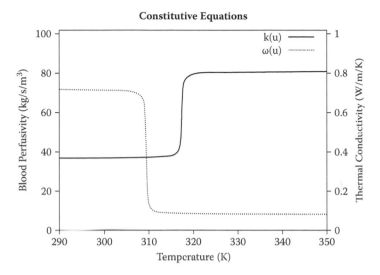

Figure 9.4   The temperature dependence of the blood perfusion and thermal conductivity material data used in the Pennes model are plotted against the left axis and right axis, respectively. The values shown were computed using inverse techniques applied to thermal imaging.

The variational form of the Pennes equation is given below:

$$C(u,\beta;v) = 0 = \int_0^T \int_\Omega \rho c_p \frac{\partial u}{\partial t} v + k(u,\mathbf{x},\beta)\nabla u \cdot \nabla v \; dxdt$$

$$+ \int_0^T \int_\Omega \omega(u,\mathbf{x},\beta)c_{blood}(u-u_a) v \; dxdt + \int_0^T \int_{\partial\Omega_N} G \, v \, dAdt \qquad (9.1)$$

$$- \int_0^T \int_\Omega Q_{laser}(\beta,\mathbf{x},t)v \; dxdt + \int_0^T \int_{\partial\Omega_C} h(u-u_\infty) v \; dxdt$$

The test function is denoted $v(\mathbf{x},t)$. Notice that each of the above integrals is well defined given that the thermal conductivity $k(u,\mathbf{x},\beta)$ and the perfusion $\omega(u,\mathbf{x},\beta)$ are bounded and assuming that $\frac{\partial u}{\partial t} \in L^2(\Omega)$.

The goal of the temperature-based optimal control and calibration problems within the control loop is to find the set of model parameters that minimize the space–time norm of the difference between the computed temperature field $u(\beta,\mathbf{x},t)$ and an ideal field $u^{ideal}(\mathbf{x},t)$. The mathematical structure is formally stated as follows.

Find the model coefficients, $\beta^* \in \mathbb{P}$, that produce the temperature field, $u^* \in \mathcal{V}$, such that

$$Q(u^*(\beta^*),\beta^*) = \frac{1}{2}\int_0^T \int_\Omega \chi(\mathbf{x})(u^*(\beta,\mathbf{x},t) - u_{ideal}(\mathbf{x},t))^2 \; dxdt + \Phi(\beta)$$

satisfies

$$Q(u^*(\beta^*),\beta^*) = \inf_{\beta \in \mathbb{P}} Q(u(\beta),\beta)$$

(9.2)

$$\mathbb{P} = \{\beta : \exists! u \text{ s.t. } C(u,\beta;v) = 0 \quad \forall v \in \mathcal{V}\}$$

The ideal field for the calibration problem is the experimentally determined temperature field $u_{exp}(\mathbf{x},t)$. The ideal field for temperature-based optimal control is given as follows:

$$u^{ideal} = \begin{cases} 37.0^\circ C & \mathbf{x} \in \Omega_H \\ 50.0^\circ C & \mathbf{x} \in \Omega_C \end{cases}$$

The function $\Phi(\beta)$ is a penalty term added to the objective function used to keep the model parameters within physically acceptable bounds.

$$\Phi(\beta) = \sum_{i=1}^{N_{params}} \exp\left(\gamma \frac{\beta_i - \beta_i^{UB}}{\beta_i^{UB} - \beta_i^{LB}}\right) + \exp\left(-\gamma \frac{\beta_i - \beta_i^{LB}}{\beta_i^{UB} - \beta_i^{LB}}\right)$$

The dimension of the parameter space is denoted N_{params}. β_i denotes a particular parameter. The physically acceptable lower and upper bounds of the parameter, β_i, are denoted by β_i^{LB} and β_i^{UB}, respectively. The penalty term is scaled by $\gamma = 1000.0$.

9.4.1 Discretization of Equations

The optimization problem in Equation (9.2) is solved using an adjoint method to compute the gradient of the quantity of interest. The following Galerkin representation of the temperature field and adjoint variable is assumed:

$$u(\mathbf{x},t) = \sum_{k=1}^{N_{step}} \sum_{j=1}^{N_{dof}} \alpha_j^k(t)\phi_j(\mathbf{x}) \qquad p(\mathbf{x},t) = \sum_{k=1}^{N_{step}} \sum_{i=1}^{N_{dof}} \lambda_i^k(t)\phi_i(\mathbf{x})$$

where N_{step} is the number of time steps, N_{dof} is the number of Galerkin coefficients, and ϕ_i's are the finite element shape functions of polynomial order $p = 1,2,3\ldots$

$$\alpha_j^k(t) = \begin{cases} \dfrac{t_k - t}{t_k - t_{k-1}}\alpha_j^{k-1} + \dfrac{t - t_{k-1}}{t_k - t_{k-1}}\alpha_j^k, & t \in [t_{k-1},t_k) \\ 0, & \text{otherwise} \end{cases} \qquad \lambda_i^k(t) = \begin{cases} \lambda_i^k, & t \in [t_{k-1},t_k) \\ 0, & \text{otherwise} \end{cases}$$

The time discretization of the power is assumed to be piecewise constant in time.

$$P(t) = \begin{cases} P_k, & t \in [t_{k-1}, t_k) \\ 0, & \text{otherwise} \end{cases}$$

The spatial variation of the parameter fields is assumed to have the following Galerkin representation:

$$k_0(\mathbf{x}) = \sum_j k_0^j \psi^j(\mathbf{x})$$

$$\omega_0(\mathbf{x}) = \sum_j \omega_0^j \psi^j(\mathbf{x})$$

where $\Psi(\mathbf{x})$ are piecewise constant across elements.

9.4.1.1 Time Stepping

Assuming that the test function is piecewise constant in time,

$$v(\mathbf{x},t) = \sum_{k=1}^{N_{step}} \sum_{i=1}^{N_{dof}} v_i^k(t)\phi_i(\mathbf{x}) \qquad v_i^k(t) = \begin{cases} v_i^k, & t \in [t_{k-1}, t_k) \\ 0, & \text{otherwise} \end{cases}$$

The governing equations (Equation 9.1) are solved with the following Crank-Nicolson time-stepping scheme.

$$\Delta t_k \int_\Omega \rho c_p \frac{u_k - u_{k-1}}{\Delta t_k} v_k + k(u_{k-\frac{1}{2}}, \mathbf{x}, \beta)\nabla u_{k-\frac{1}{2}} \cdot \nabla v_k \, dx$$

$$+ \Delta t_k \int_\Omega \omega(u_{k-\frac{1}{2}}, \mathbf{x}, \beta)c_{blood}(u_{k-\frac{1}{2}} - u_a) v_k \, dx$$

$$+ \Delta t_k \int_{\partial\Omega_C} h(u_{k-\frac{1}{2}} - u_\infty) v_k \, dA = \Delta t_k \int_\Omega Q_{laser}(\beta, \mathbf{x}, t_{k-\frac{1}{2}}) v_k \, dx \tag{9.3}$$

$$- \int_{t_{k-1}}^{t_k} \int_{\partial\Omega_N} g \, v_k \, dA \quad \forall v_k \quad k = 1, 2, ..., N_{step}$$

where (using Einstein summation notation)

$$u_k = \alpha_j^k \phi_j(\mathbf{x}) \qquad u_{k-\frac{1}{2}} = \frac{\alpha_j^{k-1} + \alpha_j^k}{2}\phi_j(\mathbf{x}) \qquad v_k = v_i^k \phi_i(\mathbf{x})$$

The discretization (Equation 9.3) is of the form

$$\text{find } \vec{\alpha}^k = \left(\alpha_1^k, \alpha_2^k, \dots \right) \text{s.t.}$$

$$\vec{f}(\vec{\alpha}^k) = \vec{0}$$

The Jacobian of this system of equations is

$$\frac{\partial f_i}{\partial \alpha_j^k} = \Delta t_k \int_\Omega \frac{\rho c_p}{\Delta t_k} \phi_j \phi_i \, dx + \Delta t_k \frac{1}{2} \int_{\partial \Omega_C} h \phi_j \, \phi_i \, dA$$

$$+ \Delta t_k \frac{1}{2} \int_\Omega \left(\frac{\partial k}{\partial u} \left(u_{k-\frac{1}{2}}, \mathbf{x}, \beta \right) \nabla u_{k-\frac{1}{2}} \phi_j + k \left(u_{k-\frac{1}{2}}, \mathbf{x}, \beta \right) \nabla \phi_j \right) \cdot \nabla \phi_i \, dx$$

$$+ \Delta t_k \frac{1}{2} \int_\Omega c_{blood} \left(\frac{\partial \omega}{\partial u} \left(u_{k-\frac{1}{2}}, \mathbf{x}, \beta \right) \left[u_{k-\frac{1}{2}} - u_a \right] + \omega \left(u_{k-\frac{1}{2}}, \mathbf{x}, \beta \right) \right) \phi_j \, \phi_i \, dx$$

9.4.1.2 Adjoint Gradient

The Adjoint Gradient of the quantity of interest is constructed from the derivative of the discretized equations with respect to a single model variable. The chain rule is used to compute the gradient of the quantity of interest for the optimization. The initial condition does not depend on the model parameters, $\frac{\partial u_0}{\partial \beta_i} = 0$.

$$\frac{\partial}{\partial \beta_i} Q(u(\beta, \mathbf{x}, t), \beta) = \sum_{k=1}^{N_{step}} \frac{\partial Q}{\partial u_k} \frac{\partial u_k}{\partial \beta_i} + \frac{\partial}{\partial \beta_i} \Phi(\beta)$$

The same Galerkin representation is used for u^{ideal} as u.

$$Q(u(\beta), \beta) = \frac{1}{2} \int_0^T \int_\Omega \rho c_p \chi(\mathbf{x}) (u(\beta, \mathbf{x}, t) - u^{ideal}(\mathbf{x}, t))^2 \, dx dt + \Phi(\beta)$$

$$= \frac{1}{2} \sum_{k=1}^{N_{step}} \int_{t_{k-1}}^{t_k} \int_\Omega \rho c_p \chi(\mathbf{x}) (u(\beta, \mathbf{x}, t) - u^{ideal}(\mathbf{x}, t))^2 \, dx dt + \Phi(\beta)$$

$$= \frac{\Delta t_k}{6} \sum_{k=1}^{N_{step}} \int_\Omega \rho c_p \chi(\mathbf{x}) \begin{pmatrix} u_{k-1}^2 - 2u_{k-1}u_{k-1}^{ideal} + u_{k-1}u_k \\ -u_{k-1}u^{ideal} + \left(u_{k-1}^{ideal} \right)^2 \\ -u_{k-1}^{ideal}u_k + u_{k-1}^{ideal}u_k^{ideal} + u_k^2 \\ -2u_k u_k^{ideal} + \left(u_k^{ideal} \right)^2 \end{pmatrix} + \Phi(\beta)$$

Derivatives are taken with respect to the numerically computed quantity of interest.

$$
\frac{\partial Q(u,\beta)}{\partial \beta_i} = \sum_{k=1}^{N_{step}} \int_{t_{k-1}}^{t_k} \int_{\Omega} \rho c_p \chi(\mathbf{x})(u(\beta,\mathbf{x},t) - u^{ideal}(\mathbf{x},t)) \frac{\partial u}{\partial \beta_i} \, dx dt + \frac{\partial}{\partial \beta_i} \Phi(\beta)
$$

$$
= \sum_{k=1}^{N_{step}} \Delta t_k \int_{\Omega} \rho c_p \chi(\mathbf{x}) \left(\begin{array}{c} \left[\dfrac{\left(u_{k-1} - u_{k-1}^{ideal}\right)}{3} + \dfrac{\left(u_k - u_k^{ideal}\right)}{6} \right] \dfrac{\partial u_{k-1}}{\partial \beta_i} \\[3ex] + \left[\dfrac{\left(u_{k-1} - u_{k-1}^{ideal}\right)}{6} + \dfrac{\left(u_k - u_k^{ideal}\right)}{3} \right] \dfrac{\partial u_k}{\partial \beta_i} \end{array} \right) dx
$$

$$
+ \frac{\gamma}{\beta_i^{UB} - \beta_i^{LB}} \left[\exp\left(\gamma \frac{\beta_i - \beta_i^{UB}}{\beta_i^{UB} - \beta_i^{LB}} \right) - \exp\left(-\gamma \frac{\beta_i - \beta_i^{LB}}{\beta_i^{UB} - \beta_i^{LB}} \right) \right]
$$

The derivative of the discretized state (Equation 9.3) with respect to a single model variable yields the following:

$$
\frac{\partial}{\partial \beta_i} C(u,\beta,v) = \Delta t_k \int_{\Omega} \frac{\rho c_p}{\Delta t_k} \left(\frac{\partial u_k}{\partial \beta} - \frac{\partial u_{k-1}}{\partial \beta} \right) v_k \, dx
$$

$$
+ \Delta t_k \int_{\Omega} \frac{\partial k}{\partial u} \left(u_{k-\frac{1}{2}}, \mathbf{x}, \beta \right) \frac{1}{2} \left[\frac{\partial u_{k-1}}{\partial \beta} + \frac{\partial u_k}{\partial \beta} \right] \nabla u_{k-\frac{1}{2}} \cdot \nabla v_k \, dx
$$

$$
+ \Delta t_k \int_{\Omega} k \left(u_{k-\frac{1}{2}}, \mathbf{x}, \beta \right) \frac{1}{2} \left[\nabla \frac{\partial u_{k-1}}{\partial \beta} + \nabla \frac{\partial u_k}{\partial \beta} \right] \cdot \nabla v_k \, dx
$$

$$
+ \Delta t_k \int_{\Omega} \frac{\partial \omega}{\partial u} \left(u_{k-\frac{1}{2}}, \mathbf{x}, \beta \right) \frac{c_{blood}}{2} \left[\frac{\partial u_{k-1}}{\partial \beta} + \frac{\partial u_k}{\partial \beta} \right] \left(u_{k-\frac{1}{2}} - u_a \right) v_k \, dx
$$

$$
+ \Delta t_k \int_{\Omega} \omega \left(u_{k-\frac{1}{2}}, \mathbf{x}, \beta \right) \frac{c_{blood}}{2} \left[\frac{\partial u_{k-1}}{\partial \beta} + \frac{\partial u_k}{\partial \beta} \right] v_k \, dx
$$

$$
+ \Delta t_k \int_{\partial \Omega_C} \frac{h}{2} \left[\frac{\partial u_{k-1}}{\partial \beta} + \frac{\partial u_k}{\partial \beta} \right] v_k \, dA
$$

$$
+ \Delta t_k \int_{\Omega} \frac{\partial k}{\partial \beta} \left(u_{k-\frac{1}{2}}, \mathbf{x}, \beta \right) \nabla u_{k-\frac{1}{2}} \cdot \nabla v_k \, dx
$$

$$
+ \Delta t_k \int_{\Omega} c_{blood} \frac{\partial \omega}{\partial \beta} \left(u_{k-\frac{1}{2}}, \mathbf{x}, \beta \right) \left(u_{k-\frac{1}{2}} - u_a \right) v_k \, dx
$$

$$
- \Delta t_k \int_{\Omega} \frac{\partial Q_{laser}}{\partial \beta} (\beta, \mathbf{x}, t_k) v_k \, dx = 0 \qquad k = 1, 2, \ldots, N_{step}
$$

Solving for the adjoint variable, P_k, such that

$$\Delta t_k \int_\Omega \frac{\rho c_p}{\Delta t_k} \hat{u} p_k + \frac{1}{2} \frac{\partial k}{\partial u}\left(u_{k-\frac{1}{2}}, \mathbf{x}, \beta\right) \hat{u} \nabla u_{k-\frac{1}{2}} \cdot \nabla p_k \, dx$$

$$+ \Delta t_k \int_\Omega \frac{1}{2} \frac{\partial \omega}{\partial u}\left(u_{k-\frac{1}{2}}, \mathbf{x}, \beta\right) \hat{u}\left(u_{k-\frac{1}{2}} - u_a\right) p_k \, dx$$

$$+ \Delta t_k \int_\Omega \frac{1}{2} k\left(u_{k-\frac{1}{2}}, \mathbf{x}, \beta\right) \nabla \hat{u} \cdot \nabla p_k \, dx + \frac{1}{2} \omega\left(u_{k-\frac{1}{2}}, \mathbf{x}, \beta\right) \hat{u} p_k \, dx$$

$$+ \Delta t_k \int_{\partial \Omega_C} \frac{h}{2} \hat{u} p_k \, dA = \Delta t_k \int_\Omega \rho c_p \chi(\mathbf{x}) \left[\frac{\left(u_{k-1} - u_{k-1}^{ideal}\right)}{6} - \frac{\left(u_k - u_k^{ideal}\right)}{3}\right] \hat{u}$$

$$\forall \hat{u}, \quad k = N_{step}$$

and

$$\Delta t_k \int_\Omega \frac{\rho c_p}{\Delta t_k} \hat{u} p_k + \frac{1}{2} \frac{\partial k}{\partial u}(u_{k-\frac{1}{2}}, \mathbf{x}, \beta) \hat{u} \nabla u_{k-\frac{1}{2}} \cdot \nabla p_k \, dx$$

$$+ \Delta t_k \int_\Omega \frac{1}{2} \frac{\partial \omega}{\partial u}\left(u_{k-\frac{1}{2}}, \mathbf{x}, \beta\right) \hat{u}\left(u_{k-\frac{1}{2}} - u_a\right) p_k \, dx + \Delta t_k \int_{\partial \Omega_C} \frac{h}{2} \hat{u} p_k \, dA$$

$$+ \Delta t_k \int_\Omega \frac{1}{2} k\left(u_{k-\frac{1}{2}}, \mathbf{x}, \beta\right) \nabla \hat{u} \cdot \nabla p_k \, dx + \frac{1}{2} \omega\left(u_{k-\frac{1}{2}}, \mathbf{x}, \beta\right) \hat{u} p_k \, dx$$

$$= \int_\Omega \rho c_p \chi(\mathbf{x}) \Delta t_k \left[\frac{1}{6}\left(u_{k-1} - u_{k-1}^{ideal}\right) + \frac{1}{3}\left(u_k - u_k^{ideal}\right)\right] \hat{u}$$

$$+ \int_\Omega \rho c_p \chi(\mathbf{x}) \Delta t_{k+1} \left[\frac{1}{3}\left(u_k - u_k^{ideal}\right) + \frac{1}{6}\left(u_{k+1} - u_{k+1}^{ideal}\right)\right] \hat{u} \, dx$$

$$- \begin{pmatrix} -\Delta t_{k+1} \int_\Omega \frac{\rho c_p}{\Delta t_{k+1}} \hat{u} p_{k+1} + \frac{1}{2} \frac{\partial k}{\partial u}\left(u_{k+\frac{1}{2}}, \beta\right) \hat{u} \nabla u_{k+\frac{1}{2}} \cdot \nabla p_{k+1} \\[2mm] + \Delta t_{k+1} \int_\Omega \frac{1}{2} \frac{\partial \omega}{\partial u}(u_{k+\frac{1}{2}}, \beta) \hat{u}\left(u_{k+\frac{1}{2}} - u_a\right) p_{k+1} \, dx \\[2mm] + \Delta t_{k+1} \int_\Omega \frac{1}{2} k\left(u_{k+\frac{1}{2}}, \beta\right) \nabla \hat{u} \cdot \nabla p_{k+1} \, dx \\[2mm] + \Delta t_{k+1} \int_\Omega \frac{1}{2} \omega\left(u_{k+\frac{1}{2}}, \beta\right) \hat{u} p_{k+1} \, dx + \Delta t_{k+1} \int_{\partial \Omega_C} \frac{h}{2} \hat{u} p_{k+1} \, dA \end{pmatrix}$$

$$\forall \hat{u}, \quad k = N_{step} - 1, N_{step} - 2, \ldots, 1$$

implies that the numerical gradient of the quantity of interest may be computed as follows:

$$
\frac{\partial Q(u,\beta)}{\partial \beta} = \sum_{k=1}^{N_{step}} \left(\begin{array}{c} -\Delta t_k \int_\Omega \frac{\partial k}{\partial \beta}(u_{k-\frac{1}{2}},\mathbf{x},\beta)\nabla u_{k-\frac{1}{2}} \cdot \nabla p_k \, dx \\[2ex] -\Delta t_k \int_\Omega c_{blood} \frac{\partial \omega}{\partial \beta}(u_{k-\frac{1}{2}},\mathbf{x},\beta)(u_{k-\frac{1}{2}} - u_a) p_k \, dx \\[2ex] +\Delta t_k \int_\Omega \frac{\partial Q_{laser}}{\partial \beta}(\beta,\mathbf{x},t_k) p_k \, dx \end{array} \right)
$$

9.4.2 Calibration Results

The computational results presented in this section revisit the thermal-imaging data studied in Feng et al. [30] and Oden et al. [4]. The calibration algorithm presented in Section 9.4.1 is applied to these data sets to invert for the constitutive equation nonlinearities as well as the heterogeneity of the biological domain.

The effect of calibrating the nonlinear perfusion, $\omega(u,\mathbf{x},\beta)$, and the thermal conductivity, $k(u,\mathbf{x},\beta)$, parameters in the Pennes model was studied. Calibration was done with respect to MRTI thermal-imaging data of *in vivo* heating of canine brain tissue. The thermal-imaging data were acquired in the form of five two-dimensional (2D) 256 × 256 pixel images every 6 seconds for 120 time steps. The spacing between images was 3.5 mm. A manual craniotomy of a canine skull was performed to allow insertion of an interstitial laser fiber to provide the heating. A template base [32] finite element mesh was generated from 36 2D 256 × 256 pixel MRI images of the canine brain. The field view was 200 mm × 200 mm, with each image spaced 1 mm apart. The FEM prediction using *CRC Handbook* [31] linear material coefficients, $k(u,\mathbf{x},\beta)=0.527[\frac{W}{mK}]$ and $\omega(u,\mathbf{x},\beta)=6.0[\frac{kg}{sm^3}]$, is shown in Figure 9.5. Figure 9.6 shows the FEM prediction using calibrated nonlinear coefficients. A plot of the material coefficients obtained is shown in Figure 9.4. The data shown in Figures 9.5 and 9.6 illustrate a particular time instance. The upper-right windows in Figures 9.5 and 9.6 each show a cutline comparison of the filtered MRTI data with the unfiltered data. The upper-left windows in Figures 9.5 and 9.6 display an overlay of the MRTI thermal image onto the anatomical MRI image. The lower-left window in Figures 9.5 and 9.6 shows a 2D temperature slice through the 3D domain. The lower-right window is a cutline comparison of the filtered MRTI data to the FEM prediction. The accuracy of the predicted FEM solution shown in Figure 9.6 was obtained by inverting for the constitutive nonlinearities. The damage of the tissue is reflected in the decrease in perfusion for high temperatures within the damaged region and hyperperfusion surrounding the damaged region. As shown in Oden et al. [4],

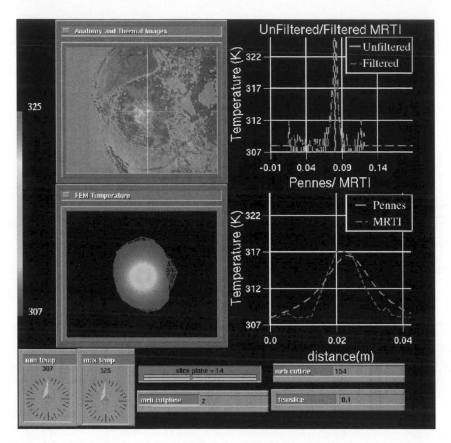

Figure 9.5   Thermal images of *in vivo* heating of a canine brain were taken every 6 seconds over a period of 12 minutes. The top left shows the anatomy with a particular time instance of the thermal images overlaid. The linear Pennes equation was solved using *CRC Handbook* [31] perfusion and thermal conductivity values for the canine brain, and the bottom left shows the linear finite element method (FEM) prediction at the same time instance for comparison. The temperature range shown is from 307 K to 325 K. The top right shows a cutline through the thermal image data. The unfiltered and filtered image data are plotted along the cutline. The bottom right compares the FEM predicted temperature and filtered thermal image along a cutline through the FEM mesh.

the linear case captures the early time heating well, but the cutlines in Figures 9.5 and 9.6 illustrate that the material nonlinearities are necessary to model the late time heat dissipation. The results presented in Figure 9.8 indicate that a spatially varying inversion for the perfusion field should provide a means to further increase the accuracy of the FEM temperature prediction of the canine brain data (Figure 9.6).

Allowing the perfusion and thermal conductivity model parameters to vary as a spatial field is seen to have a tremendous effect on the model calibrations. Inverting for the spatial variation in the parameters embeds the biological tissue

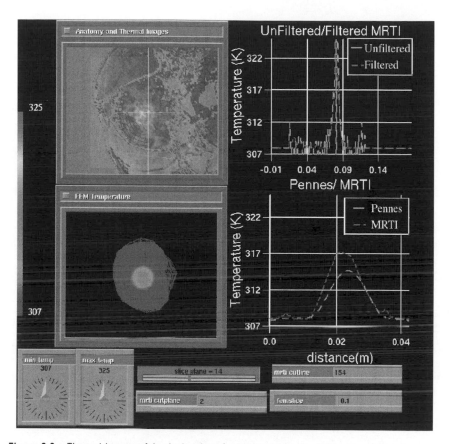

Figure 9.6   Thermal images of *in vivo* heating of a canine brain were taken every 6 seconds over a period of 12 minutes. The top left shows the anatomy with a particular time instance of the thermal images overlaid. The Pennes equation was solved with a set of nonlinear perfusion and thermal conductivity material coefficients that were calibrated to the thermal images, and the bottom left shows the nonlinear FEM prediction at the same time instance for comparison. The temperature range shown is from 307 K to 325 K. The top right shows a cutline through the thermal image data. The unfiltered and filtered image data are plotted along the cutline. The bottom right compares the FEM predicted temperature and filtered thermal image along a cutline through the FEM mesh.

heterogeneity within the Pennes model. Imaging data of an external laser applied to a tumor grown on a mouse's hind leg were used to study the effect of the parameter field inversion. Sixty thermal images were acquired at an interval of 5 seconds. A single time instance of the data is shown in Figures 9.7 and 9.8. The field of view is 4×6 cm^2, and the thickness associated with the MRI and MRTI images is 3 mm. Figures 9.7 and 9.8 compare the FEM prediction using textbook linear material coefficients to the calibrated heterogeneous material coefficients applied to the *in vivo* heating of a tumor grown on a mouse. The upper-right windows in Figures 9.7 and 9.8 each show a cutline comparison of the filtered MRTI

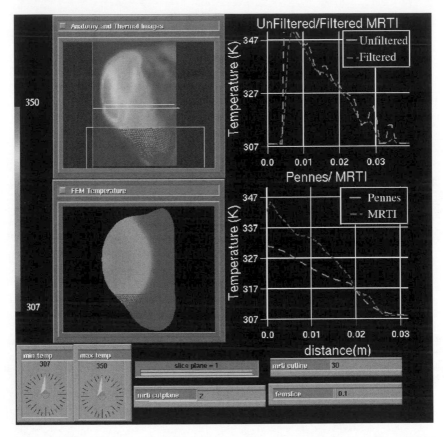

Figure 9.7 Thermal images of *in vivo* heating of a tumor grown on the hind leg of a mouse were taken every 5 seconds over a period of 5 minutes. The top left shows the anatomy with a particular time instance of the thermal images overlaid. The linear Pennes equation was solved using *CRC Handbook of Mechanical Engineering* [31] perfusion and thermal conductivity values for the tissue, and the bottom left shows the linear FEM prediction at the same time instance for comparison. The temperature range shown is from 307 K to 350 K. The top right shows a cutline through the thermal image data. The unfiltered and filtered image data are plotted along the cutline. The bottom right compares the FEM predicted temperature and filtered thermal image along a cutline through the FEM mesh.

data with the unfiltered data. The upper-left windows in Figures 9.7 and 9.8 display an overlay of the MRTI thermal image onto the anatomical MRI image. The images are 49 × 56 pixels. The lower-left window in Figures 9.7 and 9.8 shows a 2D temperature slice through the 3D domain. The lower-right window is a cutline comparison of the filtered MRTI data to the FEM prediction. The agreement between the predicted FEM solution and the MRTI thermal images shown in Figure 9.8 illustrates the importance of inverting for the field of material heterogeneity. Figure 9.8 represents a ≈4100 parameter optimization problem. Figure 9.9

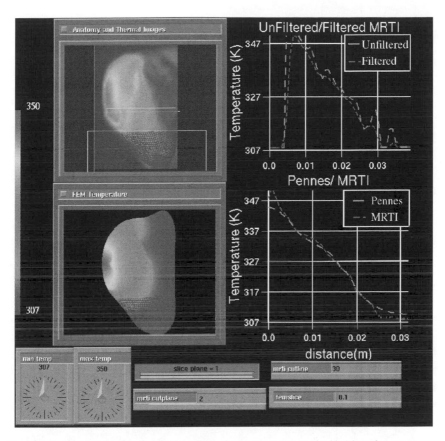

Figure 9.8   Thermal images of *in vivo* heating of a tumor grown on the hind leg of a mouse were taken every 5 seconds over a period of 5 minutes. The top left shows the anatomy with a particular time instance of the thermal images overlaid. The Pennes equation was solved with a spatially varying field of perfusion and thermal conductivity material that were calibrated to the thermal images, and the bottom left shows the nonlinear FEM prediction at the same time instance for comparison. ≈4100 model parameters were optimized to recover the material heterogeneity. The temperature range shown is from 307 K to 350 K. The top right shows a cutline through the thermal image data. The unfiltered and filtered image data are plotted along the cutline. The bottom right compares the FEM predicted temperature and filtered thermal image along a cutline through the FEM mesh.

shows the optimizer evolution of the material heterogeneity inversion for the thermal conductivity fields and blood perfusion fields. The initial guess for the material coefficients was assumed homogeneous, and the optimizer determined a sufficient field variation of the parameters that allows the Pennes model to accurately predict the temperature field seen in the thermal images. The values of the thermal conductivity field found by the optimizer are above the physical range seen in the *CRC Handbook of Mechanical Engineering* [31]. Future work includes correlating the computed inverted parameter field with the physical tissue.

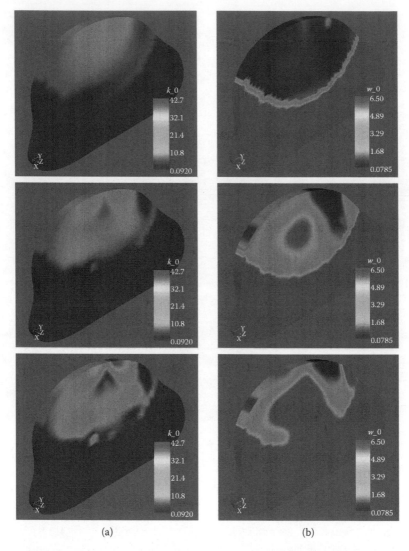

(a)   (b)

Figure 9.9 Thermal-imaging data were used to drive an inverse problem to recover the biological tissue heterogeneity. ≈ 4100 model parameters representing the blood perfusion and thermal conductivity fields of the biological domain were optimized to allow the Pennes model to accurately predict the temperature field seen in the thermal images. As shown at the top of the figure, the parameter fields are initially assumed homogeneous. The evolution of the optimizer's inversion for the parameter fields is shown. (a) and (b) show the evolution of the thermal conductivity and blood perfusion fields, respectively.

9.5 CELL STUDIES

Damage and heat shock protein (HSP) optimizations are based on *in vitro* cellular data. Due to the role of HSPs in posttraumatic cell survival, one needs to have proper data determining the changes in HSP expression induced by the thermal stress. HSPs are molecular chaperones responsible for protecting the cells from damage. They are normally present in both prokaryotic and eukaryotic cells in minimal quantities. Various stressful stimuli can lead to increased expression of HSPs, including, but not limited to, high or low temperature, acidosis, ischemia, hypoxia, and ultraviolet (UV) irradiation [33]. Denatured proteins, if bigger than 100 to 150 amino acids in size, cannot refold properly by themselves [34]. HSPs prevent improper aggregation of these damaged proteins and direct newly formed proteins for final packaging, degradation, and repair [35]. As mentioned earlier, exposure to sublethal stimuli would cause an increase in production of HSPs, which, in turn, would cause increased tolerance to further stress. This is a very important protection mechanism for cells [36]. Some members of the HSP family are inhibitors of apoptosis proteins. Therefore, an increase in expression of HSPs would lead to blocking pathways that lead to apoptosis. The prevention of apoptotic death of the damaged cells would increase cell survival rates [37]. In cancer treatment, it lowers the response of the cells to chemotherapeutic agents.

The goal of thermal therapy is to ablate malignant tissue while preserving normal tissue as much as possible. To make a correct estimate of damage in malignant and normal areas, one needs to have prior knowledge of the mechanics and kinetics of thermal damage in the appropriate cell types. This information can be acquired by conducting controlled stress experiments in model systems such as cell cultures. The subject culture can be obtained either from established cell lines or by establishing primary cultures from harvested tissue. These procedures are widely reported in the literature [38–40].

9.5.1 Heating Protocol

The heating experiments were done on primary cultures of prostate stromal cells isolated from canine prostate tissue following the protocol described by Srinivasan et al. [38]. Controlled thermal stress is applied to culture systems by immersing them in a heated liquid bath. The heating medium used in our experiments contains no L-glutamine, to prevent cellular damage due to production of ammonia secondary to breakage of L-glutamine. The heating is done in a water bath for only one culture flask at each given time to prevent a drop in temperature of the bath. In order to induce a sharp increase in the temperature of the samples, the heating medium is preheated to the same temperature as is desired for the heating experiment itself. The flasks are washed with 37°C PBS. Preheated medium is added to the flasks, which are submerged in a heated water bath for the desired duration of heating. Then the heating medium is replaced with 37°C growth medium, and the flasks are returned to the 37°C incubator. Samples are checked for viability 60 hours postheating. At the end of this time, samples are trypsinized, and the cells are removed and washed in cold PBS twice and resuspended in binding buffer at 10 g/ml concentration. Annexin V and PI (BD Biosciences) (Franklin Lakes, New Jersey) are used to detect apoptotic and dead cells, respectively. The samples are analyzed using BD FACSCalibur (BD Biosciences). This flow cytometer is used primarily for cell analysis with simultaneous acquisition of forward scatter, side scatter, and fluorescence. On the forward-scattering (FS) versus side-scattering (SC) plot, the events with a low level of SC or FS are considered to be dead cells or debris from disintegrated cells. These events are later distinguished from each other using the fluorescence detected by FL 1 (filter 1 detects fluorescence emission centered at 530 nm; this would be able to detect the emission from Annexin V) and FL 2 (detects emission centered at 585 nm; this corresponds to emission from PI). On a dot plot of FL 1 versus FL 2, events negative for both are considered to be debris, whereas the events positive for either Annexin V or PI are counted as dead cells. Furthermore, on the FS versus SC plot, the events with a high level of SC and FS are considered to be live or dead cells, which later on would be distinguished from each other using the dot plot of FL 1 versus FL 2; events negative for both are considered to be live cells, and the events positive for either Annexin V or PI (or both) would be considered dead cells. Data analysis is performed using FlowJo 8.0 (BD Biosciences). HSP expression data presented here are based on the work done by Rylander et al. [35]. To determine the relationship between HSP expression and thermal stress, after heating the cell flasks by the method described above, cells lysed were in a buffer solution 16 to 18 hours postheating, and then the supernatant was analyzed using Western blotting. Later, a spectrophotometer at 595 nm (Beckman DU 530) (Beckman Coulter, Fullerton, California) and protein dye assay (Bio-Rad 500-0002) (Bio-Rad, Hercules, California) were used, and the relative concentration of HSP to actin in the cell lysate was measured.

9.5.2 Cell Study Results

Figure 9.10 shows the viability data for heating done at 45°C, 50°C, and 56°C for time durations of 5, 10, and 15 minutes. Cell viability is defined as the percentage of live cells in the cell culture flasks 60 hours after thermal stress. As expected, the higher the temperature and duration of heating, the lower the cell viability.

Rylander measurements for HSP70 (i.e., 70 kilodalton heat shock proteins) expression at different durations and temperatures are shown in Figure 9.11 [35]. As can be seen, the HSP expression increases for each given temperature up to a certain duration of heating, after which it drops sharply. Figure 9.12 shows Rylander et al. [35] cell viability data for different time durations and temperatures.

The goal of our experiments is to find the ideal duration and temperature that maximizes trauma in tumoral cells and minimizes damage in the normal region.

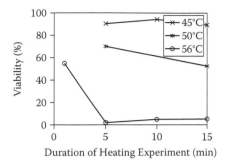

Figure 9.10   Cell viability as a function of heating time and temperature evaluated at 60 hours post-heating.

To study the effect of cell viability and HSP expression, Rylander et al. [41] have introduced a figure of merit (FOM), which is defined as follows:

$$FOM = \left(\frac{cell\ viability}{10}\right)\left(\frac{\left(\frac{HSP70}{actin}\right)_{sample}}{\left(\frac{HSP70}{actin}\right)_{control}}\right)$$

The FOM has its maximum value at 50°C and duration of 1 minute. This analysis provides an initial guide in the design of therapeutic protocols.

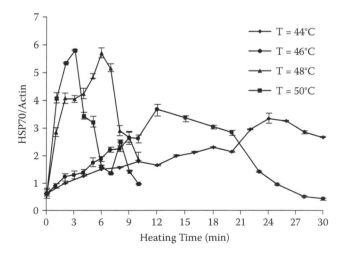

Figure 9.11   Normalized heat shock protein (HSP) and actin expression as a function of heating time and temperature evaluated at 16 to 18 hours postheating. The average error is ±0.13 mg/ml, $n = 3$ [35].

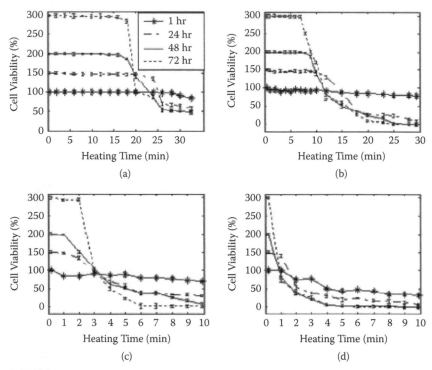

Figure 9.12   Cell viability following variable duration of thermal stress for temperatures of (a) 44°C, (b) 46°C, (c) 48°C, and (d) 50°C measured at t = 1, 24, 48, and 72 hours postheating. The average error is ±3%, and the number of samples = 3 [35].

9.6   CELL DAMAGE MODELS

Thermal damage processes in cells and tissues are usually quantified by kinetic models based on a first-order rate process to characterize pathological transformation to specific states by observable alterations such as coagulation or desiccation. While the Arrhenius law is commonly used to describe the rate of chemical reactions involving temperature [42,43], Henriques and Moritz were the first to propose a model of this form in 1947 to quantify thermal damage specifically for tissue [44,45]. The thermal injury associated with exposing cells to hyperthermia conditions is generally predicted using the Arrhenius law based on the assumption that the rate of cell damage is proportional to $\exp(-E_a/R_u)$, where E_a is the activation energy (or the heat of activation), R the universal gas constant, and u the temperature in degrees Kelvin [46], with a few exceptions (e.g., Roti Roti and Henle [47]).

Although thermal damage models based on the Arrhenius law are widely used, the model possesses some inherent limitations, which include its inability to predict cellular injury over a wide hyperthermic temperature range and throughout the entire heating process, its sensitivity to small changes in parameters, and the ambiguity in interpretation of model parameters characterizing the cell damage formulation. In fact, several basic questions can be raised. How do cells essentially respond to temperature? Why does the rate of thermal damage follow the first-order unimolecular chemical reaction? What is the biophysical interpretation of both parameters in the Arrhenius model? Some answers related to these questions may be found in Lepock [48], Pearce and Thomsen [49], and Philibert [50]. However, further investigation is warranted due to the complexity of these questions.

Experimental data [41,51] suggest that there are at least two transitional temperatures ("break points") at 43°C and 52°C in the temperature range from 39°C to 60°C. A different injury mechanism may be initiated at each of these temperatures. At the cellular or subcellular level, the sigmoidal phenomenon observed in the cell viability profile could be related to the melting of oligomers that occurs in DNA molecules [52], or due to the damage of the lipid bilayer of cell membranes. Although phospholipids are the major components of the bilayer, it is worth noting that other lipid components in the bilayer such as dodecanoic and tetradecanoic acids have melting temperatures of 44.2°C and 53.9°C, respectively [53]. Cells initially exhibit resistance to the thermal damage due to the induction of heat shock proteins by sublethal temperatures as autoregulatory mechanisms. With increases in temperature or extended exposure times, the heat shock proteins that participate in the rescue process are denatured and rendered nonfunctional. This phenomenon can be observed in measured cell viability data in which the cell damage rate is initially slow, followed by an injury rate dominated by exponential decay [41].

Usually, the Arrhenius model permits fitting of data solely within the exponential decay region of the curve where cell viability is plummeting due to extensive injury, but is not able to accommodate the "shoulder region" characterized by sustained high cell viability encountered in the initial stages of the heating process for lower temperatures. Cells exposed to $u < 54$°C experience high viability initially for a range of exposure times until a threshold thermal dose is achieved to initiate cellular injury and a corresponding decline in cell viability. In general, although it depends on cell line and temperature, cell viability profiles in this temperature range initially exhibit a shoulder region where cell viability remains high until a threshold lethal thermal dose is achieved to initiate rapid declines in cell viability. The Arrhenius model is capable of predicting the complete injury phenomena for cells exposed to $u < 54$°C, where thermal dose is substantial at short exposure times, causing rapid declines in cell viability immediately following thermal stress. However, three sets of injury parameters are required for different temperature regimes ($u < 43$°C, 43°C $\leq u \leq 54$°C, and $u > 54$°C) in order to permit accurate fitting of cellular injury data for the entire range of temperatures.

An additional problem with the thermal damage model based on the Arrhenius model is its numerical sensitivity of model parameters to small changes in measurement data. As a result, therapeutic outcomes could be compromised if the treatment planning is based on such models.

To overcome the weaknesses of the Arrhenius model described above, other types of models have been proposed for thermal damage of cells and tissues. These include models that are derived using statistical methods [54], an enzyme denaturation approach [55], and widely used kinetic theory [51,56–59]. For a discussion of various cell damage models, He and Bischof [46] provide a comprehensive review. Most of these models, however, relate thermal damage to the rate process in such a way that the rate of change with respect to temperature and time are decoupled.

In a recent study [60], a two-state model of *in vitro* cell death due to thermal insult is derived based on simple arguments motivated by classical statistical thermodynamics. This model characterizes two populations of viable (live) and damaged (dead) cells, which leads to the damaged cell population of the form $C(u,t) = \exp(-\Delta G/ku)/(1 + \exp(-\Delta G/ku))$ or, alternatively, $C(u, t) = \frac{1}{2} + \frac{1}{2}\tanh(-\Delta G/2ku)$, where k is the Boltzmann constant and ΔG is interpreted as a change in a functional analogous to a classical Gibbs free energy, depending on both temperature and exposure time. We postulate that ΔG *is a linear function of time and is inversely proportional to temperature*. To determine ΔG, we use cell viability data for human prostate cancerous (PC3) and normal (RWPE-1) cells to calibrate the two-state cell damage model derived in this study. Excellent agreement between experimental data and the derived model is obtained through least-squares regression. As compared to the Arrhenius model, the two-state model captures the damage process more accurately over a wide hyperthermic temperature range, including the beginning phase (the shoulder region) when cells are first exposed to the heat shock. Also, the model successfully characterizes the sigmoidal phenomenon of the cell response.

9.6.1 Two-State Cell Damage Model

Consider a cell population with two distinct states (i.e., a cell is either live or dead). We apply classical arguments of statistical thermodynamics to derive a two-state model for cell damage under hyperthermia conditions. In an *in vitro* system of the fixed population of total n cells, we assume that there are only two species of cells in this population: dead or dying cells (including apoptotic and necrotic cells), and live cells.

Based on the standard argument of statistical thermodynamics, the following results for a two-state population can be obtained [60]:

$$C(u,t) = \frac{e^{-\Delta G/ku}}{1 + e^{-\Delta G/ku}} \quad \text{and} \quad D(u,t) = \frac{1}{1 + e^{-\Delta G/ku}} \tag{9.4}$$

or, equivalently,

$$C(u,t) = \frac{1}{2} + \frac{1}{2} \tanh(-\Delta G/2ku) \tag{9.5}$$

$$D(u,t) = 1 - C(u,t) = \frac{1}{2} - \frac{1}{2} \tanh(-\Delta G/2ku) \tag{9.6}$$

where $C(u,t)$ is the cell viability function and $D(u,t)$ is the cell damage function. Inspired by statistical thermodynamics, we let $\Delta G = \Delta H - u\Delta S$ and postulate that ΔH is a constant and ΔS a linear function in time (i.e., $\Delta S = \alpha_o t + \beta_o$). The notions of ΔG, ΔH, and ΔS are chosen to mimic changes in Gibbs free energy, enthalpy, and entropy.

Based on the results in Equation (9.4), the cell viability function $C(u,t)$ can also be defined as a solution to the following system of partial differential equations:

$$\begin{cases} \dfrac{\partial C(u,t)}{\partial t} = \alpha \cdot \dfrac{C(u,t)}{1 + e^{-\Delta G/ku}} \\[4mm] \dfrac{\partial C(u,t)}{\partial u} = \dfrac{h}{u^2} \cdot \dfrac{C(u,t)}{1 + e^{-\Delta G/ku}} \end{cases} \tag{9.7}$$

where $h = \Delta H/k$, $\alpha = \alpha_o/k$, and $\beta = \beta_o/k$ are constants. If $e^{-\Delta G/ku}$ is very small, then $1 + e^{-\Delta G/ku} \approx 1$ and Equation (9.7) can be approximated by

$$\begin{cases} \dfrac{\partial C(u,t)}{\partial t} = \alpha \cdot C(u,t) (9) \\[4mm] \dfrac{\partial C(u,t)}{\partial u} = h \cdot \dfrac{C(u,t)}{u^2} \end{cases} \tag{9.8}$$

9.6.2 Parameter Estimation

To determine ΔG, we let $\Delta G = \Delta H - u\Delta S$ and $\Delta S = \alpha_o t + \beta_o$, where u and t are temperature and time, respectively, and ΔH, α_o, and β_o are constants. Then the first equation in Equation (9.7) can be converted to

$$\frac{\Delta G}{ku} = \frac{\Delta H - u\Delta S}{ku} = \ln\left(\frac{1 - C(u,t)}{C(u,t)}\right) \tag{9.9}$$

In other words,

$$\frac{\Delta H}{ku} - \frac{\Delta S}{k} = \left(\frac{\Delta H}{k}\right)\frac{1}{u} - \left(\frac{\alpha_o}{k}\right)t - \frac{\beta_o}{k} = \ln\left(\frac{1 - C(u,t)}{C(u,t)}\right) \tag{9.10}$$

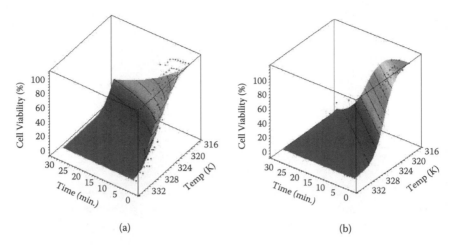

Figure 9.13   Two-dimensional projections of (a) the Arrhenius model and (b) the two-state model.

For simplicity, let $h = (\Delta H)/k$, $\alpha = \alpha_o/k$, and $\beta = \beta_o/k$. Suppose that there are $m \times n$ experimental data points for cell viability $C(u_\ell, t_\ell)$, $\ell = 1, \ldots, m \times n$ (i.e., m temperature measurement points with n exposure time for each temperature). Denote by z the function $z = \ln[(1 - C(u,t))/C(u,t)]$, then the data points (u_ℓ, t_ℓ, z_ℓ), $\ell = 1, \ldots, m \times n$ can be plotted in three-dimensional (3D) space with respect to $1/u$ and t.

At each point (u_ℓ, t_ℓ, z_ℓ), Equation (9.10) can be rewritten as

$$h\left(\frac{1}{u_\ell}\right) - \alpha t_\ell - \beta = z_\ell, \qquad \ell = 1, \ldots, m \times n \tag{9.11}$$

where parameters h, α, and β are to be determined by the standard least-squares regression using measurement data. Figure 9.13 illustrates that the transformation by introducing the z-variable converts a curved surface representing cell viability into a flat plane in 3D space. The 2D projections on the $C - 1/u$ and $C - t$ planes are also presented on Figure 9.13.

We summarize the algorithmic steps of parameter estimation for the two-state models as follows:

1. Compute $z_\ell = \ln[(1 - C(u_i, t_j))/C(u_i, t_j)], i = 1, \ldots, m; j = 1, \ldots, n; \ell = 1, \ldots, m \times n$.
2. Plot data points z_ℓ versus time t and $1/u$ on a 3D graph.
3. Use bilinear regression to find the coefficients: h, α, and β.
4. The resulting coefficients are used directly in the model (Equation 9.4).

Since the range of $C(u,t) \in [0,1]$, we need to exclude initial points $C(u_i, 0) = 1$, $i = 1, \ldots, m$, in the least-squares regression process. This will not affect the final results because the initial conditions in the original form (Equation 9.4) are automatically satisfied.

As a comparison, we also list the major steps to establish the Arrhenius model:

1. Plot data points $C(u_i, t_j)$ versus time for all $i = 1, \ldots, m; j = 1, \ldots, n$.
2. Determine time τ_i such that $C(u_i, \tau_i) = 36.8\%(\Omega = 1), i = 1, \cdots, m$.
3. Plot data points $(\ln \tau_i, 1/u_i)$ on a 2D graph, $i = 1, \ldots, m$.
4. Use linear regression to find the slope and $(\ln \tau)$-intercept.
5. The resulting slope is E_a/R, and the $(\ln \tau)$-intercept is $\ln A$.

Next, we discuss the major differences and similarities between two cell damage models.

9.6.3 Model Comparison between the Arrhenius Model and the Two-State Model

The *cell damage index* Ω is defined as usual. When the cell viability function $C(u,t)$ is normalized and $C(u,0)$ is set to one, we have $\Omega = -\ln C(u,t)$. Recall that the Arrhenius model assumes that the cell damage rate is proportional to the rate of reaction $k(u) = e^{-\frac{E_a}{Ru}}$. Thus, the cell damage index based on the Arrhenius model is

$$\Omega = \int_0^t A e^{\frac{-E_a}{Ru(\tau)}} d\tau \tag{9.12}$$

where t is the total exposure time and A is a constant that is often referred to as the frequency factor [49]. If temperature is kept constant during the entire exposure time t, then

$$\Omega = A t\, e^{\frac{-E_a}{Ru}} \tag{9.13}$$

In order to compare the Arrhenius model with the two-state model, we rewrite Equations (9.15) and (9.12) in terms of cell viability in a differential form:

$$\begin{cases} \dfrac{\partial C(u,t)}{\partial t} = -k(u) \cdot C(u,t) \\[2mm] \dfrac{\partial C(u,t)}{\partial u} = -\left(t k(u) \dfrac{E_a}{R} \right) \cdot \dfrac{C(u,t)}{u^2} \end{cases} \tag{9.14}$$

Therefore, the Arrhenius model is an approximation to the two-state model when $e^{-\Delta G/Ru} \ll 1$ with different choices of parameters.

Note that both functions $C(u,t)$ and $D(u,t)$ depend on ΔG, which is a function of temperature and time involving parameters that will be determined by experiments. In addition, it is important to recognize that the variable t is treated simply as a label for a continuous sequence of quasistatic states during heating.

9.7 SENSITIVITY STUDY

When creating a high-fidelity bioheat transfer model meant for prediction, a crucial element is capturing the heating source accurately. In the case of laser therapy, this mainly involves modeling the laser fluence in the tissue correctly. While there are two standard ways of modeling this—analytically or by using a Monte Carlo method—they both depend on three optical parameters of interest: the absorption coefficient, μ_a; the scattering coefficient, μ_s; and the anisotropic factor, g. Respectively, these give the average number of photons that are absorbed per unit length, the average number of photons scattered per unit length, and the expected value of the cosine of the scattering angle. In living tissue, each of these parameters is truly a function of space, light wavelength, and temperature. However, the way in which they are obtained experimentally usually limits them to functions of wavelength only, though the temperature can sometimes be accounted for. The model being presented in this chapter is currently implementing these parameters as functions of wavelength, and thus (since a single wavelength is used during treatment) leaving them constant throughout the entire simulation. Experiments have been done by Nau et al., however, showing that as the temperature increases, these parameters do not remain constant [61]. As the treatment being considered is meant to increase the temperature in the modeling region, it seems that this effect should most definitely be captured. However, the Pennes bioheat transfer model highly diffuses the heat in the domain. Thus, there is a question of whether the change in the heat source parameters will affect the overall heating profile enough to warrant modeling them as functions of temperature instead of constants. A sensitivity study was conducted to answer this question, the results of which are given here.

$$\Omega = \ln\left(\frac{C(u,0)}{C(u,t)}\right) \tag{9.15}$$

Laser fluence terms are generally obtained in one of two ways: analytically or via a Monte Carlo method. When capturing fluence analytically, a diffusion theory must be used, and generally many simplifying assumptions are made regarding boundary conditions and the source geometry. However, once the equation form is found, it is differentiable and easily calculated for different sets of parameters. Generally, the diffusion theory assumption, which states that the light radiance is mostly isotropic but for a small perturbation in one direction, holds when the source is not collimated and far from a boundary of two layers with very different refraction indices or optical parameters (i.e., tissue and air). In contrast, the Monte Carlo method allows for a more accurate handling of boundary conditions and source geometries.

However, once the fluence is found for one set of parameters, there is no easy way to convert it into the fluence for another set of parameters; the entire simulation must be run again. Their differences make them appropriate or necessary for different situations; analytic solutions are appropriate for interstitial diffuse lasers, while Monte Carlo solutions are necessary, for high accuracy, when modeling laser beams incident to a flat surface.

In the experiments done for this research, despite the specific interest in interstitial tumors, both topical and interstitial lasers have been used. Since different source terms should be used for these situations, the sensitivity study presented here considers both an analytic solution as well as a Monte Carlo solution. In particular, the analytic fluence term considered is

$$\Phi(\mathbf{r}) = 3P(t)\mu_{tr} \frac{e^{-\mu_{eff}\mathbf{r}}}{4\pi\mathbf{r}}$$

where P is the power, $\mu_{tr} = \mu_a + (1 \quad g)\mu_s$, and $\mu_{eff} = \sqrt{3\mu_a\mu_{tr}}$. This is derived as a solution to the transport equation as a spherically isotropic point source. The Monte Carlo fluence used follows the algorithm in Welch and van Gemert [62] incorporating a Gaussian initialization profile of a $1/r^2$ radius of 2 cm with 3 million photons for each simulation. For each of the source terms, the sensitivity was analyzed by running many simulations with each, varying one parameter while holding the others constant. Original values of parameters were $\mu_a = 2.15$ [1/cm], $\mu_s = 14.2$ [1/cm], and $g = 0.7$. The squared L^2 space–time norm of the temperature field was then calculated as the quantity of interest and plotted. The squared L^2 norm used is given here:

$$\int_0^T \int_\Omega |u(\mathbf{x},t)|^2 \, d\mathbf{x} \, dt$$

The different simulations were all run on a mesh representing a tumor on the leg of a mouse, a setup that is admittedly more appropriate for the Monte Carlo approach due to the boundary interface. Use of the isotropic source term, however, is not inappropriate for the purposes of this study since the general changes in the model's predictions are what are of interest. Figures 9.14 and 9.15 each show representative snapshots of the temperature profiles. The snapshots are each at time step 40, equivalent to 3 minutes and 20 seconds of heating.

The general effects of changing these parameters, which can be seen in Figures 9.14 and 9.15, are the same for both source terms. Higher values of μ_a and μ_s are associated with an increase in heating, whereas an increase in g is associated with a decrease in heating. However, the relative increase or decrease

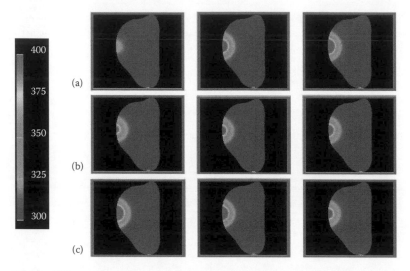

Figure 9.14 All images here show the temperature field generated with a Monte Carlo source term at time step 40, equivalent to 3 minutes and 20 seconds of heating. Line (a) contains images from simulations using values of μ_a = 0.44, 3.14, and 5.0 [1/cm], respectively. Line (b) contains images from simulations using values of μ_s = 1, 11, and 25 [1/cm], respectively. And line (c) contains images from simulations using values of g = 0, 0.5, and 0.99, respectively. Original values of the parameters are μ_a = 2.15, μ_s = 14.2, and g = 0.7. Temperature is given in degrees of Kelvin.

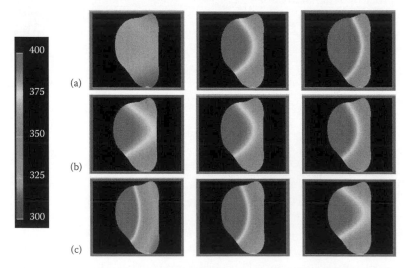

Figure 9.15 All images here show the temperature field generated with an isotropic source term at time step 40, equivalent to 3 minutes and 20 seconds of heating. Line (a) contains images from simulations using values of μ_a = 0.44, 3.14, and 5.0 [1/cm], respectively. Line (b) contains images from simulations using values of μ_s = 1, 11, and 25 [1/cm], respectively. And line (c) contains images from simulations using values of g = 0, 0.5, and 0.99, respectively. Original values of the parameters are μ_a = 2.15, μ_s = 14.2, and g = 0.7. Temperature is given in degrees Kelvin.

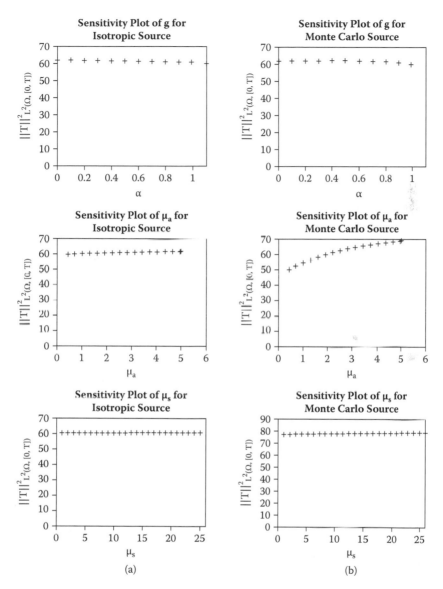

Figure 9.16   (a) Sensitivity plots for the isotropic source term simulations. (b) Sensitivity plots for the Monte Carlo source term simulations.

in temperature does appear to be different for the two source terms. The graphs in Figure 9.16a,b show the values of the quantity of interest, the L^2 norm, generated in each simulation and graphically show the general effects mentioned above. As a remark, the Monte Carlo simulations have produced more heating than the analogous isotropic simulations because the source is assumed to be 2 cm in diameter, whereas the isotropic source was merely at a point.

Table 9.1   Linear Regression Slopes of Plotted Sensitivities

	Monte Carlo	Isotropic
μ_a	4.95	0.398
μ_s	.071	0.017
g	−1.55	−1.09

These graphs indicate that the model's sensitivity to the parameters, for both source terms, is for the most part linear. For the range tested here, the linear regression slopes are given in Table 9.1. It can be argued from these results that in the isotropic case, both the parameters μ_s and μ_a may be treated as functions of wavelength only, but that g may need to be a function of temperature as well. However, in the Monte Carlo case, while μ_s can still remain just a function of wavelength, both μ_a and g should be dependent on temperature.

9.8 CONCLUSIONS AND FUTURE DIRECTIONS

Every aspect of the control system is operational and has been tested on a 1% agar phantom material. This testing represents a project milestone. The phantom material has provided an animal-free method of testing and debugging the entire control system. The phantom is meshed, it is registered, the computers in Austin control the heating, thermal images of the heating are acquired and sent to HP3d, HP3d calibrates and optimizes the model parameters, and, finally, a visualization of the entire process is provided in Houston. Figure 9.17a shows a particular time instance of heating of the phantom with an external collimated source. For

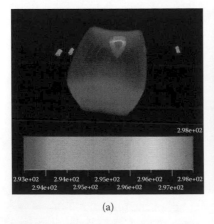
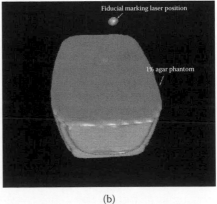

(a)   (b)

Figure 9.17   (a) 3D volume rendering of MRI and MRTI images at a selected time instance. The phantom geometry is shown in grayscale, and the thermal image is overlaid. The color bar illustrates the temperature range. (b) Isosurface visualization of MRI images of the geometry of the phantom material. The fiducial used to mark the laser probe for the FEM calculations are also shown.

this particular test, the thermal images were mainly used to periodically update the initial condition of the FEM computation in real time. After each update, the calculations proceeded to predict all the way to the end of the treatment. On current parallel computing architectures, the prediction capabilities seen are that 10 seconds of computation time at high-performance computing facilities provide ≈40 to 50 seconds' worth of prediction. Figure 9.17b conveys the geometry of the phantom, as seen by an isosurface visualization of the MRI images. A fiducial marking the external laser position is shown above the phantom. This is used to obtain the coordinates of the laser tip in the FEM calculations.

The next milestone on the horizon for this multidisciplinary effort is to perform an *in vivo* trial of the control system. Current results indicate that it is indeed feasible to accurately control the bioheat transfer through real-time imaging and computational prediction. The culmination of adaptive hp finite element technology implemented on parallel computing architectures, modern data transfer and visualization infrastructure, thermal-imaging modalities, and cellular damage mechanisms as cancer treatment tools will provide a very powerful methodology for planning and optimizing thermal therapy delivery for cancer treatments.

ACKNOWLEDGMENTS

The research in this chapter was supported in part by the National Science Foundation (NSF) under grants CNS-0540033 and IIS-0325550, and National Institutes of Health (NIH) Contracts P20RR0206475 and GM074258. The authors also acknowledge the important support of dynamic data-driven application system (DDDAS) research by Dr. Frederica Darema of NSF.

REFERENCES

1. K. Shinohara, Thermal ablation of prostate diseases: advantages and limitations, *Int. J. Hyperthermia*, vol. 20, no. 7, pp. 679–697, 2004.
2. R. Salomir et al., Hyperthermia by MR-guided focused ultrasound: accurate temperature control based on fast MRI and a physical model of local energy deposition and heat conduction, *Magn. Reson. Med.*, vol. 43, no. 3, pp. 342–347, 2000.
3. F. C. Vimeux et al., Real-time control of focused ultrasound heating based on rapid MR thermometry, *Invest. Radiol.*, vol. 34, no. 3, pp. 190–193, 1999.
4. J. T. Oden, K. R. Diller, C. Bajaj, J. C. Browne, J. Hazle, I. Babuška, J. Bass, L. Demkowicz, Y. Feng, D. Fuentes, S. Prudhomme, M. N. Rylander, R. J. Stafford, and Y. Zhang, Dynamic data-driven finite element models for laser treatment of prostate cancer, *Num. Meth. PDE*, vol. 23, no. 4, pp. 904–922, 2007.
5. J. T. Oden and S. Prudhomme, Goal-oriented error estimation and adaptivity for the finite element method, *Comput. Math. App.*, vol. 41, nos. 5–6, pp. 735–756, 2001.
6. S. Balay, W. D. Gropp, L. C. McInnes, and B. F. Smith, *Petsc users manual*, Tech. Rep. ANL-95/11—Revision 2.1.5, Argonne, Ill.: Argonne National Laboratory, 2003.
7. S. J. Benson, L. C. McInnes, J. Moré, and J. Sarich, *TAO user manual* (revision 1.8), Tech. Rep. ANL/MCS-TM-242, Argonne, Ill.: Argonne National Laboratory, Mathematics and Computer Science Division, 2005. http://www.mcs.anl.gov/tao.

8. A. V. S. Inc., AVS user's guide, May 1992, National Congress on Computational Mechanics, San Francisco, conference presentation.

9. Z. Yu and C. Bajaj, A fast and adaptive algorithm for image contrast enhancement, in *Proceedings of IEEE International Conference on Image Processing*, pp. 1001–1004, 2004.

10. C. Bajaj, Q. Wu, and G. Xu, Level set based volumetric anisotropic diffusion, in *ICES Technical Report 301*, Austin: University of Texas at Austin, 2003.

11. J. Sethian, A marching level set method for monotonically advancing fronts, *Proc. Natl. Acad. Sci.*, vol. 93, no. 4, pp. 1591–1595, 1996.

12. Z. Yu and C. Bajaj, Normalized gradient vector diffusion and image segmentation, in *Proceedings of European Conference on Computer Vision*, pp. 517–530, 2002.

13. Z. Yu and C. Bajaj, A segmentation-free approach for skeletonization of gray-scale images via anisotropic vector diffusion, in *IEEE International Conference on Computer Vision and Pattern Recognition (CVPR'04)*, vol. 1, pp. 415–420, 2004.

14. S. M. Park, G. W. Gladish, and C. L. Bajaj, Artery-vein separation from thoracic CTA scans, *IEEE Trans. Med. Imag.*, pp. 23–30, 2006.

15. R. Araiza, M. Averill, G. Keller, S. Starks, and C. Bajaj, 3D image registration using Fast Fourier Transform, with potential applications to geoinformatics and bioinformatics, in *Proceedings of the International Conference on Information Processing and Management of Uncertainty in Knowledge-Based Systems IPMU06*, pp. 817–824, 2006.

16. W. Lorensen and H. Cline, Marching cubes: a high resolution 3D surface construction algorithm, in *Siggraph*, pp. 163–169, 1987.

17. T. Ju, F. Losasso, S. Schaefer, and J. Warren, Dual contouring of hermite data, in *Siggraph 2002, Computer Graphics Proceedings*, pp. 339–346, Reading, Mass.: ACM Press/ACM SIGGRAPH/Addison Wesley Longman, 2002.

18. T. K. Dey and S. Goswami, Tight cocone: a water-tight surface reconstructor, in *Proceedings of the 8th ACM Symposium on Solid Modeling and Applications*, pp. 127–134, 2003.

19. T. K. Dey and S. Goswami, Provable surface reconstruction from noisy samples, in *Proceedings of the 20th ACM-SIAM Symposium on Computational Geometry*, pp. 330–339, 2004.

20. C. Bajaj, G. Xu, and X. Zhang, Smooth surface constructions via a higher-order level-set method, in *Computer-Aided Design and Computer Graphics, 2007 10th IEEE International Conference*, Beijing, China, October 15–18, 2007.

21. T. K. Dey, J. Giesen, S. Goswami, and W. Zhao, Shape dimension and approximation from samples, *Discr. Comput. Geom.*, vol. 29, pp. 419–434, 2003.

22. T. K. Dey, J. Giesen, and S. Goswami, Shape segmentation and matching with flow discretization, in *Proceedings of the Workshop Algorithms Data Structures (WADS 03)* (F. Dehne, J.-R. Sack, and M. Smid, eds.), LNCS 2748, Berlin, Germany, pp. 25–36, 2003.

23. C. Bajaj, A. Gillette, and S. Goswami, Topology-based selection and curation of level sets, in *Topology-in-Visualization, Topo-in-Vis'07* (A. Wiebel, H. Hege, K. Polthier, and G. Scheuermann, eds.), pp. 45–58, Leipzig, Germany: Springer Verlag, 2009.

24. Y. Zhang, Y. Bazilevs, S. Goswami, C. L. Bajaj, and T. J. R. Hughes, Patient-specific vascular NURBS modeling for isogeometric analysis of blood flow, *Comp. Met. Appl. Mech. Eng. (CMAME)*, vol. 196, nos. 29–30, pp. 2943–2959, 2007.

25. S. Goswami, T. K. Dey, and C. L. Bajaj, Identifying flat and tubular regions of a shape by unstable manifolds, in *Proceedings of the 11th Symposium on Solid and Physical Modeling*, pp. 27–37, 2006.

26. Y. Zhang and C. Bajaj, Adaptive and quality quadrilateral/hexahedral meshing from volumetric data, *Computer Methods in Applied Mechanics and Engineering (CMAME)*, vol. 195, nos. 9–12, pp. 942–960, 2006.

27. C. Bajaj, J. Chen, and G. Xu, Modeling with cubic A-patches, *ACM Transactions on Graphics*, vol. 14, no. 2, pp. 103–133, 1995.

28. Y. Zhang, C. Bajaj, and B.-S. Sohn, 3D finite element meshing from imaging data, *CMAME Unstruct. Mesh Gen.*, vol. 194, nos. 48–49, pp. 5083–5106, 2005.

29. C. Bajaj and G. Xu, Smooth shell construction with mixed prism fat surfaces, *Geom. Model. Comp. Suppl.*, vol. 14, pp. 19–35, 2001.

30. Y. Feng, D. Fuentes, A. Hawkins, J. Bass, M. N. Rylander, A. Elliott, A. Shetty, R. J. Stafford, and J. T. Oden, Nanoshell-mediated laser surgery simulation for prostate cancer treatment, *Eng. Comp.*, DOI: 10.1007/s00366-008-0109-y, 2007.

31. K. R. Diller, J. W. Valvano, and J. A. Pearce, Bioheat transfer, in *The CRC Handbook of Mechanical Engineering*, 2nd ed. (F. Kreith and Y. Goswami, eds.), pp. 4-278–4-357, Boca Raton, Fla.: CRC Press, 2005.

32. Y. Zhang, Y. Bazilevs, S. Goswami, C. Bajaj, and T. J. R. Hughes, Patient-specific vascular NURBS modeling for isogeometric analysis of blood flow, *CMAME*, vol. 196, pp. 2943–2959, 2007.

33. K. R. Diller, Stress protein expression kinetics, *Annu. Rev. Biomed. Eng.*, vol. 8, pp. 403–424, 2006.

34. A. Horwich, Protein aggregation in disease: a role for folding intermediates forming specific multimeric interactions, *J. Clin. Invest.*, vol. 110, pp. 1221–1232, 2002.

35. M. N. Rylander, S. Wang, S. Aggarwal, and K. R. Diller, Correlation of HSP 70 expression and cell viability following thermal stimulation of bovine aortic endothelial cells, *J. Biomech. Eng.*, vol. 127, pp. 751–757, 2005.

36. A. Peper, C. A. Grimbergen, J. A. Spaan, J. E. Souren, and R. van Wijk, A mathematical model of the HSP70 regulation in the cell, *Int. J. Hypertherm.*, vol. 14, pp. 97–124, 1998.

37. R. A. Coss, Inhibiting induction of heat shock proteins as a strategy to enhance cancer therapy, *Int. J. Hypertherm.*, vol. 21, pp. 695–701, 2005.

38. D. Srinivasan, L. R Burbach, D. V. Daniels, A. P. D. W. Ford, and A. Bhattacharya, Pharmacological characterization of canine bradykinin receptors in prostatic culture and in isolated prostate, *Brit. J. Pharmacol.*, vol. 142, pp. 297–304, 2004.

39. P. D. Walden, M. Ittmann, M. E. Monaco, and H. Lepor, Endothelin-1 production and agonist activities in cultured prostate-derived cells: Implications for regulation of endothelin bioactivity and bioavailability in prostatic hyperplasia, *Prostate*, vol. 34, pp. 241–250, 1998.

40. D. M. Peehl, G. K. Leung, and S. T. Wong, Keratin expression: a measure of phenotypic modulation of human prostatic epithelial cells by growth inhibitory factors, *Cell Tiss. Res.*, vol. 277, pp. 11–18, 1994.

41. M. N. Rylander, K. R. Diller, S. Wang, and S. Aggarwal, Correlation of HSP70 expression and cell viability following thermal stimulation of bovine aortic endothelial cells, *J. Biomech. Eng.*, vol. 127, pp. 751–757, 2005.

42. S. Arrhenius, On the reaction velocity of the inversion of cane sugar by acids, *Zeitschrift für Physikalische Chemie*, vol. 4, p. 226, 1889.

43. K. J. Laidler, *Chemical Kinetics*, 3rd ed., New York: Harper & Row, 1987.

44. F. Henriques and A. Moritz, Studies of thermal injury. i. the conduction of heat to and through skin and the temperatures attained therein. A theoretical and an experimental investigation, *Am. J. Pathol.*, vol. 23, no. 53, pp. 1–549, 1947.

45. A. Moritz and F. Henriques, Studies of thermal injury. ii. The relative importance of time and surface temperature in the causation of cutaneous burns, *Am. J. Pathol.*, vol. 23, pp. 695–720, 1947.

46. X. He and J. Bischof, Quantification of temperature and injury response in thermal therapy and cryosurgery, *J. Biomech. Eng.*, vol. 31, pp. 355–422, 2003.

47. J. Roti Roti and K. Henle, Comparison of two mathematical models for describing heat-induced cell killing, *Radiat. Res.*, vol. 81, no. 3, pp. 374–383, 1980.

48. J. Lepock, How do cells respond to their thermal environment? *Int. J. Hypertherm.*, vol. 21, no. 8, pp. 681–687, 2005.

49. J. Pearce and S. Thomsen, Rate process analysis of thermal damage, in *Optical-Thermal Response of Laser-Irradiated Tissue* (A. J. Welch and M. J. C. van Gemert, eds.), pp. 561–606, New York: Plenum, 1995.

50. J. Philibert, Some thoughts and/or questions about activation energy and pre-exponential factor, *Def. Diffus. Forum*, vol. 249, pp. 61–72, 2006.

51. W. C. Dewey, L. Hopwood, S. Sapareto, and L. Gerweck, Cellular responses to combinations of hyperthermia and radiation, *Radiat. Biol.*, vol. 123, no. 2, pp. 463–474, 1977.

52. J. Bayer, J. Rädler, and R. Blossey, Chains, dimers, and sandwiches: melting behavior of DNA nanoassemblies, *Nano Lett.*, vol. 5, no. 3, pp. 497–501, 2005.

53. D. Nelson and M. Cox, *Lehninger Principles of Biochemistry*, 4th ed., New York: Worth, 2004.

54. H. Jung, A generalized concept for cell killing by heat, *Radiat. Res.*, vol. 106, no. 1, pp. 56–72, 1986.

55. W. Tsang, V. Bedanov, and M. Zachariah, Master equation analysis of thermal activation reactions: energy-transfer constraints on falloff behavior in the decomposition of reactive intermediates with low thresholds, *J. Phys. Chem.*, vol. 100, pp. 4011–4018, 1996.

56. K. Dill, Theory for the folding and stability of globular proteins, *Biochemistry*, vol. 24, no. 6, pp. 1501–1509, 1985.

57. X. He and J. Bischof, The kinetics of thermal injury in human renal carcinoma cells, *Annal. Biomed. Eng.*, vol. 33, no. 4, pp. 502–510, 2005.

58. C. Merlo, K. Dill, and T. Weikl, Φ values in protein-folding kinetics have energetic and structural components, *Proc. Natl. Acad. Sci.*, vol. 102, no. 29, pp. 10171–10175, 2005.

59. T. Weikl, M. Palassini, and K. Dill, Cooperativity in two-state protein folding kinetics, *Prot. Sci.*, vol. 13, no. 3, pp. 822–829, 2004.

60. Y. Feng, J. Oden, and M. Rylander, A statistical thermodynamics based cell damage models and its validation *in vitro*, *J. Biomech. Eng.*, vol. 130, no. 041016, pp. 1–10, 2008.

61. W. H. Nau, R. J. Roselli, and D. F. Milam, Measurement of thermal effects on the optical properties of prostate tissue at wavelengths of 1,064 and 633 nm, *Laser Surg. Med.*, vol. 24, no. 1, pp. 38–47, 1999.

62. A. J. Welch and M. J. C. van Gemert, *Optical-Thermal Response of Laser-Irradiated Tissue*, New York: Plenum, 1995.

10

A Mathematical Model to Predict Tissue Temperatures and Necrosis during Microwave Thermal Ablation of the Prostate

S. Ramadhyani, J. P. Abraham, and E. M. Sparrow

CONTENTS

10.1   INTRODUCTION

Benign prostatic hyperplasia (BPH) is a condition commonly encountered in elderly men. It involves proliferative (but benign) growth of tissue in the prostate leading to a complex of bothersome urinary symptoms. Transurethral ablation of prostatic tissue by microwave radiation is now an accepted modality for the treatment of BPH and is offered as an alternative to a lifelong regimen of drugs or invasive surgery [1]. The procedure involves heating prostatic tissue by means of a microwave delivery catheter inserted into the prostate via the urethra. Figure 10.1 shows a cross-sectional view of the Targis microwave catheter

345

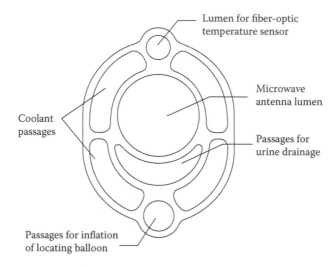

Figure 10.1   Cross section of the Targis microwave delivery catheter (Urologix, Inc.). The catheter includes a fiber-optic sensor to measure the temperature of the surface of the urethra.

(Urologix, Inc., Minneapolis, Minnesota), one of several commonly used types. The catheter contains a microwave antenna surrounded by coolant channels, the latter serving to reduce the temperature elevation of the urethral surface. The catheter is positioned in the prostatic urethra, as shown in Figure 10.2. A balloon

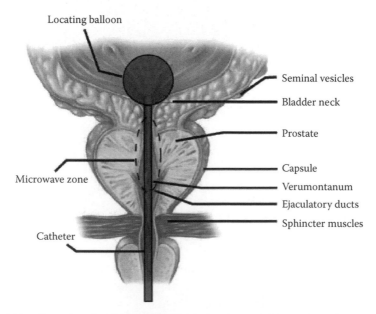

Figure 10.2   Illustration of catheter and locating balloon inserted through the urethra and into the bladder cavity.

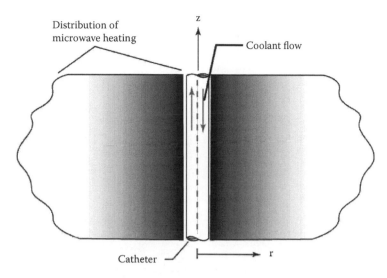

Figure 10.3   Diagram of the cooling catheter passing through prostate tissue that is heated by microwave radiation.

at the tip of the catheter is inflated once the catheter is passed into the prostate. The catheter is then gently retracted to seat the balloon at the bladder neck, thus accurately locating the microwave antenna within the prostate. The flow of coolant through the catheter as well as the distribution of microwave heating within the prostate are depicted pictorially in Figure 10.3.

Thermal energy generation occurs volumetrically within the prostate, resulting in temperature elevation to between 50°C and 85°C within the gland, while urethral surface temperatures typically remain under 40°C due to the circulating cooling water. Since the urethra is not heated significantly, the procedure may be performed with oral analgesia and topical anesthetic with little discomfort to the patient [2]. The high intraprostatic temperatures achieved during therapy result in necrosis of a sizeable volume of prostatic tissue. Over a period of weeks, much of this necrosed tissue is removed by the body's natural processes, while the remainder resolves into a compact fibrotic mass. The concomitant reduction in prostate volume leads to alleviation of the mechanical obstruction of the urethra. In addition, destruction of nerve endings in the zone of necrosis leads to a resolution of the irritative symptoms of BPH (i.e., frequency, urgency, and nocturia).

A satisfactory response to the treatment hinges on accurate control of the microwave power, the coolant temperature, and the duration of the therapy. The objective is to achieve a sizeable zone of tissue necrosis while maintaining the integrity of the urethra, the prostate capsule, and the adjoining section of the bowel. Intrinsic factors affecting the temperature field within the gland include the size of the gland, the magnitude of the blood perfusion and its distribution within the gland, and the composition of the tissue itself. While the size of the gland can be measured with good accuracy by noninvasive ultrasound imaging, the blood perfusion and the tissue composition are not easily determinable. Nevertheless, it is important to be able to predict the spatial and time-wise variations of temperature within the gland in order to control the therapy parameters and achieve the desired volume of tissue necrosis.

This chapter reports on progress in developing a mathematical model capable of predicting intraprostatic temperatures and necrosis zones during microwave treatment for BPH. The details of the model are presented, and comparisons between model predictions and measurements in human patients are provided.

10.1.1   Previous Modeling Efforts

Significant strides have been made by previous investigators in developing mathematical models for prediction of the temperature distribution in the prostate during microwave therapy. Xu and coworkers presented a one-dimensional, transient, finite difference model of the prostate and compared its predictions to temperature fields measured in a nonperfused, tissue-equivalent phantom gel as well as temperatures measured in several canine prostates [3]. A Urologix Targis catheter was employed for microwave delivery. The Pennes bioheat equation was used in cylindrical coordinates to predict the radial temperature distributions in the cross sections of the canine prostates. Two different, temporally constant values of perfusion were used in the calculation region. The periurethral region was assumed to be much more highly perfused than the parenchymal region of the gland. While qualitative justification for the use of two different values of perfusion was provided on the basis of the vasculature of the canine prostate [4], the perfusion values themselves were inferred from fitting the model predictions to one of the data sets. In general, the measured and predicted temperatures in the phantom gel were in very good agreement. The level of agreement of model predictions with the canine prostate data cannot be assessed from the reported results, but we have observed that better agreement with the data could be obtained by allowing the perfusion to increase in response to increases in local tissue temperature.

A two-dimensional, transient, finite difference model was developed and presented by Yuan et al. [5]. This model, also formulated in cylindrical coordinates, was capable of predicting the variations of temperature both radially and circumferentially in the cross section of the prostate. The model predictions were compared to experimental data obtained in a nonperfused, tissue-equivalent phantom heated with a Urologix T3 catheter. The model was used to predict temperature fields with various assumed distributions of the perfusion, but no comparisons were made against data measured *in vivo*. In a subsequent clinical study, the predictions of this model were compared to intraprostatic tissue temperatures measured in 22 human subjects. The measured steady-state temperature distributions were found to be in qualitative agreement with those predicted by the model [6]. However, comparisons of the measured and predicted transient temperature variations during the initial part of the therapy were not provided. Thus, the ability of the model to predict the temperature rise in the prostate during the initial heating period was not tested.

Bolmsjo and coworkers have presented a qualitative description of a mathematical model developed to predict the temperature distribution produced by the Prostalund ProSitex microwave catheter (Lund Instruments, Sweden) [7]. No mathematical details are presented, but it appears that their model is similar to that developed by Xu et al. [3].

10.1.2   Factors Influencing Model Accuracy

As noted earlier, spatial and temporal variations of perfusion have an important bearing on the temperature distribution in the prostate [8]. Prior modeling attempts have hinted at the nature of these variations, but no comparisons have yet been made between model predictions and measured tissue temperatures during the entire course of a microwave treatment. Therefore, although good agreement has been obtained in prior studies by adjusting perfusion levels to match the steady-state temperatures, there has been no attempt to incorporate the thermoregulatory perfusion response into a predictive model of the tissue temperatures.

Other important factors in developing an accurate predictive model include accurate thermophysical property data on prostatic tissue, accurate knowledge of the specific absorption rate (SAR) of microwave radiation within the volume of the gland, and accurate treatment of the thermal effects of blood flow in the tissue volume. Among these factors, the SAR distribution is, perhaps, the most readily quantified on the basis of measured temperature elevations in a volume of phantom gel. Analytical expressions for the SAR distribution produced by the Urologix catheter have been developed in this manner by both Xu et al. [3] and Zhu et al. [9]. The thermophysical properties of human prostatic tissue are not well established, but the properties of canine prostatic tissue have been measured by Yuan et al. [10]. Those values are used to guide the choice of values in the current model.

The rate of blood perfusion and its thermal effects are the least well understood of the various factors influencing the model. Most prior modeling efforts have been based on the Pennes bioheat equation, which employs a special term to account for the energy inflow and outflow produced by the flow of blood through a volume of tissue. This term depends on the local "rate of blood perfusion" as well as the difference in temperature between the body core and the local tissue. Perfusion is a nebulous concept, and its measurement is difficult. In humans, blood perfusion in the prostate is known to vary widely from person to person. It is estimated that the range of perfusion values encountered in the elderly male population is between 8 ml/min/100 ml and 50 ml/min/100 ml. Various alternatives to the Pennes model have been developed, but it continues to be used because of its relative simplicity and reasonable accuracy compared to more elaborate formulations [11]. In addition to the energy inflow and outflow, a more subtle effect of blood perfusion is an enhancement of the apparent thermal conductivity of the tissue. This enhancement, which is the result of the flow of blood in the microvasculature, has received little attention to date, and little information is available to guide model development.

10.2 TEMPERATURE MEASUREMENTS IN HUMAN SUBJECTS

Interstitial temperature-mapping studies were conducted on nine volunteer human subjects to obtain the necessary experimental database for model calibration. In all cases, the patients were administered a spinal anesthetic, and several fiber-optic temperature probes (Luxtron Corp., Mountain View, California) were inserted into the prostate through nylon needle-tipped cannulae introduced percutaneously through the perineum. The cannula insertion points were guided by a template that contained an array of holes for insertion of interstitial needles. Insertion through the perineum results in the probe axis being oriented roughly parallel to the axis of the prostatic urethra. The locations of the cannulae were confirmed by inserting metal needles through the cannulae and placing the Urologix catheter in the urethra. Since these objects are radio-opaque, fluoroscopy could be used to ascertain the needle positions relative to the urethra.

Each fiber-optic probe contained either three or four sensors equally spaced along the fiber at either 0.5 cm or 1.0 cm intervals, starting at the tip of the fiber. Since it is extremely difficult to precisely ascertain the axial position of each sensor in the gland, the peak temperature reading from each multisensor probe was assumed to correspond to the reading at the axial midplane of the microwave antenna. The sensors were connected to two Luxtron thermometry units (Model 3000), each of which had the capacity of monitoring eight sensors. Thus, a total of 16 interstitial temperatures could be monitored by the setup. The temperature sensors were scanned once every 10 seconds, and the data were automatically logged into a laptop computer. Additional details of the temperature-mapping technique are available in a prior publication by Larson et al. [6].

Patients were treated using the Urologix Targis microwave catheter. After insertion of the catheter, transrectal ultrasound imaging was used to confirm its proper positioning in the prostatic urethra. Rectal temperatures were monitored with a rectal thermometer unit equipped with five resistance temperature detectors (RTDs). Coolant was circulated through the channels of the catheter at a rate of 100 ± 10 ml/min at a set temperature of $8°C$. Microwave power was applied in increments until the target urethral temperature of $40 \pm 1°C$ was achieved. Treatment was continued for between 45 and 60 minutes after attainment of the target temperature, with appropriate adjustments being made to the power input to maintain the urethral temperature within the specified tolerance range. At the end of the treatment period, microwave power was discontinued and the coolant circulation was maintained for an additional 5 minutes before withdrawal of the catheter.

10.3 THE MATHEMATICAL MODEL

The present model is closely related to the one-dimensional model in cylindrical coordinates developed by Xu et al. [3]. The computational domain is taken to be a long tissue cylinder of radius R that encompasses the entire prostate as well as a layer of fatty connective tissue surrounding the capsule of the gland (see Figure 10.4). The innermost layer of tissue, 5 mm in radial extent, is taken to be the periurethral zone. In this section of the chapter, which deals with model development, the temperature field is assumed to be purely one-dimensional (i.e., only radial variations in temperature are considered), while axial and circumferential variations are ignored. In Section 10.7, the model is extended to include radial and axial variations.

The model is based on the thermodynamic energy balance embodied in the Pennes bioheat equation [12]:

$$\rho_t c_t \frac{\partial T}{\partial t} = div(k\, grad\, T) - \omega \rho_b c_b (T - T_a) + Q + Q_m \tag{10.1}$$

In this equation, T is the tissue temperature, t stands for time, k is the thermal conductivity of tissue, c is the specific heat, Q is the volumetric rate of heating produced by the microwave radiation (i.e., the volumetric SAR), Q_m is the rate of heat generation due to metabolism, ρ is the density, ω is the blood perfusion rate, and T_a is the arterial blood temperature. The subscripts t and b stand for tissue and blood, respectively.

Equation (10.1) reveals that the time-wise variation of the tissue temperature during microwave treatment is dependent on the properties of the tissue (density, conductivity, and specific heat), the rate of heating produced by the microwaves, the metabolic heat generation rate, and the rate of heat absorbed by the circulating blood flow. In microwave treatments for BPH, the metabolic heat generation, Q_m, is negligibly small in comparison to the heat generated by the microwaves, and it may be discarded from the equation.

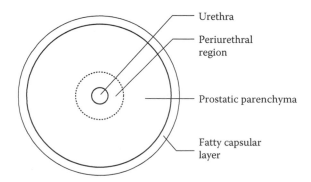

Figure 10.4   Computational domain for model development.

The microwave antenna is assumed to radiate as a uniform line source located at the axis of the catheter. The electromagnetic power flux is, accordingly, taken to have the form

$$q'' = \frac{(A_t P)e^{(-2\beta r)}}{r^2} \tag{10.2}$$

where A_t is a constant of proportionality, β is a constant related to the rate of absorption of microwaves in the tissue, and P is the electrical power delivered to the catheter. The electromagnetic power flux is maximum at the surface of the catheter and diminishes somewhat faster than the inverse square of the radial distance. The value of A_t is determined from experiments in tissue-equivalent phantom gel, as described later. The divergence of the power flux yields the volumetric rate of absorption of microwave radiation per unit volume of tissue:

$$Q = \left(2\beta + \frac{1}{r} \right) q'' \tag{10.3}$$

The radial variation of Q is depicted in Figure 10.5. The minor radius of the Urologix Targis catheter is 2.5 mm. At this location, the tissue-heating rate is close to 10^7 W/m³. In comparison, at a distance of 2.5 mm from the surface of the catheter (5 mm from the axis), the heating rate is only about 10^6 W/m³. Thus, the heating is seen to be extremely intense close to the surface of the catheter but to diminish rapidly with increasing radial distance.

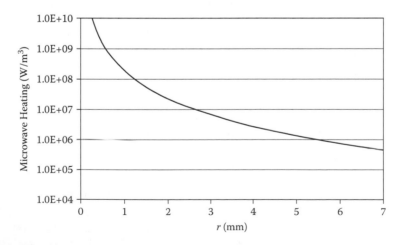

Figure 10.5   Radial variation of the function representing microwave heating (Equation 10.3).

To predict the temperature distribution within the prostate, the Pennes bio-heat equation is solved subject to the appropriate boundary conditions. At the surface of the catheter (which is the inner boundary of the calculation region), the boundary condition is specified by the temperature of the coolant and the overall heat transfer coefficient, U, between the tissue and the coolant. For the present calculations, the value of U is taken to be 750 W/m² K. This value was calculated based on standard correlations for convective heat transfer in ducts and the known thermal resistance of the catheter wall. In addition, a simple calo-rimetric experiment was conducted to confirm the calculated value. At the outer boundary of the calculation region, which is taken to be so far from the catheter that the temperature elevation there is small, the conduction heat flow rate is assumed to be zero. The outer boundary of the calculation domain is taken to be at a distance of 33 mm from the axis.

10.3.1 Thermophysical Properties

Thermophysical properties k, ρ, and c were obtained from various literature sources [3,10,13] and were assigned as follows:

In the prostate:

$$\rho_b c_b = \rho_t c_t = 3.9 * 10^6 \text{ J/m}^3 \text{ K}$$

and

$$k = 0.55 \text{ W/m K} \quad \text{for} \quad T < 49°C$$

$$k = 0.42 \text{ W/m K} \quad \text{for} \quad T > 60°C$$

$$k = \{0.55 - 0.13 * (T - 49)/11\} \text{ W/m K} \quad \text{for} \quad 49°C < T < 60°C$$

In the capsular layer:

$$\rho_b c_b = \rho_t c_t = 3.9 * 10^6 \text{ J/m}^3 \text{ K}$$

and

$$k = 0.25 \text{ W/m K}$$

10.3.2 Blood Perfusion Values

Observations of both canine and human prostates show that the blood supply to the periurethral region is distinct from that to the rest of the gland [4]. The large blood vessels around the urethra run parallel to the urethral axis. These blood vessels serve as a significant heat sink in the periurethral region. Xu and cowork-ers measured the blood perfusion in canine prostates and reported higher values in the periurethral zone than in the parenchyma of the gland [14]. In addition, the perfusion values were found to be dependent on the temperature of the local tissue. Using the data from Xu et al. [14] as a guide, perfusion values in the current model are assigned according to the following expressions:

Periurethral Zone

$$\omega = 1.5 * \omega_{base} \quad \text{for} \quad T < 40°C \tag{10.4a}$$

$$\omega = 2.25 * \omega_{base} \quad \text{for} \quad 40°C < T < 43°C \tag{10.4b}$$

$$\omega = 3.75 * \omega_{base} \quad \text{for} \quad 43°C < T < 48°C \tag{10.4c}$$

$$\omega = 5.25 * \omega_{base} \quad \text{for} \quad T > 48°C \tag{10.4d}$$

Parenchyma

$$\omega = \omega_{base} \quad \text{for} \quad T < 40°C \tag{10.5a}$$

$$\omega = 1.20 * \omega_{base} \quad \text{for} \quad 40°C < T < 43°C \tag{10.5b}$$

$$\omega = 1.35 * \omega_{base} \quad \text{for} \quad 43°C < T < 48°C \tag{10.5c}$$

$$\omega = 1.6 * \omega_{base} \quad \text{for} \quad T > 48°C \tag{10.5d}$$

Capsular Layer

$$\omega = 0.0012 \ s^{-1} \tag{10.6}$$

The quantity ω_{base} appearing in the foregoing expressions is a baseline value prevailing at the beginning of treatment. This baseline value has to be determined either by measurement just prior to treatment or by inference from measured intraprostatic temperatures during treatment. Equations (10.4) and (10.5) attempt to mimic the vasodilatation observed in canine prostates by Xu et al. [14]. As is evident, the baseline blood perfusion values are incremented by various factors depending on the extent of the local tissue temperature elevation. Although the vasodilatation model is largely based on the measurements of Xu et al. [14], the numerical values of the various factors featured in the two equations have been adjusted by trial and error to obtain the best overall agreement with the measured intraprostatic temperatures in human subjects.

The perfusion adjustment rules described by Equations (10.4) and (10.5) are invoked as long as the extent of local "thermal damage" of the tissue remains below a threshold level. Once the local thermal damage exceeds a threshold value, the local blood perfusion is taken to diminish abruptly as a consequence of shutdown of circulation in the local microvasculature. It was found necessary to incorporate vascular shutdown in the model to replicate certain features of the measured temperature variations in human subjects. Further discussion of this issue is presented in Section 10.5. The phenomenon of local vascular shutdown resulting from thermal damage to the tissue has been noticed by prior investigators [15,16].

The extent of local thermal damage is computed by evaluating the "damage integral,"

$$\Omega = \int \kappa \, dt \tag{10.7}$$

as suggested originally by Henriques [17]. The rate constant, κ, was taken to vary with temperature according to the expression $\kappa = \kappa_0 * 10^{(T - T_0)/z}$, where κ_0 is the value of κ at the reference temperature T_0 (taken to be 47.5°C), and $z = 6.5$°C. This computationally convenient expression for κ is commonly used for cell death calculations. It is derivable from the standard Arrhenius expression for the variation of the chemical rate constant with temperature. The value of κ_0 was assigned to be $4.2 * 10^{-4}$ in the parenchyma and capsular layer and $1.5 * 10^{-5}$ in the periurethral region. The threshold value of the damage integral was taken to be 1.0. Once the local damage integral exceeded this threshold, the local values of blood perfusion were set as follows:
Periurethral Zone

$$\omega = 1.70 * \omega_{base} \tag{10.8a}$$

Parenchyma

$$\omega = 0.2 * \omega_{base} \tag{10.8b}$$

As with Equations (10.4) and (10.5), the numerical values featured in Equation (10.8a,b) were obtained by trial and error until model predictions closely matched measured interstitial temperatures in human patients undergoing microwave treatment. The difference in the values of κ_0 in the periurethral region and the parenchyma suggests that the periurethral vasculature is much less susceptible to shutdown when subjected to heating. The perfusion adjustment rules presented in Equations (10.4) through (10.8) have previously been described in a patent application by Ramadhyani et al. [18].

The essential details of the model are now complete. Obviously, the complications associated with variations in the thermophysical properties and blood perfusion preclude a closed-form analytical solution of the bioheat equation. Instead, a standard, fully implicit, finite-volume, numerical solution scheme is employed [19]. The computational domain is discretized into 180 uniform radial increments, and time steps of 1-second duration are taken. Extensive numerical studies established that the spatial and temporal discretizations were sufficiently fine to yield grid-independent solutions.

10.4 STUDIES IN TISSUE-EQUIVALENT PHANTOM GEL

As noted previously, the proportionality constant A_t featured in Equation (10.2) was determined through experiments in tissue-equivalent phantom gel. The phantom gel is a mixture of water, ethylene glycol, and sodium chloride mixed in appropriate proportions to closely replicate the real and imaginary parts of the complex dielectric constant of prostatic tissue ($\varepsilon' = 55$, $\varepsilon'' = 24$). The solution is cross-linked by the addition of formaldehyde to create a gel.

A Urologix Targis catheter was placed vertically within a large mass of phantom gel equilibrated to laboratory temperature (24°C). The gel was instrumented with several fiber-optic temperature sensors to monitor temperatures at various spatial locations. The catheter was operated at an input microwave power of 20 W with coolant circulating at 8°C. According to the test protocol, the coolant was circulated for the first 5 minutes with microwave power at zero. At the 5-minute time point, power was switched on and maintained at 20 W for a period of 25 minutes.

To determine the value of A_t, the mathematical model was run with the actual coolant temperatures and power values employed in the phantom-gel test, and with the blood perfusion rate set to zero. The value of A_t was adjusted until a close match was obtained between the measured and predicted temperatures at several locations in the phantom gel. Figure 10.6 displays a comparison of the measured and predicted temperatures at four different locations within the gel. The measurement locations are in the middle horizontal plane of the vertically positioned antenna, and are at radial distances of 5, 10, 20, and 30 mm from the surface of the catheter. In each case, the continuous dotted curve represents the model prediction, while the discrete data symbols represent the measurements.

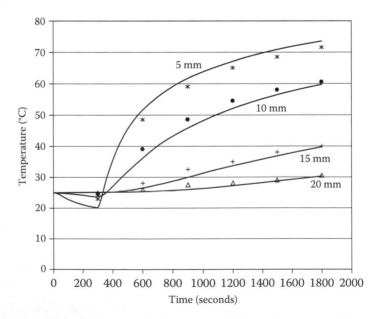

Figure 10.6 Comparison of predicted and measured temperatures over time at several locations in phantom gel; $A_t = 0.0095$ for the best match.

At the 5-minute time point (300 seconds), the predicted and measured temperatures begin to rise as a result of the microwave heating. To achieve the level of agreement shown in Figure 10.6, the constant A_t was adjusted to the value of 0.0095. This value of A_t was then employed in predicting the interstitial temperatures in human test subjects.

10.5 PREDICTION OF INTERSTITIAL TEMPERATURES IN HUMAN SUBJECTS

To facilitate an understanding of the Targis treatment protocol, Figure 10.7 presents the time-wise variations of the microwave power and the coolant temperature during an actual Targis procedure, along with the measured urethral temperature during the treatment. The coolant is automatically maintained at the recommended set point of 8°C by the control system through on-off regulation of a thermoelectric refrigerator. The small oscillations in the coolant temperature indicate this regulation. The power is gradually incremented by the physician to drive the urethral temperature to its recommended set point of 40 ± 1°C. Once the urethral temperature is at its set point, the physician makes small power adjustments to maintain that temperature. In Figure 10.7, it is observed that the physician employed larger power increments at the beginning of the power ramp and finer increments as the target urethral temperature was approached. Toward the latter stages of the treatment, it is often necessary to decrement the power as the tissue undergoes progressive necrosis and the vasculature shuts down. Such a reduction in power is seen in Figure 10.7 at the 27-minute time point.

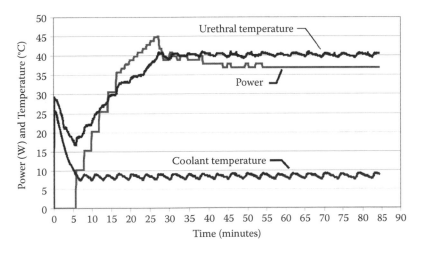

Figure 10.7 Variation of microwave power, coolant temperature, and urethral temperature during a typical Targis treatment.

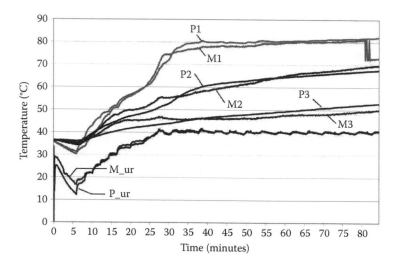

Figure 10.8 Comparison of measured and predicted interstitial temperatures in a human subject: P1, M1 at 4 mm from urethra; P2, M2 at 9 mm from urethra; and P3, M3 at 14 mm from urethra. P_ur and M_ur are predicted and measured urethral temperatures.

Figures 10.8, 10.9, and 10.10 present direct comparisons of measured and predicted interstitial temperatures in three of the nine human subjects in the interstitial mapping study. In each case, the model was run retrospectively, using the actual coolant and power values used in the treatment. With one exception, none

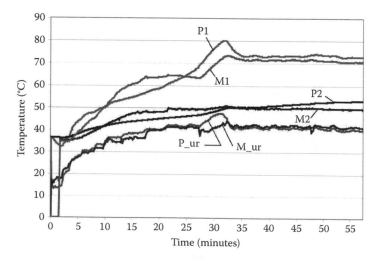

Figure 10.9 Comparison of measured and predicted interstitial temperatures in a human subject: P1, M1 at 5 mm from urethra; and P2, M2 at 12 mm from urethra. P_ur and M_ur are predicted and measured urethral temperatures.

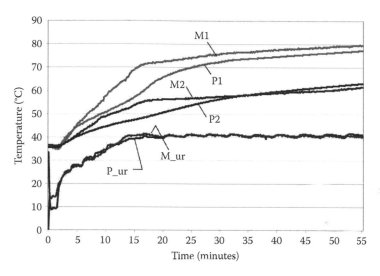

Figure 10.10 Comparison of measured and predicted interstitial temperatures in a human subject: P1, M1 at 7 mm from urethra; and P2, M2 at 11 mm from urethra. P_ur and M_ur are predicted and measured urethral temperatures.

of the model constants was adjusted to improve the level of agreement between the measurements and the predictions. The exception was the value of the baseline blood perfusion, ω_{base}. In each case, the value of ω_{base} was adjusted by trial and error to obtain a close fit between the predicted and measured urethral surface temperatures (labeled as P_ur and M_ur, respectively, in each of the figures). Once satisfactory agreement between P_ur and M_ur was achieved, the model-predicted interstitial temperatures were compared to the corresponding measured values.

A comment on the adjustment of the value of ω_{base} may be helpful. The Targis catheter contains a fiber-optic temperature sensor to monitor the temperature of the urethra. Accordingly, it is feasible to adjust the value of the baseline perfusion, during an actual treatment, to obtain a match between the predicted and the measured temperature history of the urethra. The value of the predictive model lies in being able to accurately predict interstitial temperatures and zones of necrosis, once the baseline blood perfusion has been established partway through the treatment.

Attention is directed to the pairs of curves P1, M1 in Figures 10.8 and 10.9. In Figure 10.8, both curves display a rapid rise between 22.5 minutes and 30 minutes (the concave-upward shapes of the curves are noteworthy). This rapid rise is associated with the local shutdown of the vasculature at a distance of approximately 4 mm from the urethra. The model correctly captures this event and accurately predicts the resulting sharp temperature increase. An even more clearly defined spike in the measured temperature is observed in Figure 10.9. Again, the model accurately predicts the time at which the event occurs and realistically reproduces the sharp temperature rise due to vascular shutdown.

By appropriate choice of the blood perfusion values, it was possible to obtain extremely close agreement between the measured and predicted urethral temperatures in Figures 10.8 and 10.10. In Figure 10.9, there are noticeable differences between the two curves, especially in the time range between 27.5 minutes and 32.5 minutes. This discrepancy is due to a sudden loss of contact between the temperature sensor and the urethral wall in that period. Such events may occur due to patient movement or muscular spasms in the urethra.

In general, the level of agreement between the measurements and the predictions in Figures 10.8, 10.9, and 10.10 is satisfactory. Particularly toward the end of each of the treatments, the differences between the predicted and measured interstitial temperatures are less than 5°C in all three cases. It must be noted, however, that larger differences are observed earlier in the treatments, with the measured temperatures generally exceeding the predicted values. These differences are most probably due to a shortcoming in the method used for adjusting the perfusion with rises in temperature. As described in Section 10.3, the perfusion is assumed to rise abruptly when the local temperature reaches a specified threshold. This approach makes no allowance for time delays or oscillations in the vasodilatory response [14]. As additional clinical data become available in the future, it may be possible to incorporate a more accurate vasodilatation model.

To provide further perspective on the temperature distribution and the distribution of necrotic tissue in the gland, attention is directed to Figures 10.11 and 10.12. Both of these figures present model predictions for the patient whose treatment parameters were presented in Figures 10.7 and 10.8. Figure 10.11 shows the calculated radial variation of temperature, from the surface of the catheter to the outer edge of the calculation domain, at the end of power application

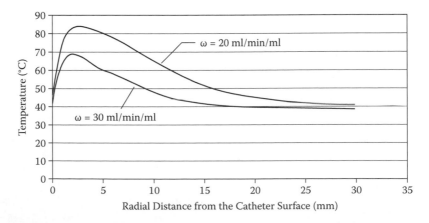

Figure 10.11   Radial variation of tissue temperature near the end of a Targis treatment—comparison of predictions with two different values of blood perfusion.

(approximately the 80-minute mark on Figure 10.7). The two curves correspond to two different assumed values of the baseline blood perfusion, ω_{base}. With both curves, it is observed that the temperature at the surface of the catheter (the temperature of the urethral mucosa) is low (approximately 42°C) due to the cooling effect of the circulating water. With progressive increase in the radial location, the temperature initially increases extremely steeply as the cooling effect of the water diminishes, while the microwave energy dissipation (SAR) remains very strong. For ω_{base} = 20 ml/min/100 ml, the temperature reaches a peak value of about 83°C at a distance of approximately 3 mm from the surface of the catheter. Beyond that point, the temperature diminishes steadily with increasing radial distance because of the rapid decrease in the SAR function. For ω_{base} = 30 ml/min/100 ml, the temperature reaches a peak value of only about 70°C before decreasing. For this patient, an accurate match between the measurements and predictions was obtained for ω_{base} = 20 ml/min/100 ml, as seen in Figure 10.5. The difference between the two curves shows the great sensitivity of the predictions to the assumed value of ω_{base}. For accurate predictions, it is essential to correctly establish the value of the baseline blood perfusion by matching model predictions to the measured temperature of the urethral surface.

The shape of the temperature distribution displayed in Figure 10.11 is characteristic of the Targis treatment system. With this temperature distribution, tissue thermal damage is expected to be greatest at about 3 to 5 mm from the surface of the catheter. Correspondingly, vascular shutdown is expected to commence at about 3 to 5 mm from the surface of the catheter and progressively expand (both radially inward and outward) with time. Therefore, a temperature sensor located around 3 to 5 mm from the surface would be expected to register a sharp rise in temperature, signaling the abrupt shutdown of the microvasculature near that location.

Figure 10.12 displays the radial variation of the computed thermal damage at three different instants during the treatment. At 20 minutes, the value of the damage integral is less than 0.5 in the periurethral region. At 5 mm from the

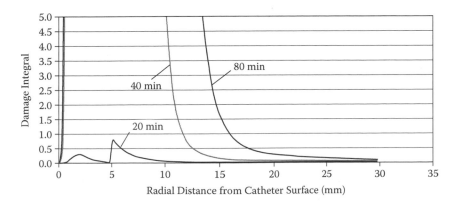

Figure 10.12   Radial distribution of thermal damage at various times during a Targis treatment.

catheter surface, the computed damage integral rises sharply to about 0.8 and then declines with increasing radial distance. Thus, none of the tissue in the computational domain has reached the damage threshold of 1.0 up to this time. At 40 minutes, the picture is dramatically different. A significant portion of the computational domain (between 0.37 mm and 11.7 mm) is now composed of tissue in which the damage integral exceeds 1.0. At the end of treatment, 80 minutes, the zone of necrosis has widened as additional thermal damage has accrued in the intervening time. The figure clearly illustrates the growth of the necrotic zone over time as the treatment progresses. Note that the bumpy shape of the damage integral curve at 20 minutes is an outcome of the different values of κ_0 assigned to the periurethral region and the prostatic parenchyma.

10.6 COMPARISON OF OBSERVED AND PREDICTED ZONES OF TISSUE NECROSIS

An independent data set was used to assess the accuracy of the model. A separate clinical study had previously been conducted by Urologix Inc. to measure the size of the necrotic lesion created by the Targis treatment. In a randomized trial, patients had been treated for two different durations—28.5 minutes and 60 minutes [20]. Magnetic resonance imaging (MRI) of the prostate was done on all patients one week after the treatment. Images were obtained with gadolinium contrast agent to delineate the boundary of the zone of necrosis. It has been shown in previous studies that this zone correlates well with the histologically determined zone of cell destruction [21]. One such MR image is presented in Figure 10.13. The zone of necrosis is the irregular (approximately circular) dark region. In the middle of this region is a small black circle with a lighter center. This small circle marks the location of the urethra. The prostate itself is the gray zone surrounding the necrotic region. The boundary of the prostate is clearly delineated, particularly near the top.

Figure 10.13 shows one particular transverse section of the prostate. In each patient, MR images were obtained at multiple transverse sections, the distance between adjacent slices being fixed at approximately 2 mm. By electronically processing all the image slices from a given patient, it was possible to reconstruct a full, three-dimensional image of the whole prostate as well as the zone of ischemic necrosis within the gland. Through such reconstruction, it was possible to determine the volume of the whole gland, and the ischemic zone within the gland, for each patient. The electronic analysis of the MR images was done at the Mayo Clinic (Rochester, Minnesota) on all study patients, and the results reported here are from that analysis.

10.6.1 Prostate

Figure 10.14 presents a comparison of the numerical model predictions with the data from the Mayo Clinic analysis of the MR images. The abscissa plots the necrosis area predicted by the one-dimensional model, while the ordinate plots

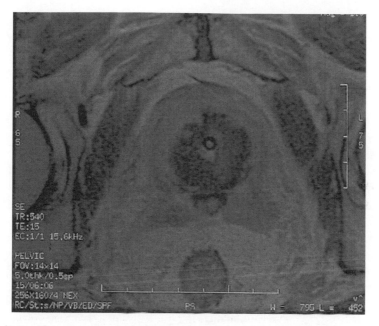

Figure 10.13 Magnetic resonance image showing zone of necrosis in a human subject.

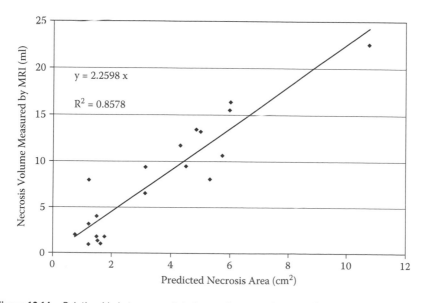

Figure 10.14 Relationship between predicted necrosis area and measured necrosis volume.

the necrosis volume measured by MRI. The necrosis area is computed as $\pi\,(r_2^2 - r_1^2)$, where r_2 and r_1 are the radii of the outer and inner boundaries of the necrotic zone. The best-fit line as well as the individual data points from 21 patients are shown. The plot shows that there is an approximately linear relationship between the calculated necrosis area and the measured necrosis volume. The standard deviation, σ, of the difference between the best-fit straight line and the measured necrosis volumes is 2.3 ml.

A strictly linear relationship between the two variables would be expected if the SAR distribution of the antenna were invariant in the axial direction and if all the tissue properties were perfectly invariant in the direction parallel to the urethra. Although these conditions are not strictly satisfied, the generally good agreement between the predictions and the measurements provides confidence in the validity of the simplifications invoked in the model.

10.7 EXTENSION OF THE MODEL TO ACCOUNT FOR PHASE CHANGE

Examination of Figures 10.8, 10.9, and 10.10 reveals that peak interstitial temperatures during Targis treatments can approach or even exceed 80°C. An excessive application of microwave power might result in temperatures reaching 100°C, with the resulting generation of water vapor in the heated zone. Such a scenario can have disastrous consequences to the patient as the pressure generated by the vapor drives heated liquid deeper into the tissue, thus greatly extending the zone of necrosis. Current microwave delivery systems have numerous safety features to forestall the occurrence of such a situation. Nevertheless, there is one documented instance of patient injury occurring as a result of excessive treatment with an earlier-generation microwave system [22].

To account for vapor evolution at 100°C, the Pennes model described by Equation (10.1) is modified to an enthalpy form as follows:

$$\frac{\partial h}{\partial t} = div(k\,gradT) - \omega\rho_b c_b(T - T_a) + Q + Q_m \tag{10.9}$$

The quantity h in the foregoing equation represents the specific enthalpy of the tissue. It is defined by the following equations:

$$h = (\rho c)^{pre}(T - 37°C), \quad 37 < T < 100°C \tag{10.10a}$$

$$h = h(100) + h_{fg}C_{liq}\frac{(T - 100°C)}{(101 - 100°C)}, \quad 100 < T < 101°C \tag{10.10b}$$

$$h = h(101) + (\rho c)^{post}(T - 101°C), \quad T > 101°C \tag{10.10c}$$

According to Equation (10.10a), the datum for enthalpy is set at 37°C. Between 37°C and 100°C, there is no phase change, and the enthalpy increases linearly with temperature. The quantity $(\rho c)^{pre}$ is the product of the density and the specific heat of tissue prior to phase change. According to Equation (10.10b), phase change is assumed to occur between 100°C and 101°C. The quantity C_{liq} appearing in Equation (10.10b) takes the value 0.8, which represents the concentration of liquid in normal tissue. In the temperature range from 100°C to 101°C, the concentration of liquid drops from 80% to 0% [23]. For temperatures greater than 101°C, it is assumed that no liquid is present. According to Equation (10.10c), the enthalpy increases linearly with temperature. The quantity $(\rho c)^{post}$ appearing in Equation (10.10c) is the product of the density and the specific heat of dry tissue. The superscripts "pre" and "post" refer, respectively, to temperatures less than 100°C and greater than 101°C. For temperatures less than or equal to 100°C, the thermal conductivity and specific heat of prostatic tissue are specified in Section 10.3.1. For temperatures greater than 101°C, the thermal conductivity and the quantity $(\rho c)^{post}$ take on the values 0.2 W/m K and 2.2×10^6 W/m³ K, respectively.

The enthalpy-based formulation enables prediction of tissue temperatures up to 100°C with good accuracy. Once phase change begins, the accuracy of the predictions is degraded by the fact that liquid water in the tissues is moved by the pressure generated by the evolving vapor. This effect is not reflected in the model. Nevertheless, it is interesting to employ the enthalpy-based model to examine the consequences of an excessive microwave treatment delivered to one particular patient. These computations, reported by Abraham et al. [22], were performed in axisymmetric cylindrical coordinates. Thus, temperature variations both radially and longitudinally were captured by the computations. Since the prostate is not axisymmetric but approximately ellipsoidal, Abraham et al. [22] considered two axisymmetric geometries to obtain upper and lower bounds on the solution. In one case, the diameter of the computational domain is based on the anterior–posterior dimension of the prostate, while in the other, it is based on the lateral dimension of the gland.

The computational domains used are depicted in Figure 10.15. Figure 10.15a depicts the lateral view, while Figure 10.15b depicts the anterior–posterior view. In each case, it is seen that the computational domain encompasses the prostate, the bladder, a layer of connective tissue around both the prostate and the bladder, and the urinary sphincter at the pelvic floor. The computational domain is closed by a layer of "highly perfused" tissue, taken to be 10 mm thick, within which the perfusion is set to a value of 0.063⁻¹ sec. This layer, in reality, is mainly composed of intestine. The temperature is set to the body core value of 37°C at the outer boundary of the highly perfused layer. The possible presence of a pool of urine within the bladder is also considered in the model.

The microwave power delivered to the treatment catheter is depicted as a function of time in Figure 10.16. It can be seen that microwave power delivery was maintained for a period of about 110 minutes during this particular treatment. In contrast, current protocols limit the treatment duration to about 45 minutes.

As described earlier in Section 10.5, the value of ω_{base} was iteratively adjusted until a good match was obtained between the measured and model-predicted

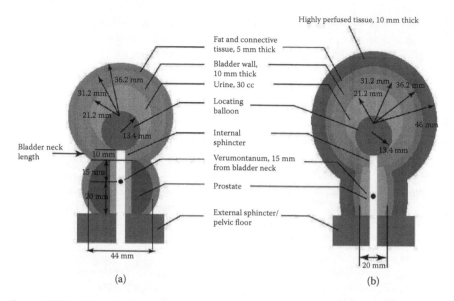

Figure 10.15   Schematic diagrams of the computational domain: (a) lateral view; and (b) anterior–posterior view.

catheter temperatures. The comparison is shown in Figure 10.17, and it is seen that the model replicates well the time-wise variations registered by the catheter temperature sensor.

The computed temperature distributions are depicted at two different time points in Figures 10.18 and 10.19. As seen in Figure 10.18, the peak tissue

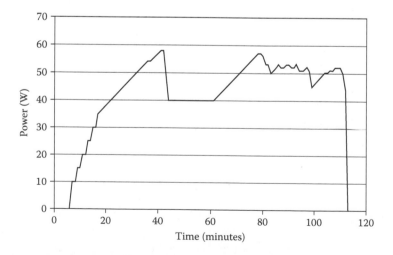

Figure 10.16   Microwave power levels used during the treatment.

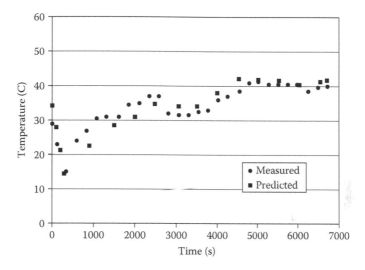

Figure 10.17 Comparison of measured and predicted catheter temperatures used for the determination of blood perfusion rate; ω_{base} = 38 ml/min/100 ml for best fit.

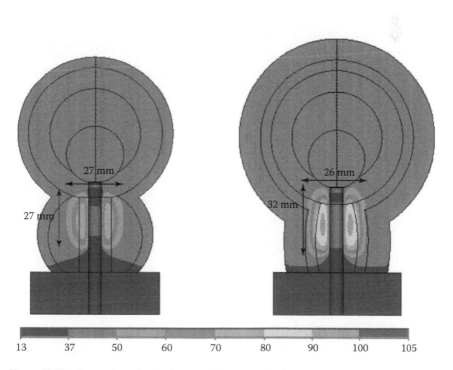

Figure 10.18 Temperature distributions at 3500 seconds in the prostate and adjacent anatomical structures. At left, the lateral view; and at right, the anterior–posterior view.

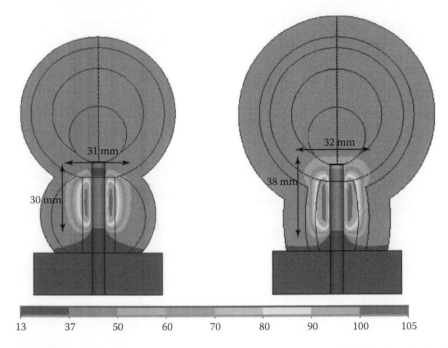

Figure 10.19   Temperature distributions at 4800 seconds in the prostate and adjacent anatomical structures. At left, the lateral view; and at right, the anterior–posterior view.

temperature in the prostate is already around 90°C after 3500 seconds of treatment. Although treatments are typically terminated at around this duration, in this particular instance microwave delivery was continued for much longer. As seen in Figure 10.19, peak temperatures within the prostate eventually reached and exceeded 100°C. The extent of the injury to the surrounding tissues is described in detail by Abraham et al. [22].

10.8 CONCLUSIONS

This chapter describes a mathematical model for the calculation of tissue temperatures and zones of tissue destruction during transurethral microwave ablation of the prostate. The model, which is based on the Pennes bioheat equation, accounts for increases in the blood perfusion with increases in temperature. The model also accounts for an abrupt reduction in perfusion when the thermal damage exceeds a critical threshold. The perfusion adjustment rules as well as the various model constants have been derived by matching model predictions to bench-top measurements as well as *in vivo* measurements in human patients.

Comparisons have been presented between model-predicted tissue temperatures and measurements in human patients undergoing microwave ablation treatment. Comparisons have also been presented between model-predicted necrosis volumes and necrosis volumes measured in human patients by MRI. These comparisons show that the model is able to predict necrosis volumes within +/– 3.5 ml with 90% confidence. This ability of the model is clinically valuable. In Targis treatments of 60 minutes duration, necrosis volumes have been observed to range between 3 ml and 30 ml, with the lower limit corresponding to the most highly vascularized gland and the upper limit to the least vascularized gland. The model provides the physician with the ability to adjust the treatment duration on a patient-specific basis to achieve a more uniform volume of necrosis, optimally around 12 to 15 ml for the average 45 ml prostate.

It is shown that during normal microwave ablation treatments, peak tissue temperatures can approach 90°C. In unusual circumstances, tissue temperatures may approach, or even exceed, 100°C. A method has been described for extending the model to account for the possibility of phase change in these rare circumstances

NOMENCLATURE

A_i: constant of proportionality in Equation (10.2)
c: specific heat
h: specific enthalpy
k: thermal conductivity of tissue
P: electrical power input to the catheter
Q: heat generation rate per unit volume in tissue by microwaves
q'':   microwave power flux from the catheter
r: radial coordinate
T: temperature
T_a: arterial blood temperature
t: time

GREEK SYMBOLS

β: attenuation constant of microwave radiation in tissue
κ: thermal damage rate constant
κ_0:   rate constant at the reference temperature of 47.5°C
Ω: damage integral
ω: blood perfusion rate
ω_{base}:   baseline blood perfusion rate before the start of treatment

REFERENCES

1. D. Simopoulos and M. Blute, Assessing the Value of Transurethral Microwave Thermotherapy, *Contemporary Urology*, vol. 12, pp. 30–46, 2000.
2. B. Djavan, S. Shahrokh, B. Schafer, and M. Marberger, Tolerability of High Energy Transurethral Microwave Thermotherapy with Topical Urethral Anesthesia: Results of a Prospective, Randomized, Single-Blinded Clinical Trail, *Journal of Urology*, vol. 160, pp. 772–776, 1998.
3. L. Xu, E. Rudie, and K. Holmes, Transurethral Thermal Therapy (T3) for the Treatment of Benign Prostatic Hyperplasia (BPH) in the Canine: Analysis Using the Pennes Bioheat Equation, in *Advances in Bioheat and Mass Transfer: Microscale Analysis of Thermal Injury Processes, Instrumentation, Modeling, and Clinical Applications, HTD-Vol. 268*, pp. 31–35, New York: American Society of Mechanical Engineers, 1993.
4. D. Yuan, K. Holmes, and J. Valvano, Morphometry of the Canine Prostate Vasculature, *Microvascular Research*, vol. 59, pp. 115–121, 2000.
5. D. Yuan, J. Valvano, E. Rudie, and L. Xu, 2-D Finite Difference Modeling of Microwave Heating in the Prostate, in *Advances in Heat and Mass Transfer in Biotechnology, HTD-Vol. 322/BED-Vol. 32*, pp. 107–115, New York: American Society of Mechanical Engineers, 1995.
6. T. Larson, J. Collins, and A. Corica, Detailed Interstitial Temperature Mapping during Treatment with a Novel Transurethral Microwave Thermoablation System in Patients with Benign Prostatic Hyperplasia, *Journal of Urology*, vol. 159, pp. 258–264, 1998.
7. M. Bolmsjo, C. Sturesson, L. Wagrell, A. Andersson-Engels, and A. Mattiasson, Optimizing Transurethral Microwave Thermotherapy: A Model for Studying Power, Blood Flow, Temperature Variations and Tissue Destruction, *British Journal of Urology*, vol. 81, pp. 811–816, 1998.
8. M. Devonec, N. Berger, J. Fendler, P. Joubert, M. Nasser, and P. Perrin, Thermoregulation during Transurethral Microwave Thermotherapy: Experimental and Clinical Fundamentals, *European Urology*, vol. 23 (Suppl. 1), pp. 63–67, 1993.
9. L. Zhu, L. Xu, and N. Chencinski, Quantification of the 3-D Electromagnetic Power Absorption Rate in Tissue during Transurethral Prostatic Microwave Thermotherapy Using Heat Transfer Model, *IEEE Transactions on Biomedical Engineering*, vol. 45, pp. 1163–1172, 1998.
10. D. Yuan, J. Valvano, and G. Anderson, Measurement of Thermal Conductivity, Thermal Diffusivity, and Perfusion, *Biomedical Scientific Instrumentation*, vol. 29, pp. 435–442, 1993.
11. H. Arkin, L. Xu, and K. Holmes, Recent Developments in Modeling Heat Transfer in Blood Perfused Tissues, *IEEE Transactions on Biomedical Engineering*, vol. 41, pp. 97–107, 1994.
12. H. Pennes, Analysis of Tissue and Arterial Blood Temperatures in the Resting Human Forearm, *Journal of Applied Physiology*, vol. 1, pp. 93–122, 1948.
13. J. Chato, *Fundamentals of Bioheat Transfer, Thermal Dosimetry and Treatment Planning* (ed. M. Gautherie), Berlin: Springer-Verlag, 1991.
14. L. Xu, L. Zhu, and K. Holmes, Thermoregulation in the Canine Prostate during Transurethral Microwave Hyperthermia, Part I: Temperature Response, *International Journal of Hyperthermia*, vol. 14, pp. 29–37, 1998.
15. C. Song, Effect of Local Hyperthermia on Blood Flow and Microenvironment: A Review, *Cancer Research*, vol. 44 (Suppl.), pp. S4721–4730, 1984.

16. R. Roemer, J. Olesen, and T. Cetas, Oscillatory Temperature Response to Constant Power Applied to Canine Muscle, *American Journal of Physiology*, vol. 249, pp. R153–R158, 1985.

17. F. Henriques, Studies of Thermal Injury V: The Predictability and the Significance of Thermally Induced Rate Processes Leading to Irreversible Epidermal Injury, *Archives of Pathology*, vol. 43, pp. 489–502, 1947.

18. S. Ramadhyani, J. Flachman, and E. Rudie, Thermodynamic Modeling of Tissue Treatment Procedure, U.S. Patent No. 6,312,391, 2001.

19. S. Patankar, *Numerical Heat Transfer and Fluid Flow*, Washington, D.C.: Hemisphere Publishing, 1980.

20. A. Partin, C. Roehrborn, M. Blute, and H. Hezmall, Shortened Treatment Time Using Targeted Microwave Thermotherapy for the Treatment of BPH, Abstract 4198, *AUA Ninety-Fifth Annual Meeting*, Atlanta, GA, 2000.

21. T. Larson, J. Collins, D. Bostwick, and P. de Geeter, MRI's GAD Defect Can Quantify Coagulative Necrosis in Thermotherapy: A New Comparative Model, *Journal of Urology*, vol. 157, p. 1705A, 1997.

22. J. Abraham, E. Sparrow, and S. Ramadhyani, Numerical Simulation of a Necrosis-Causing Thermal Therapy: A Case Study Involving TUMT, *Journal of Biomedical Engineering*, vol. 129, pp. 549–557, 2007.

23. F. Duck, *Physical Properties of Tissue: A Comprehensive Reference Book*, London: Academic Press, 1990.

Index